GREEN CHEMISTRY USING LIQUID AND SUPERCRITICAL CARBON DIOXIDE

GREEN CHEMISTRY USING LIQUID AND SUPERCRITICAL CARBON DIOXIDE

Edited by

Joseph M. DeSimone

William Tumas

OXFORD
UNIVERSITY PRESS

2003

OXFORD
UNIVERSITY PRESS

Oxford New York
Auckland Bangkok Buenos Aires Cape Town Chennai
Dar es Salaam Delhi Hong Kong Istanbul Karachi Kolkata
Kuala Lumpur Madrid Melbourne Mexico City Mumbai Nairobi
São Paulo Shanghai Taipei Tokyo Toronto

Published by Oxford University Press, Inc.
198 Madison Avenue, New York, New York 10016

www.oup.com

Library of Congress Cataloging-in-Publication Data
Green chemistry using liquid and supercritical carbon dioxide / [edited by]
Joseph M. DeSimone, William Tumas.
p. cm.
Includes bibliographical references and index.
ISBN 978-0-19-515483-2
1. Liquid carbon dioxide—Industrial applications. 2. Environmental chemistry—Industrial applications. I. DeSimone, Joseph M. II. Tumas, William.

TP244.C1 G74 2002
660—dc21 2002071531

Printed in the United States of America

Contents

Contributors vii
Introduction xi
J. Young, J. M. DeSimone, and W. Tumas

Part I: Catalysis and Chemical Synthesis in Carbon Dioxide

1. Phase Behavior and Its Effects on Reactions in Liquid and Supercritical Carbon Dioxide 3
   L. A. Blanchard, G. Xu, M. A. Stadtherr, and J. F. Brennecke
2. Advances in Homogeneous, Heterogeneous, and Biphasic Metal-Catalyzed Reactions in Dense-Phase Carbon Dioxide 17
   T. Ikariya, R. Noyori, M. B. Abrams, C. A. G. Carter, G. B. Jacobson, F. Liu, D. R. Pesiri, and W. Tumas
3. Carbon Dioxide as a Reactant and Solvent in Catalysis 48
   T. Ikariya and R. Noyori
4. Free-Radical Chemistry in Supercritical Carbon Dioxide 64
   J. M. Tanko
5. Fluorous Phases and Compressed Carbon Dioxide as Alternative Solvents for Chemical Synthesis: A Comparison 81
   W. Leitner
6. Enzyme Chemistry in Carbon Dioxide 103
   R. L. Rodney and A. J. Russell

Part II: Polymers in Carbon Dioxide

7. Solubility of Polymers in Supercritical Carbon Dioxide 125
M. A. McHugh

8. Interfacial Phenomena with Carbon Dioxide Soluble Surfactants 134
K. P. Johnston, S. R. P. da Rocha, J. D. Holmes, G. B. Jacobson, C. T. Lee, and M. Z. Yates

9. Synthesis and Characterization of Polymers: From Polymeric Micelles to Step-Growth Polymerizations 149
J. L. Young and J. M. DeSimone

10. Preparation and Studies of Polymer/Polymer Composites Prepared Using Supercritical Carbon Dioxide 164
E. Kung, A. J. Lesser, and T. J. McCarthy

11. Rheological Properties of Polymers Modified with Carbon Dioxide 174
C. W. Manke and E. Gulari

Part III: Industrial Processes and Applications Utilizing Carbon Dioxide

12. Coatings from Liquid and Supercritical Carbon Dioxide 193
Y. Chernyak, F. Henon, E. Hoggan, B. Novick, J. M. DeSimone, and R. Carbonell

13. Dry Cleaning with Liquid Carbon Dioxide 215
G. Stewart

14. Selective and Complete Hydrogenation of Vegetable Oils and Free Fatty Acids in Supercritical Fluids 228
T. Tacke, S. Wieland, and P. Panster

15. Enhancing the Properties of Portland Cements Using Supercritical Carbon Dioxide 241
J. B. Rubin, C. M. V. Taylor, T. Hartmann, and P. Paviet-Hartmann

Index 257

Contributors

Michael B. Abrams
Chemistry Division
Los Alamos National Laboratory
Los Alamos, NM 87545

Lynnette A. Blanchard
Department of Chemical Engineering
University of Notre Dame
Notre Dame, IN 46556
lblancha@darwin.cc.nd.edu

Joan F. Brennecke
Department of Chemical Engineering
University of Notre Dame
Notre Dame, IN 46556
Joan.Brennecke.1@nd.edu

Ruben Carbonell
Department of Chemical Engineering
P.O. Box 7006
North Carolina State University
Raleigh, NC 27695
ruben@ncsu.edu

Charles A. Carter
Chemistry Division
Los Alamos National Laboratory
Los Alamos, NM 87545

Yuri Chernyak
Huntsman Chemical Company
Somerville Road
West Footscray
Victoria 3012
Australia

S. R. P. da Rocha
Department of Chemical Engineering & Materials Science
Wayne State University
Detroit, MI 48202

Joseph M. DeSimone
Department of Chemistry
CB 3290, Venable Hall
University of North Carolina, Chapel Hill
Chapel Hill, NC 27599
desimone@unc.edu

Esin Gulari
Department of Chemical
Engineering & Materials Science
5050 Anthony Wayne Drive
Wayne State University
Detroit, MI 48202
egulari@chem1.eng.wayne.edu

Thomas Hartmann
Environmental Technology Group
Environmental Science & Waste
Technology Division
Los Alamos National Laboratory
Los Alamos, NM 87545
hartmann@lanl.gov

Florence Henon
Atofina
900 First Avenue, P.O. Box 1536
King of Prussia, PA 19406-6970
fhenon@ato.com

Erik Hoggan
Department of Chemical
Engineering
P.O. Box 7006
North Carolina State University
Raleigh, NC 27695
enhoggan@eos.ncsu.edu

J. D. Holmes
Texas Materials Institute
Center of Nano & Molecular
Science & Technology
University of Texas
Austin, TX 78712

Takao Ikariya
Graduate School of Science &
Engineering
Department of Applied Chemistry
Tokyo Institute of Technology
Meguro Ku, 2-12-1
Ookayama, Tokyo 1528552
Japan

Gunilla B. Jacobson
Chemistry Division
Los Alamos National
Laboratory
Los Alamos, NM 87545

Keith P. Johnston
Department of Chemical
Engineering
University of Texas at Austin
Austin, TX 78712
kpj@che.utexas.edu

Edward Kung
Polymer Science & Engineering
Department
University of Massachusetts
Amherst, MA 01003

C. T. Lee
MIT Chemical Engineering
77 Massachusetts Avenue, 66-350
Cambridge, MA 02139-4307

Walter Leitner
Max-Planck-Institut für
Kohlenforschung
Kaiser-Wilhelm-Platz 1
45470 Mülheim an der Ruhr
Germany
leitner@mpi-muelheim.mpg.de

Alan J. Lesser
Polymer Science & Engineering
Department
University of Massachusetts
Amherst, MA 01003

Fuchen Liu
Chemistry Division
Los Alamos National
Laboratory
Los Alamos, NM 87545

Charles W. Manke
Department of Chemical
Engineering & Materials Science
5050 Anthony Wayne Drive
Wayne State University
Detroit, MI 48202
cmanke@che.eng.wayne.edu

Thomas J. McCarthy
Polymer Science & Engineering
Department
University of Massachusetts
Amherst, MA 01003
tmccarthy@polysci.umass.edu

Mark A. McHugh
Department of Chemical Engineering
601 West Main Street
P.O. Box 843028
Virginia Commonwealth University
Richmond, VA 23284-3028
mmchugh@saturn.vcu.edu

Brian Novick
Department of Chemical
Engineering
P.O. Box 7006
North Carolina State University
Raleigh, NC 27695
bjnovick@unity.ncsu.edu

Ryoji Noyori
Department of Chemistry
Nagoya University
Chikusa, Nagoya 464-01
Japan
noyori@chem3.chem.nagoya-u.ac.jp

Peter Panster
Applied Research & Development
Catalysts
Degussa AG
P.O. Box 1345
D-63403 Hanau
Germany

Patricia Paviet-Hartmann
Environmental Technology Group
Environmental Science & Waste
Technology Division
Los Alamos National Laboratory
Los Alamos, NM 87545

David R. Pesiri
Chemistry Division
Los Alamos National Laboratory
Los Alamos, NM 87545

Rebecca L. Rodney
1249 Benedum Engineering Hall
Department of Chemical &
Petroleum Engineering
University of Pittsburgh
Pittsburgh, PA 15261

James B. Rubin
Nuclear Materials Technology
Division
Los Alamos National Laboratory
Los Alamos, NM 87545

Alan J. Russell
1249 Benedum Engineering Hall
Department of Chemical &
Petroleum Engineering
University of Pittsburgh
Pittsburgh, PA 15261
russel@civengl.engrng.pitt.edu

Mark A. Stadtherr
Department of Chemical Engineering
University of Notre Dame
Notre Dame, IN 46556
markst@nd.edu

Gina Stewart
207 Bebington Drive
Cary, NC 27513
gstewart@nc.rr.com

Thomas Tacke
Applied Research & Development Catalysts
Degussa AG
P.O. Box 1345
D-63403 Hanau
Germany
thomas.tacke@degussa.com

J. M. Tanko
107 Davidson Hall
Department of Chemistry
Virgina Polytechnic Institute & State University
Blacksberg, VA 24061-0212
jtanko@vt.edu

Craig M. V. Taylor
Chemistry Division
Los Alamos National Laboratory
Los Alamos, NM 87545
eggus_taylor@lanl.gov

William Tumas
Chemistry Division
Los Alamos National Laboratory
C-SIC, MS J514
Los Alamos, NM 87545
tumas@lanl.gov

Stefan Wieland
Applied Research & Development Catalysts
Degussa AG
P.O. Box 1345
D-63403 Hanau
Germany

Gang Xu
Department of Chemical Engineering
University of Notre Dame
Notre Dame, IN 46556

M. Z. Yates
Department of Chemical Engineering
University of Rochester
Rochester, NY 14627

Jennifer L. Young
Dupont Experimental Station
P.O. Box 80352
Wilmington, DE 19880-0357
Jennifer.L.Young.USA@Dupont.com

Introduction

JENNIFER YOUNG
JOSEPH M. DESIMONE
WILLIAM TUMAS

One of the most exciting challenges that scientists and engineers face today involves using their skills and creativity to design, develop, and implement processes that enable environmental and economic sustainability. Green chemistry involves "carrying out chemical activities—including chemical design, manufacture, use, and disposal—such that hazardous substances will not be used and generated" (Anastas and Williamson, 1998). With increasing emphasis placed on environmental responsibility, there is a clear need for reactions and processes that eliminate or significantly reduce harmful residues such as used organic solvents, salt streams, and dangerous organic or inorganic by-products. The energy required for thermal control, mixing, separation, and purification makes the development of more efficient processes fundamental to designing greener processes. Green chemistry represents a new way of thinking about chemical synthesis and processing by incorporating environmentally responsible principles throughout the product life cycle (Anastas, 1994; Anastas and Williamson, 1998). Green chemistry approaches strive to eliminate or significantly reduce hazards, environmental impacts, waste, and energy usage. A key aspect of green chemistry is pollution prevention, which obviates the need for future costly environmental remediation and becomes increasingly important as environmental regulations become more stringent. Anastas (Anastas and Williamson, 1996) has categorized advances in green chemistry into four general areas: alternative feedstocks or starting materials, alternative reagents or transformations, alternative reaction conditions, and alternative final products or target molecules.

This book describes green chemistry approaches that employ liquid or supercritical carbon dioxide. Supercritical fluids have been investigated for more than a century; however, applications to chemical synthesis and materials processing are relatively new. Since the 1990s, we have witnessed a veritable explosion of increased research activity in the area of chemical reactions and processes in liquid or supercritical carbon dioxide with the development of a number of new concepts and a myriad of emerging applications.

Carbon dioxide, when compressed to a liquid or, above its critical point ($T_c = 31.1\,^\circ\text{C}$, $P_c = 73.8$ bar), to a supercritical fluid, represents an environmentally benign solvent that can have a number of positive impacts on green chemistry. These include solvent replacement, energy efficiency, better separations, or even entirely new processes. Since it is nonflammable, nontoxic, and inexpensive, CO_2 may replace solvents that are carcinogenic, environmentally hazardous, volatile, or legislatively regulated. Since liquid CO_2 has a low heat of vaporization and supercritical CO_2 has zero heat of vaporization, this solvent can be easily and completely removed by depressurization, eliminating the energy-intensive drying process generally used to remove traditional solvents including water. Depressurization and recompression schemes allow for facile separation from other chemical components and therefore recycling. Although CO_2 is a "greenhouse" gas, it can be acquired from natural sources or as a by-product from industrial processes.

The properties and physical chemistry of liquid and supercritical carbon dioxide have been extensively reviewed (Kiran and Brennecke, 1992), as have many fundamentals and applications for separation, chromatography, and extraction (McHugh and Krukonis, 1994). The phase diagram for pure CO_2 is illustrated in Figure I.1. Due to its relatively low critical point, CO_2 is frequently used in the supercritical state. Other common supercritical fluids require higher temperatures and pressures, such as water with $T_c = 374.2\,^\circ\text{C}$ and $P_c = 220.5$ bar, while propane ($T_c = 96.7\,^\circ\text{C}$ and $P_c = 42.5$ bar) and ethane ($T_c = 32.2\,^\circ\text{C}$ and $P_c = 48.8$ bar) have lower critical pressures but are flammable (McHugh and Krukonis, 1994).

Due to its compressibility in the liquid (near the critical point) and in the supercritical fluid state, the dielectric constant and density, and thus the solvent quality of CO_2, are tunable with pressure and temperature (Keyes and Kirkwood, 1930). As illustrated in Figure I.2, this compressibility provides for control of the density and therefore solvent-dependent properties such as dielectric constant and overall solvent strength (Giddings et al., 1968). While supercritical CO_2 can have high liquidlike densities, it shares many of the

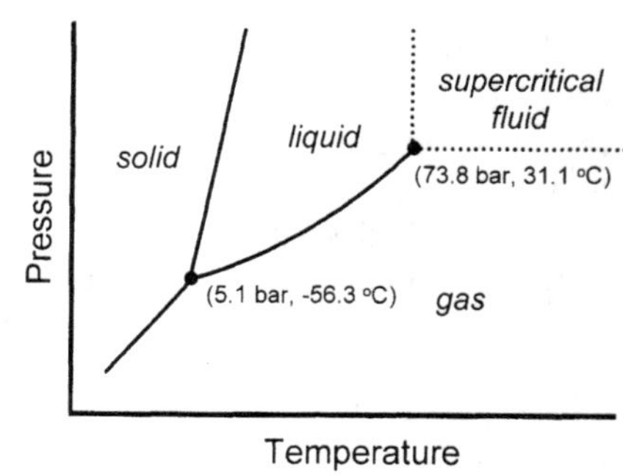

Figure I.1. Phase diagram of pure carbon dioxide.

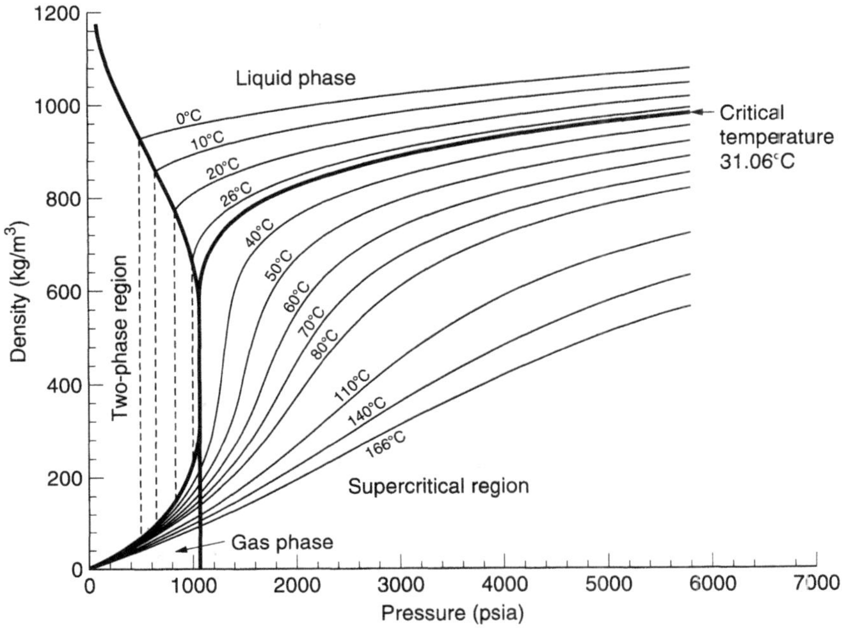

Figure I.2. Plot of density for neat supercritical carbon dioxide as a function of pressure at various temperatures.

processing advantages of gases (Kiran and Brennecke, 1992; McHugh and Krukonis, 1994) including lower viscosity, higher diffusivity, and miscibility with gases including oxygen, carbon monoxide, and hydrogen.

As shown in Figure I.2, the solvent strength of supercritical carbon dioxide approaches that of hydrocarbons or halocarbons. As a solvent, CO_2 is often compared to fluorinated solvents. In general, most nonpolar molecules are soluble in CO_2, while most polar compounds and polymers are insoluble (Hyatt, 1984). High vapor pressure fluids (e.g., acetone, methanol, ethers), many vinyl monomers (e.g., acrylates, styrenics, and olefins), free-radical initiators (e.g., azo- and peroxy-based initiators), and fluorocarbons are soluble in liquid and supercritical CO_2. Water and highly ionic compounds, however, are fairly insoluble in CO_2 (King et al., 1992; Lowry and Erickson, 1927). Only two classes of polymers, siloxane-based polymers and amorphous fluoropolymers, are soluble in CO_2 at relatively mild conditions ($T < 100\,^{\circ}C$ and $P < 350$ bar) (DeSimone et al., 1992, 1994; McHugh and Krukonis, 1994). Accordingly, a number of reactions may be homogeneous, while most polymerizations become heterogeneous as polymer forms. Considerable effort has been placed on both enhancing the solubility of materials in carbon dioxide and developing multiphase processing concepts.

The aim of this book is to illustrate how liquid and supercritical carbon dioxide have provided a powerful tool for the development of green chemical processes. We also intended to capture the excitement of the research activity and new technology in chemistry, chemical engineering, and polymer science, where there are now many examples of using CO_2 as an alternative solvent, reagent, and processing aid. The contributors are leading researchers in the field and their chapters are grouped into three sections: Part I, Catalysis and Chemical Synthesis in Carbon Dioxide; Part II, Polymers in Carbon Dioxide; and Part III, Industrial Processes and Applications Utilizing Carbon Dioxide. Most of the chapters not only highlight the environmental benefits of using CO_2 but also call out other advantages resulting from the low viscosity of CO_2, its miscibility with gases, its inertness to the reaction conditions, and the plasticization of polymers by CO_2. The first part of the book describes the considerable amount of recent research focused on chemical reactions, including catalytic transformations (metal catalysis, catalytic activation for use of CO_2 as a reactant and solvent, and enzymatic transformations) and free-radical reactions.

In Chapter 1, the complex nature of phase behavior of species in CO_2 is introduced by Blanchard et al. Two specific examples of catalyzed reactions are used to illustrate the importance of understanding and modeling phase behavior for determining the feasibility of the reaction and the suitable pressure range for a single-phase system. Chapter 2 reviews advances in homogeneous, heterogeneous, and biphasic catalysis in carbon dioxide. Four general reaction classes are discussed: acid catalysis, reduction by hydrogenation, selective oxidation, and catalytic carbon–carbon bond formation. Chapter 3 continues the catalysis theme by incorporating CO_2 both as a reactant and a solvent. Ikariya and Noyori highlight the catalytic activation of CO_2 in supercritical carbon dioxide, focusing on the hydrogenation of CO_2 and copolymerization of CO_2 with epoxides. In Chapter 4, Tanko covers the generation of free radicals, free-radical chain reactions, termination reactions, inhibition, and electron paramagnetic resonance studies of radicals. Due to the inert nature of CO_2 and the solubility of small organic molecules in it, CO_2 has proven to be an excellent replacement solvent for free-radical reactions. This chapter also helps set the stage for free-radical polymerizations presented in Chapter 9 by Young and DeSimone. Due to the similarities between reactions in fluorous solvents and CO_2, Leitner critically evaluates these two reaction media for similar chemical reactions in Chapter 5. Rodney and Russell finish Part I by reviewing enzyme chemistry and applications in CO_2. The catalytic performance of enzymatic reactions including hydrolysis, esterification, transesterification, and oxidation is dependent on the enzyme, temperature, pressure, and water content.

Part II focuses on polymer science and technology in CO_2. In Chapter 7, McHugh delves deeper into the study of polymer solubility in CO_2, with a particular emphasis on particular acrylates and highly fluorinated polymers at extreme pressures and temperatures. Johnston et al., Chapter 8, review the exciting area of interfacially active surfactants in CO_2 and their use for the formation of emulsions, microemulsions, and dispersion polymerizations. In Chapter 9, Young and DeSimone describe the polymerizations in CO_2 of fluorinated polymers and nonfluorinated polymers, as well as polymer characterization in CO_2. Composites of high-density polyethylene and polystyrene are described in detail in Chapter 10, including their synthesis in CO_2, characterization, and mechanical testing. In Chapter 11, Manke and Gulari examine the rheological properties of poly(dimethyl siloxane) and polystyrene in CO_2. Since many polymers are formed into final products by various melt processing techniques, it is important to study the rheology of polymer melts in the presence of CO_2.

Part III describes specific applications and industrial processes that employ CO_2. In Chapter 12, the topic of coating applications for CO_2 is reviewed, including spray, spin, free meniscus, and impregnation coatings. The spray coating processes have been investigated in the most detail, with the UNICARB process illustrating the industrial feasibility of CO_2 for coatings. Another industrial example of CO_2 is dry cleaning. Stewart, in Chapter 13, elaborates on the CO_2 dry cleaning process that was developed by Micell Technologies and is put into practice by the franchise of Hangers Cleaners. The cleaning is enhanced by detergents designed for CO_2. In Chapter 14, the catalytic hydrogenation of vegetable oils and fatty acids in CO_2 is described by Tacke et al. Finally, the use of CO_2 for enhancing cemented materials for building or encapsulating radioactive waste is explained in Chapter 15 by Rubin et al. One application is a process being commercialized industrially by Supramics that involves the combination of CO_2 and fly ash to modify cement for low-cost building materials. Many of the chapters also describe future directions and other potential applications.

The future of CO_2 technology appears to be bright. We have seen a number of new developments for chemical and materials processing to the point of emerging industrial and commercial applications. As the number of laboratories in academia, government, and industry that are investigating the utilization of CO_2 continues to increase, we expect more applications that lead to greener processes.

References

Anastas, P. A. In *Benign by Design: Alternative Synthetic Design for Pollution Prevention*; Anastas, P. A., Farris, C. A., Eds.; American Chemical Society: Washington, DC, 1994, pp. 2–22.

Anastas, P. A.; Williamson, T. C. In *Green Chemistry: Designing Chemistry for the Environment*; Anastas, P. A., Williamson, T. C., Eds.; ACS Symposium Series 626; American Chemical Society: Washington, DC, 1996, pp. 1–17.

Anastas, P. A.; Williamson, T. C. In *Green Chemistry: Frontiers in Benign Chemical Syntheses and Processes*; Anastas, P. A., Williamson, T. C., Eds.; Oxford University Press: New York, 1998, pp. 1–26.

DeSimone, J. M.; Guan, Z.; Eisbernd, C. S. *Science* **1992**, *257*, 945–947.

DeSimone, J. M.; Maury, E. E.; Menceloglu, Y. Z.; McClain, J. B.; Romack, T. R.; Combes, J. R. *Science* **1994**, *265*, 356–359.

Giddings, J. C.; Myers, M. N.; McLaren, L.; Keller, R. A. *Science* **1968**, *162* (67).

Hyatt, J. A. *J. Org. Chem.* **1984**, *49*, 5097–5101.

Keyes, F. G.; Kirkwood, J. G. *Phys Rev.* **1930**, *36*, 754.

King, M. B.; Mubarak, A.; Kim, J. D.; Bott, T. R. *J. Supercrit. Fluids* **1992**, *5*, 296.

Kiran, E., Brennecke, J. F., Eds. *Supercritical Fluid Engineering Science: Fundamentals and Applications*; ACS Symposium Series 514; American Chemical Society: Washington, DC, 1992.

Lowry, H. H.; Erickson, W. R. *J. Am. Chem. Soc.* **1927**, *49*, 2729.

McHugh, M. A.; Krukonis, V. J. *Supercritical Fluids Extraction: Principles and Practice*, 2nd ed.; Butterworth-Heinemann: Boston, MA, 1994.

PART I

CATALYSIS AND CHEMICAL SYNTHESIS IN CARBON DIOXIDE

1

Phase Behavior and Its Effects on Reactions in Liquid and Supercritical Carbon Dioxide

LYNNETTE A. BLANCHARD

GANG XU

MARK A. STADTHERR

JOAN F. BRENNECKE

Carbon dioxide, either as an expanded liquid or as a supercritical fluid, may be a viable replacement for a variety of conventional organic solvents in reaction systems. Numerous studies have shown that many reactions can be conducted in liquid or supercritical CO_2 (sc CO_2) and, in some cases, rates and selectivities can be achieved that are greater than those possible in normal liquid- or gas-phase reactions (other chapters in this book; Noyori, 1999; Savage et al., 1995). Nonetheless, commercial exploitation of this technology has been limited.

One factor that contributes to this reluctance is the extremely complex phase behavior that can be encountered with high-pressure multicomponent systems. Even for simple binary systems, one can observe multiple fluid phases, as shown in Figure 1.1. The figure shows the pressure–temperature (PT) projection of the phase diagram of a binary system, where the vapor pressure curve of the light component (e.g., CO_2) is the solid line shown at temperatures below T_B. It is terminated by its critical point, which is shown as a solid circle. The sublimation curve, melting curve, and vapor pressure curve of the pure component 2 (say, a reactant that is a solid at ambient conditions) are the solid lines shown at higher temperatures on the right side of the diagram; that is, the triple point of this compound is above T_E. The solid might experience a significant melting point depression when exposed to CO_2 pressure [the dashed–dotted solid/liquid/vapor (SLV) line, which terminates in an upper critical end point (UCEP)]. For instance, naphthalene melts at 60.1 °C under CO_2 pressure (i.e., one might observe a three-phase solid/liquid/vapor system), even though the normal melting point is 80.5 °C (McHugh and Yogan, 1984). To complicate things even further, there will be a region close to the critical point of pure CO_2 where one

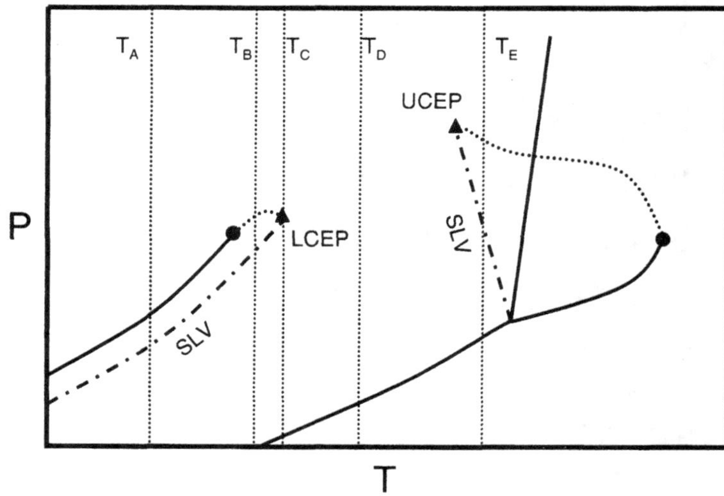

Figure 1.1. The pressure–temperature projection of a typical binary solvent–solute system. See text for discussion. SLV, solid/liquid/vapor; LCEP, lower critical end point; UCEP, upper critical end point.

will observe three phases as well, as indicated by the dashed–dotted SLV line that terminates at the lower critical end point (LCEP). The dotted line connecting the critical point of the light component and the LCEP is a vapor/liquid critical point locus. A much more detailed discussion of systems of the type shown in Figure 1.1 can be found in Xu et al. (2000). Systems of compounds that are liquid at room temperature can also be complicated by the formation of additional liquid phases when placed in equilibrium with CO_2 (see, for example, the binary phase diagrams as classified by van Konynenburg and Scott, 1980).

Nonetheless, understanding high-pressure phase behavior is vitally important to evaluating CO_2 as a potential replacement solvent for reactions. Certainly, commercial reactions are not run at dilute conditions, so the solubility of the reactants, products, and catalysts in the CO_2 (if one desires to run the reaction as a single-phase system) will frequently be the key factor in determining the economic viability of the CO_2-based reaction system.

An additional difficulty in evaluating CO_2 as a potential replacement solvent for reactions is caused by the problems that one may encounter with conventional flash algorithms when attempting to calculate high-pressure phase behavior for any particular equation-of-state model. Conventional algorithms, even in commercial implementations, may fail to converge or may converge to an incorrect solution (Stradi et al., 1999a). Although conventional algorithms may have difficulties with the modeling of normal liquid solutions, this is especially a problem for any supercritical fluid system. This is because convergence near critical points, where there is very little difference in the density and compositions of the two phases, and near three-phase lines is especially challenging. This can be a serious impediment to the design, optimization, and evaluation of a high-pressure reaction system that uses CO_2 as the solvent.

To address these limitations to the commercial evaluation and implementation of CO_2 as a substitute solvent, we (1) present a methodology to measure and model high-pressure phase behavior of CO_2-based reaction systems using minimal experimental data and (2) present a new computational technique for high-pressure phase equilibrium calculations that provides a guarantee of the correct solution to the flash problem.

Systems Studied

We have applied the methodology and newly developed computational techniques to two model reaction systems. Both examples represent classes of reactions that are of significant commercial interest and are reactions that have been shown to occur with good rates and selectivities in CO_2 by researchers at Los Alamos National Laboratories (Pesiri et al., 1998). The first reaction is the epoxidation of *trans*-2-hexen-1-ol, in which the double bond of the allylic alcohol is converted into an epoxy group by addition of an oxygen atom. This is an example of Sharpless chemistry, which produces a high-value, stereospecific product. It constitutes a good target for solvent substitution by carbon dioxide since epoxidation reactions are traditionally performed in organic solvents. The second reaction is an acylation, which is an example of a class of commercially important Friedel–Crafts reactions. The naphthalene and acetyl chloride react to form isomers of acetonaphthone. The full reaction would also require the presence of stoichiometric amounts of $AlCl_3$ or other appropriate catalyst. These reactions are shown in Figure 1.2.

Methodology

As mentioned above, we will present a two-pronged approach to understanding the multicomponent high-pressure phase behavior of potential reaction systems. First, we seek to model the multicomponent phase behavior using limited experimental data. For this effort, we have chosen to use the Peng–Robinson equation of state (Peng and Robinson, 1976) with conventional van der Waals mixing rules as described below. There are many different models that can be used to correlate and predict high-pressure phase behavior, including more complex and more fundamental models. These include other cubic equations of state like the Soave–Redlich–Kwong equation and models like the statistical associating fluid theory

Epoxidation Reaction

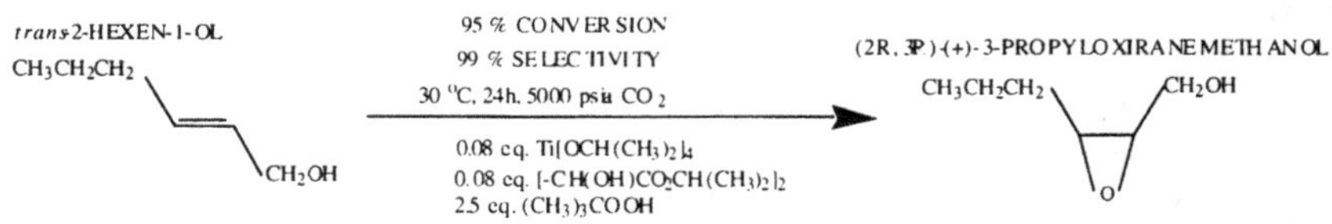

Friedel-Crafts Acylation Reaction

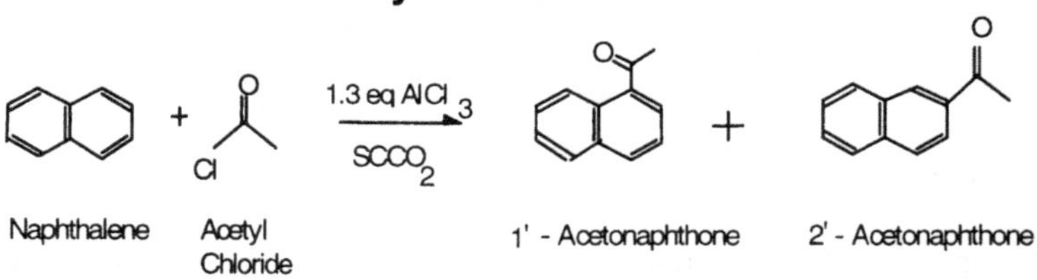

Figure 1.2. Two reactions for which the high-pressure phase behavior is studied.

(SAFT). While some may work better for certain types of systems, in general, the more adjustable parameters, the better the fit one will achieve. However, larger numbers of fit parameters tend to decrease the predictive power of the model. A discussion of the benefits and drawbacks of various models can be found in several reviews (Brennecke and Eckert, 1989; Johnston et al., 1989). We have chosen the Peng–Robinson equation since it is known to give reasonably good representations of solubilities in sc CO_2 (Brennecke and Eckert, 1989; Johnston et al., 1989), because it is relatively simple to use, and because it is readily available for use in industry. The Peng–Robinson equation requires inputs of critical temperatures and pressures, acentric factors (w) and binary interaction parameters, k_{ij}. The pure component properties can be taken from the literature or estimated (Reid et al., 1987; Stradi et al., 1998, 2001a, 2001b) but the binary interaction parameters must be fit to experimental data. Thus, our methodology is to measure binary phase behavior if it is not already available in the literature, to find the best-fit k_{ij} values to that data, and to use those binary interaction parameters to estimate the multicomponent phase equilibria.

The completely reliable computational technique that we have developed is based on interval analysis. The interval Newton/generalized bisection technique can guarantee the identification of a global optimum of a nonlinear objective function, or can identify all solutions to a set of nonlinear equations. Since the phase equilibrium problem (i.e., particularly the phase stability problem) can be formulated in either fashion, we can guarantee the correct solution to the high-pressure flash calculation. A detailed description of the interval Newton/generalized bisection technique and its application to thermodynamic systems described by cubic equations of state can be found in a variety of publications (Hua et al., 1996, 1998a, 1998b, 1999; Xu et al., 2000). In these same publications can be found a discussion of work by other researchers who apply global optimization techniques to phase stability and phase split calculations, in an effort to guarantee correct solutions to the thermodynamic models. However, thus far the method described here is the only general purpose method that can be applied to any equation-of-state model, i.e., the types of models that are needed to described high-pressure phase equilibria of CO_2-based systems.

Experimental Techniques

A discussion of a variety of techniques to measure high-pressure phase equilibria can be found in a review by Dohrn and Brunner (1995). Here, we use two different apparatuses. The first is a static high-pressure equilibrium apparatus, which has been described previously (Stradi et al., 1998). It consists of a high-pressure glass tube, into which a liquid sample can be loaded. Carbon dioxide is carefully metered into the glass cell and, by assuming that the gas phase is essentially pure CO_2, one can determine the composition of the liquid phase as a function of pressure and temperature. The apparatus can also be used to investigate vapor/liquid/liquid equilibria and solid/liquid/vapor equilibria. The second apparatus is a standard dynamic-flow ISCO 220 SX extractor (ISCO, Inc., Lincoln, NE), which is used to determine compositions in the CO_2-rich vapor or fluid phase. Either a solid or liquid solute can be loaded into the extractor, through which CO_2 is passed from an ISCO 260 syringe pump (ISCO, Inc., Lincoln, NE). The saturated solution is passed through a restrictor and the precipitated solute collected, usually in a liquid collection solvent that is analyzed by UV-visible spectrometry. Thus, these two apparatuses are used in a complementary fashion to obtain full information about the high-pressure phase behavior. The chemicals used were obtained from Aldrich Chemicals (Milwaukee, WI) and were used as received (purities: acetyl chloride, 99+%; 1′-acetonaphthone, 98%; 2′-acetonaphthone, 99%). The CO_2 was Coleman Instrument Grade with a minimum purity of 99.99% and was obtained from Mittler Gas Supply (South Bend, IN).

Modeling

As mentioned earlier, we have chosen to model the high-pressure phase behavior using the Peng–Robinson equation (Peng and Robinson. 1976) with standard van der Waals mixing rules:

$$P = \frac{RT}{(v-b)} - \frac{a}{[v(v+b)+b(v-b)]}$$

where

$$a = \frac{0.45724 R^2 T_c^2}{P_c}[1 + (0.3764 + 1.54226w - 0.2699w^2)(1 - T_r^{0.5})]^2$$

$$b = 0.07780 \frac{RT_c}{P_c}$$

and w is the acentric factor. T_c and P_c are the critical temperature and pressure of the compound, respectively, and $T_r = T/T_c$. To extend this equation to mixtures, the conventional van der Waals mixing rules were used:

$$a = \sum_{i=1}^{n}\sum_{j=1}^{n} x_i x_j a_{ij}, \qquad b = \sum_{i=1}^{n} x_i b_i, \qquad \text{and} \qquad a_{ij} = (a_{ii}a_{jj})^{0.5}(1 - k_{ij})$$

where the sums extend over all components, and a_{ii} and b_i indicate the pure component values for component i.

The key to obtaining good representation of experimental data is fitting a single binary interaction parameter, k_{ij}, to each set of binary data. In most of the modeling described here, our main concern lies with the binary interaction parameters between each of the components and CO_2, since CO_2 introduces the most asymmetry (i.e., difference in size and energy parameters) into the system. In a few cases, we include nonzero binary interaction parameters for some of the other components.

Computational Techniques

The phase equilibrium problem consists of two parts: the phase stability calculation and the phase split calculation. For a particular total mixture composition, the phase stability calculation determines if that "feed" will split into two or more phases. If it is determined that multiple phases are present, then one performs the phase split calculation, assuming some specified number of phases. One must then calculate the stability of the solutions to the phase split to ascertain that the assumed number of phases was correct. The key to this procedure is performing the phase stability calculation reliably. Unfortunately, this problem—which can be formulated as an optimization problem (or the equivalent set of nonlinear equations)—frequently has multiple minima and maxima. As a result, conventional phase equilibrium algorithms may fail to converge or may converge to the wrong solution.

We have applied a global optimization technique, based on interval analysis, to the high-pressure phase equilibrium problem (INTFLASH). It does not require any initial guesses and is guaranteed, both mathematically and computationally, to converge to the correct solution. The interval analysis method and its application to phase equilibria using equation-of-state models has been described elsewhere (Hua et al., 1996, 1998a, 1998b, 1999; Xu et al., 2000). It is a general-purpose technique that can be applied to any equation-of-state or excess Gibbs free energy model, and it guarantees correct solution to the phase equilibrium problem.

In addition to the interval method developed, we also used standard modeling tools from Aspen Plus (Aspen Technology, Inc., Williamsville, NY), including a routine to fit binary interaction parameters and two flash algorithms—the FLASH3 module and the RGIBBS module. We also used the two-phase flash routine LNGFLASH, which employs Michelsen's well-known approach, from the IVC-SEP package (Hytoft and Gani, 1996).

Results and Discussion

Epoxidation Reaction

For the epoxidation of *trans*-2-hexen-1-ol to (2*R*,3*R*)-(+)-3-propyloxiranemethanol, we have measured the high-pressure phase behavior of each of the reactants, products, and catalysts in CO_2 and modeled them quite well with the Peng–Robinson equation, even in the cases where we observed vapor/liquid/liquid equilibria (Stradi et al., 1998).

Unfortunately, even in modeling these binary systems, we experienced some computational difficulties when attempting to use Aspen Plus and LNGFLASH (Stradi et al., 2001a). This was especially pronounced at conditions close to the formation of three phases (vapor/liquid/liquid). For instance, for the *trans*-2-hexen-1-ol/CO_2 system at a feed composition of 0.8 mole fraction CO_2, a temperature of 303.15 K, and a pressure of 71.00725 bar, LNGFLASH and both Aspen Plus modules converged, but to incorrect solutions. At just a slightly higher pressure (71.00826 bar), LNGFLASH converged correctly but the Aspen Plus modules still converged to the wrong answer. At a feed composition of 0.7 mole fraction CO_2, $T = 303.15$ K, and $P = 70.09$ bar, LNGFLASH and both Aspen Plus modules indicate no phase split when, in fact, correct solution to the model yields two phases. In all cases, the newly developed routine, INTFLASH, identified the correct solutions to the model without any difficulty.

Based on the binary measurements and modeling, we estimated the multicomponent high-pressure phase behavior for the reaction mixture, extending the entire way from all reactants to full conversion (Stradi et al., 2001a). These calculations suggested that at 40 °C and 150 bar, the reaction mixture would remain single phase. However, if the reaction were run at 100 bar the reactant mixture would be two phase, and at 50 bar the system would be two phase all the way from reactants to full conversion. Thus, by maintaining the pressure above just 125 bar, one would expect the reaction mixture to remain single phase throughout the reaction. This pressure is significantly below the 346 bar used originally for this reaction. Subsequent investigations by researchers at Los Alamos showed high conversions and selectivities at lower pressures. Thus, by using modeling tools to interactively guide experimental work, an improved design was achieved that uses a much lower pressure than originally proposed.

Friedel–Crafts Reaction

Friedel–Crafts alkylation and acylation reactions are ubiquitous in industrial practice and are conducted in a variety of volatile organic solvents, since solvent polarity can be used to control product distribution. Tumas and co-workers (W. Tumas, personal communication, 1998) conducted successful preliminary investigations of the acylation of naphthalene with acetyl chloride in sc CO_2. We have measured the binary phase behavior of acetyl chloride and each of the products, 1′-acetonaphthone and 2′-acetonaphthone, with CO_2 over a wide range of pressures at temperatures of 40 °C and 50 °C. Data for naphthalene is available in the literature (McHugh and Paulaitis, 1980; McHugh and Yogan, 1984; Najour and King, 1966; Tsekhanskaya et al., 1964). For this demonstration, we have not included the catalyst, $AlCl_3$; however, its impact on the mixture phase behavior will be discussed later.

Naphthalene is a solid at room temperature [melting point of 80.5 °C (Weast, 1983)] so it is likely to exhibit solid/fluid equilibria at 40 °C and 50 °C; that is, these are likely to be temperatures like T_D shown in Figure 1.1, which are above the LCEP and below the UCEP. In fact, the UCEP for naphthalene/CO_2 is 60.1 °C (Lamb et al., 1986; McHugh and Yogan, 1984) and this is adequately modeled by the Peng–Robinson equation of state using a k_{ij} of 0.0974. This is the value that provides the best fit to the solid/fluid equilibria data at 55 °C. A detailed description of the modeling of this particular system with the Peng–Robinson equation can be found in Xu et al. (2000).

Acetyl chloride is a liquid at room temperature so we measured the solubility of CO_2 in the acetyl-chloride-rich liquid phase as a function of pressure at both 40 °C and 50 °C and these results are shown in Figure 1.3. Also shown in the figure is the modeling done with the Peng–Robinson equation of state. Using binary interaction parameters of 0.0196 at 40 °C and −0.0138 at 50 °C, we are able to obtain very good representation of the liquid-phase compositions. Also shown on the graph are the predicted values of the vapor-phase compositions that would be in equilibrium with the liquid phases measured. As shown in the graph, a binary mixture of CO_2 and acetyl chloride of any composition at 50 °C would produce a single-phase mixture at pressures above around 90 bar.

1′-Acetonaphthone is a liquid at room temperature (melting point 10.5 °C) and our measurements of the solubility of CO_2 in 1′-acetonaphthone liquid at 40 °C and 50 °C are shown in Figure 1.4. As with acetyl

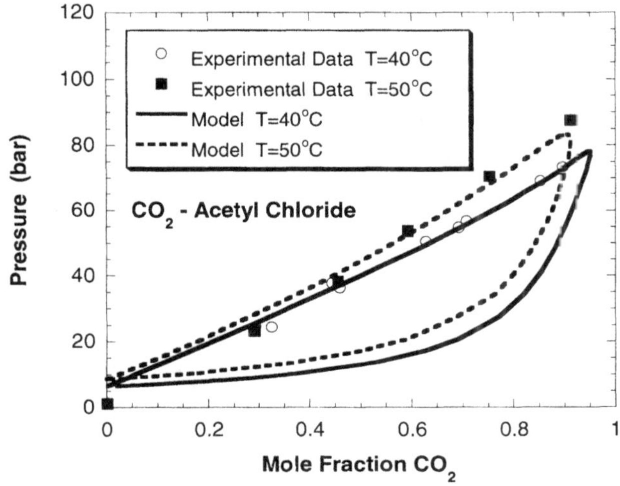

Figure 1.3. Vapor/liquid equilibrium for acetyl chloride and CO_2 at 40 °C and 50 °C. The symbols indicate experimental measurements and the lines indicate the results of modeling with the Peng–Robinson equation of state.

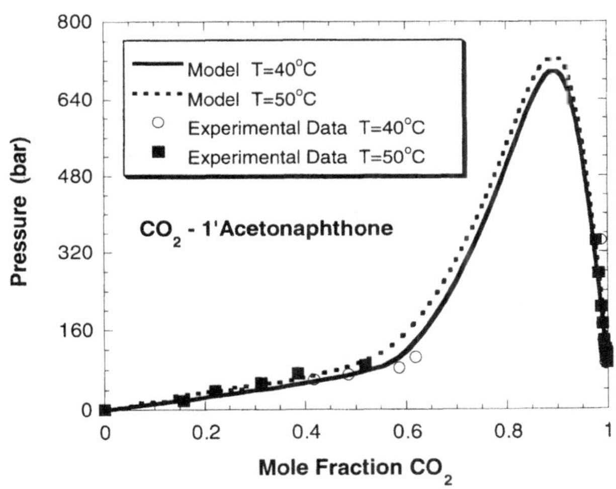

Figure 1.4. Vapor/liquid equilibrium for 1′-acetonaphthone and CO_2 at 40 °C and 50 °C. The symbols indicate experimental measurements and the lines indicate the results of modeling with the Peng–Robinson equation of state.

chloride, these measurements were taken with the static high-pressure phase equilibrium apparatus. However, for this system, we also measured the solubility of the 1′-acetonaphthone in the CO_2-rich vapor phase using the dynamic extraction apparatus at pressures up to 340 bar. Also shown in the figure is the Peng–Robinson equation-of-state modeling using a binary interaction parameter of 0.0687 at 40 °C and 0.0715 at 50 °C. These values gave the best fit to the liquid-phase composition data; that is, we did not use the vapor-phase measurements in obtaining this value. Yet, the Peng–Robinson equation, with a k_{ij} determined from liquid-phase compositions, gives remarkably good estimates of the vapor-phase compositions, as shown in the figure. The model suggests, and the experimental measurements support, that there is an extremely large two-phase envelope that extends to very high pressure. In other words, to obtain a single-phase system for CO_2/1′-acetonaphthone mixtures over the whole composition range at 40 °C or 50 °C, one would have to operate at prohibitively high pressures of greater than about 700 bar.

2′-Acetonaphthone is a solid at room temperature, with a melting point of 53–55 °C. However, at 40 °C or 50 °C it is very likely to melt under CO_2 pressure; that is, these temperatures are likely to be above the UCEP, as shown by T_E in Figure 1.1. Moreover, equimolar mixtures of 1′- and 2′-acetonaphthone are liquid even at room temperature. Thus, we sought to measure the vapor/liquid equilibrium of 2′-acetonaphthone with CO_2. This was easily achieved using both the static and dynamic apparatuses. For instance, 2′-acetonaphthone was melted to introduce it into the glass cell of the static apparatus. It did not resolidify when the cell was cooled to 50 °C; this type of subcooling is quite common. We then added CO_2 and measured its solubility in the liquid at 50 °C and these data are shown in Figure 1.5. Also shown are the solubilities of 2′-acetonaphthone in the CO_2-rich vapor-phase. Once again, the Peng–Robinson equation was fit to just the liquid-phase compositions ($k_{ij} = 0.0695$), yet it provided excellent estimates of the vapor-phase compositions. Once again, the model indicates the existence of a large two-phase envelope; pressures of 680 bar would be required to achieve single-phase mixtures across the whole composition range.

Based on these binary measurements (and the k_{ij} values determined from them), we have performed phase equilibrium computations for the multicomponent mixtures that would result in the course of this reaction system. Assuming feed compositions consisting of 90 mol % CO_2, 5 mol % naphthalene, and 5 mol % acetyl chloride, we estimate that pressures greater than about 165.5 bar would be required to achieve a single-phase reactant mixture at 50 °C. However, if the reaction proceeded to 100% conversion at this pressure and temperature, assuming formation of equal amounts of the two isomers, the system would clearly be two-phase. In fact, modeling suggests that pressures greater than 603.3 bar would be required to solubilize a mixture of 10 mol % 1′- and 2′-acetonaphthone in CO_2. Phase equilibrium calculations indicate that the mixture would have split into two phases

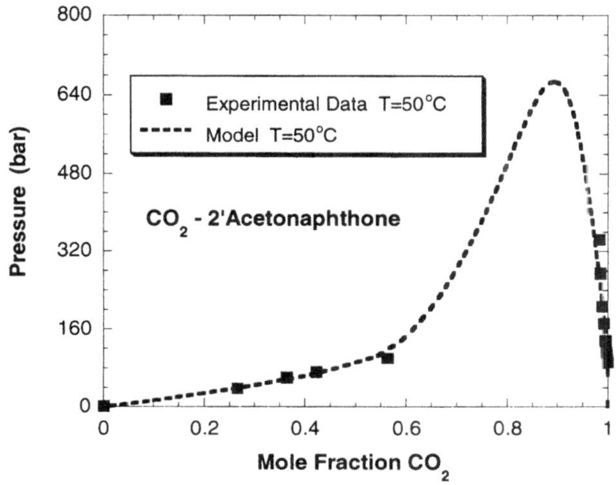

Figure 1.5. Vapor/liquid equilibrium for 2′-acetonaphthone and CO_2 at 50 °C. The symbols indicate experimental measurements and the lines indicate the results of modeling with the Peng–Robinson equation of state.

(vapor and liquid), even at 50% conversion, if one were to operate at 50 °C and 306.8 bar and start with 5 mol % of each of the reactants. The critical temperatures, pressures, and acentric factors for all of the components, as well as their binary interaction parameters with CO_2, that were used in these calculations, are shown in Table 1.1. Note that in performing the multi-component modeling, we used binary interaction parameters for each component with CO_2 but assumed that all other binary interaction parameters were zero, except the one for the 1′-acetonaphthone/2′- acetonaphthone pair. A value of 0.2013 provided a best fit to ternary 1′-acetonaphthone/2′-acetonaphthone/CO_2 vapor/liquid equilibrium data (not shown here).

Thus, the reaction of naphthalene with acetyl chloride to form acetonaphthone in CO_2 would require either prohibitively high pressures (>600 bar) or extremely dilute (well less than 1 mol % reactants and

Table 1.1. Physical Properties Used in Equation-of-State Modeling

	T_c (K)	P_c (bar)	w	k_{ij} with CO_2, T = 50 °C
Carbon dioxide	304.2	73.80	0.225	—
Acetyl chloride	409.6	58.68	0.344	−0.0138
Naphthalene	748.4	40.49	0.302	0.0974
1′-Acetonaphthone[a]	812.5	32.80	0.572	0.0715
2′-Acetonaphthone[a]	812.5	32.80	0.572	0.0695

[a]Nonzero k_{ij} of 0.2013 for 1′-acetonaphthone/2′-acetonaphthone also used in modeling.

products) concentrations if single-phase operation was desired. Since both extremely high-pressure and very large moderate-pressure vessels would require significant capital investment, these options are not likely to be economically attractive. Two-phase operation would require excellent mixing to eliminate mass-transfer resistances. It should be noted that in the real reaction system, two-phase operation would probably be unavoidable. In this example, we have neglected to include the $AlCl_3$ catalyst. The $AlCl_3$ actually forms a complex with the acetyl chloride and is required to be present in greater than stoichiometric amounts. We attempted to measure the phase behavior of the acetyl chloride/$AlCl_3$ complex with CO_2. Preliminary investigations revealed that the complex is highly insoluble in CO_2 and, when a 1.3:1 $AlCl_3$/acetyl chloride ratio is used, actually forms two liquid phases when exposed to CO_2 pressure. Since all of the components of this reaction can be solubilized in various liquid solvents, this work suggests that the Friedel–Crafts acylation of naphthalene may not be a good candidate for solvent substitution with CO_2 due to the complex, multiphase system that would result.

Conclusions

Phase behavior is an extremely important issue in designing and evaluating processes that use CO_2 as a replacement solvent. Evaluation of systems that might exhibit complex high-pressure phase behavior is further frustrated by the inability of conventional flash algorithms and process modeling tools to reliably compute the phase behavior given a particular model. Here, we have presented a methodology to model and compute complex high-pressure phase behavior for reaction systems, taking a limited amount of binary experimental data. We have developed a completely reliable computational technique for performing the flash calculations, based on interval mathematics. We have demonstrated this methodology for two systems: the allylic epoxidation of *trans*-2-hexen-1-ol and the Friedel–Crafts acylation of naphthalene. Based on model calculations, we recommend that the epoxidation reaction could be run efficiently as a single-phase system in CO_2 at pressures as low as 125 bar. Conversely, it appears that the Friedel–Crafts reaction would have to be performed under multiphase conditions, which is likely to make it a poor candidate for solvent substitution with CO_2.

References

Brennecke, J. F.; Eckert, C. A. Phase Equilibria for Supercritical Fluid Process Design. *AIChE J.* **1989**, *35*, 1409–1427.

Dohrn, R.; Brunner, G. High-Pressure Fluid-Phase Equilibria—Experimental Methods and Systems. *Fluid Phase Equilib.* **1995**, *106*, 213–282.

Hua, Z.; Brennecke, J. F.; Stadtherr, M. A. Reliable Prediction of Phase Stability Using an Interval-Newton Method. *Fluid Phase Equilib.* **1996**, *116*, 52–59.

Hua, J. Z.; Brennecke, J. F.; Stadtherr, M. A. Reliable Computation of Phase Stability Using Interval Analysis: Cubic Equation of State Models. *Comput. Chem. Eng.* **1998a**, *22*, 1207–1214.

Hua, J. Z.; Brennecke, J. F.; Stadtherr, M. A. Enhanced Interval Analysis for Phase Stability: Cubic Equation of State Models. *Ind. Eng. Chem. Res.* **1998b**, *37*, 1519–1527.

Hua, J. Z.; Maier, R. W.; Tessier, S. R.; Brennecke, J. F.; Stadtherr, M. A. Interval Analysis for Thermodynamic Calculations in Process Design: A Novel and Completely Reliable Approach. *Fluid Phase Equilib.* **1999**, *158–160*, 607–615.

Hytoft, G.; Gani, R. IVC-SEP Program Package, Danmarks Tekniske Universitet, Lyngby, Denmark, 1996.

Johnston, K. P., Peck, D. G.; Kim, S. Modeling Supercritical Mixtures—How Predictive Is It? *Ind. Eng. Chem. Res.* **1989**, *28*, 1115–1125.

Lamb, D. M., Barbara, T. M.; Jonas, J. NMR Study of Solid Naphthalene Solubilities in Supercritical Carbon Dioxide Near the Upper Critical End Point. *J. Phys. Chem.* **1986**, *90*, 4210–4215.

McHugh, M.; Paulaitis, M. E. Solid Solubilities of Naphthalene and Biphenyl in Supercritical Carbon Dioxide. *J. Chem. Eng. Data* **1980**, *25*, 326–329.

McHugh, M. A.; Yogan, T. J. Three-Phase Solid–Liquid–Gas Equilibria for Three Carbon Dioxide–Hydrocarbon Solid Systems, Two Ethane–Hydrocarbon Solid Systems, and Two Ethylene–Hydrocarbon Solid Systems. *J. Chem. Eng. Data* **1984**, *29*, 112–115.

Najour, G. C.; King, A. D. Solubility of Naphthalene in Compressed Methane, Ethylene, and Carbon Dioxide. Evidence for a Gas-Phase Complex Between Naphthalene and Carbon Dioxide. *J. Chem. Phys.* **1966**, *45*, 1915–1921.

Noyori, R., Ed. Special Issue on Supercritical Fluids. *Chem. Rev.* **1999**, *99*, 353–634.

Peng, S. Y.; Robinson, D. B. A New Two Constant Equation of State. *Ind. Eng. Chem. Fundam.* **1976**, *8*, 59–64.

Pesiri, D. R., Morita, D. K., Glaze, W.; Tumas, W. Selective Epoxidation in Dense Phase Carbon Dioxide. *Chem. Commun.* **1998**, *9*, 1015–1016.

Reid, R. C.; Prausnitz, J. M.; Poling, B. E. *The Properties of Gases and Liquids*, 4th ed., McGraw-Hill: New York, 1987.

Savage, P. E.; Gopalan, S.; Mizan, T. I.; Martino, C. J.; Brock, E. E. Reactions at Supercritical Conditions—Applications and Fundamentals. *AIChE J.* **1995**, *41*, 1723–1778.

Stradi, B. A.; Kohn, J. P.; Stadtherr, M. A.; Brennecke, J. F. Phase Behavior of the Reactants, Products and Catalysts Involved in the Allylic Epoxidation of *trans*-2-Hexen-1-ol to (2*R*,3*R*)-(+)-3-Propyloxiramethanol in High Pressure Carbon Dioxide. *J. Supercrit. Fluids* **1998**, *12*, 109–122.

Stradi, B. A.; Brennecke, J. F.; John, J. P.; Stadtherr, M. A. Reliable Computation of Mixture Critical Points. *AIChE J.*, **2001a**, *47* (1), 212–221.

Stradi, B. A.; Stadtherr, M. A.; Brennecke, J. F. Multicomponent Phase Equilibrium Measurements and Modeling for the Allylic Epoxidation of *trans*-2-hexen-1-ol to (2*R*,3*R*)-(+)-3-Propyloxiranemethanol in High-Pressure Carbon Dioxide. *J. Supercrit. Fluid* **2001b**, *20* (1), 1–13.

Tsekhanskaya, Yu. V.; Iomtev, M. B.; Mushkina, E. V. Solubility of Naphthalene in Ethylene and Carbon Dioxide Under Pressure. *Russ. J. Phys. Chem.* **1964**, *38*, 1173–1176.

van Konynenburg, P. H.; Scott, R. L. Critical Lines and Phase Equilibria in Binary van der Waals Mixtures. *Philos. Trans. R. Soc. London, Ser. A* **1980**, *298*, 495–540.

Weast, R. D., Ed. *CRC Handbook of Chemistry and Physics*. 64th ed. CRC Press; Boca Raton, FL, 1983.

Xu, G.; Scurto, A. M.; Castier, M.; Brennecke, J. F.; Stadtherr, M. A. Reliable Computation of High Pressure Solid-Fluid Equilibrium. *Ind. Eng. Chem. Res.* **2000**, *39* (6), 1624–1636.

2

Advances in Homogeneous, Heterogeneous, and Biphasic Metal-Catalyzed Reactions in Dense-Phase Carbon Dioxide

---

TAKAO IKARIYA

RYOJI NOYORI

MICHAEL B. ABRAMS

CHARLES A. G. CARTER

GUNILLA B. JACOBSON

FUCHEN LIU

DAVID R. PESIRI

WILLIAM TUMAS

The use of compressed carbon dioxide as a reaction medium, either as a liquid or a supercritical fluid (sc CO_2), offers the opportunity not only to replace conventional hazardous organic solvents but also to optimize and potentially control the effect of solvent on chemical synthesis. Although synthetic chemists, particularly those employing catalysis, may be relative latecomers to the area of supercritical fluids, the area of catalysis in carbon dioxide has grown significantly since around 1975 to the point that a number of excellent reviews have appeared (Baiker et al., 1999; Buelow et al., 1998; Jessop and Leitner, 1999; Jessop et al., 1995c, 1999; Morgenstern et al., 1996). Developing and understanding catalytic processes in dense-phase carbon dioxide could lead to "greener" processing at three levels: (1) solvent replacement, (2) improved chemistry (e.g., higher reactivity, selectivity, less energy), and (3) new chemistry (e.g., use of CO_2 as a C-1 source). In this chapter, we will highlight a number of examples from the literature in homogeneous and heterogeneous transition-metal catalysis, as well as the emerging area of biphasic catalysis in H_2O/sc CO_2 mixtures. The intent is to provide an illustrative rather than a comprehensive overview to four classes of catalytic transformations: acid catalysis, reduction via hydrogenation, selective oxidation catalysis, and catalytic carbon–carbon bond-forming reactions. The reader is referred to other chapters in this book and other reviews (King and Bott, 1993) for discussion of uncatalyzed reactions, phase-transfer catalysis, polymerization, and radical reactions in sc CO_2.

From a synthetic chemist's viewpoint, sc CO_2 has a number of potential advantages that one would like to capitalize upon.

- Solvent Replacement Carbon dioxide is a nontoxic, nonflammable, inexpensive alternative to hazardous organic solvents. Simple solvent replacement will not be a sufficient driver for all chemical reactions; however, as described below, the use of carbon dioxide could lead to better chemistry for certain reactions.
- Gas Miscibility Gases such as H_2, O_2, and CO are sparingly soluble in liquid solvents but they are highly miscible with sc CO_2. For example, the concentration of H_2 in a supercritical mixture of 85 bar of H_2 and 120 bar of CO_2 at 50 °C is 3.2 M, while the concentration of H_2 in tetrahydrofuran (THF) under the same pressure is merely 0.4 M (Jessop et al., 1996). The high miscibility of these gases could enhance catalytic hydrogenation, oxidation, or carbonylation, particularly for those reactions that are nonzero order in the concentration of the gaseous reagent.
- Enhanced Mass Transfer, Diffusivity Supercritical fluids share many of the advantages of gases, including lower viscosities and higher diffusivities relative to liquid solvents, thereby potentially providing the opportunity for faster rates, particularly for diffusion-limited reactions.
- Selectivity Enhancement and Tunable Solvent Properties The high compressibility of supercritical fluids allows for control of their densities (and therefore their density-dependent properties, including the dielectric constant, viscosity, and overall solvent strength) through changes in temperature and pressure. The ability to "tune" solvent properties could lead to higher reaction rates and/or selectivity with potentially small changes in operating conditions.
- Oxidative Stability The fact that CO_2 cannot be further oxidized makes it an ideal candidate for carrying out catalytic oxidation chemistry. The enhanced thermal conductivity of sc CO_2 relative to organic solvents suggests that it could also act as an efficient solvent for buffering heat transfer (especially relative to gas-phase reactions), even for highly exothermic reactions.
- Enhanced Separations/Recovery Supercritical fluids also offer the ability to enhance separations through pressure or temperature "tuning" of the solvent properties, as demonstrated by the vast literature on extraction and supercritical fluid chromatography (King and Bott, 1993; McHugh and Krukonis, 1994). Running reactions in sc CO_2 may help solve the often-encountered problem of recovery of the expensive transition-metal-containing species and products in solution. After completing the reaction, controlled pressure reduction could be used to selectively precipitate the catalyst precursor/resting catalyst, a product, or unreacted starting material (Kainz et al., 1999).
- Catalyst Lifetime The ability of dense-phase fluids to dissolve organic substrates and coke precursors can lead to longer catalyst lifetimes relative to gas-phase heterogeneous catalysis.

Experimental Considerations/Limitations

Despite its potential advantages as a solvent for synthetic chemistry, the use of CO_2 has some inherent limitations. Even at relatively high densities, dense-phase CO_2 has a low dielectric constant and a correspondingly limited range of the Hildebrand solubility parameter (<9) (Morgenstern et al., 1996). It therefore exhibits solvent features similar to conventional nonpolar solvents, such as hydrocarbons, fluorocarbons, or other halocarbons. Many substrates and reagents, particularly polar or high-molecular-weight species, may have limited solubility. For catalytic applications, careful attention must also be paid to the solubility of ligands and transition-metal complexes. A number of approaches have been employed to enhance catalyst solubility, including attaching highly CO_2-philic moieties such as fluoroalkyl, polysiloxane, or alkoxy groups to ligands (see Chapter 5, this book; Kayaki et al., 1999) or using lipophilic counterions for charged species (Burk et al., 1995; Chandler et al., 1998). The electrophilic nature of CO_2 may render it a non-innocent solvent. For example, it can react with Lewis or Brønsted bases with a $pK_b > 9$, such as organic amines.

The typical operating conditions for reactions in CO_2 (pressures between 30 and 700 bar, and temperatures up to ca. 130 °C) require the use of pressure equipment and safety protocols. A number of commercial autoclaves or custom high-pressure stainless steel or specialty steel reactors with sapphire view windows have been developed to investigate reactions in dense-phase CO_2 (King and Bott, 1993). This equipment allows for the ability to sample reaction mixtures at kinetically relevant time-scales, accurate control of reaction conditions (i.e., temperature, pressure, residence time, mixing intensity), and direct visual observation of the reaction, to insure system homogeneity and to monitor either visually or spectroscopically the phase behavior and reaction progress. A number of high-pressure cells have been used for in situ spectroscopy, including Fourier transform infrared (FTIR), UV-vis, Raman, NMR, neutron scattering, and time-resolved laser photolysis (Jessop and Leitner, 1999).

The phase behavior of reaction media involving sc CO_2 can be complex (see Chapter 1 in this book), particularly when using gaseous reagents and high concentrations of substrate, conditions that may be required for useful productivities. Typically, the reaction mixtures can be viewed through a sapphire window of the reactor to give an indication of a homogeneous solution, phase separation, or precipitation of catalysts, substrates, or products. Although complete phase diagrams are available for only a few binary mixtures of solutes and sc CO_2, at least some experimental data concerning the phase behavior of sc CO_2 solutions of many organic compounds is available.

Homogeneous Catalysis

Homogeneous catalysts, where the catalyst is in the same solution phase as the reactants, can easily be modified by ligand design and their structure and reaction pathways can be characterized in detail by a range of spectroscopic techniques. The ability to control the stereoelectronic environment around transition-metal centers in homogeneous catalysts through ligand design and control has led to remarkable developments over the last 30 years in homogeneously catalyzed reactions (Collman et al., 1987; Cornils and Herrmann, 1996; Parshall and Ittel, 1992). New ligands, including chiral ligands, have led to unprecendented regio-, stereo- and even enantio-selectivities. Table 2.1 summarizes homogeneous reactions in dense-phase fluids.

Acid Catalysis

Acid-catalyzed transformations, such as hydrocarbon alkylation and isomerization of organic compounds, are very important industrial processes. In a 1975 U.S. patent, Kramer and Leder describe a clear benefit of the use of sc CO_2 for the Lewis acid catalyzed isomerization of *n*-hexane using $AlBr_3$, $AlCl_3$, or BF_3 in the presence of HCl, HBr, or H_2O (eq. 2.1). The addition of hydrogen gas allows for more isomerization relative to cracking to lower alkanes. In neat hexane, cracking to lower-molecular-weight alkanes competed with isomerization (selectivity ca. 1:1), whereas isomerization was five times faster than cracking in sc CO_2. This study represents the first report of the beneficial effects on a homogeneous catalytic reaction due to higher solubilities of gases in sc CO_2. Although the conversion level is essentially the same as those achieved in heterogeneous liquid reactant/solid catalyst conditions, the ease of separation of the reaction products by pressure or temperature change gives the supercritical process an additional important advantage.

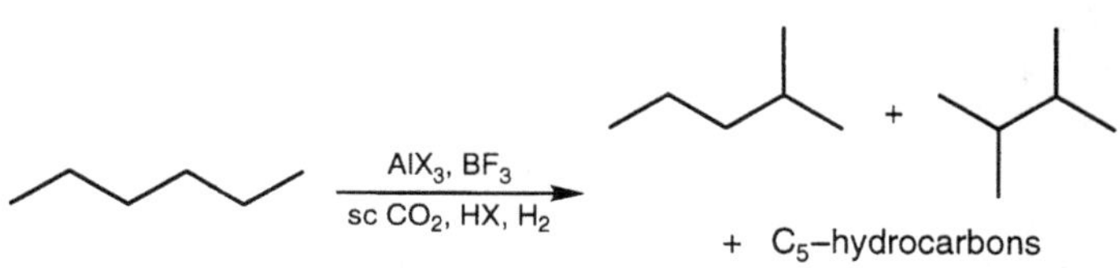

(2.1)

Two reports demonstrate that Brønsted (Olah et al., 1999) or Lewis acids (Oakes et al., 1999) may be more reactive in sc CO_2 than in organic solvents. The isobutane–isobutylene alkylation (eq. 2.2), which is used commercially to increase the octane numbers in automotive fuels, has been demonstrated to occur in sc CO_2 with several liquid acid catalysts (Olah et al., 1999). Increased selectivity for C_8-alkylates is observed in sc CO_2 relative to neat acid media when anhydrous HF, pyridine poly(hydrogen fluoride) (PPHF),

Table 2.1. Homogeneous Catalysis in Liquid or Supercritical Carbon Dioxide

Reaction	Catalyst/Ligand	Advantage(s)	Reference(s)
Acid Catalysis			
n-Hexane isomerization to methylpentanes	$AlBr_3$, $AlCl_3$, BF_3	Selectivity enhancement Gas miscibility	Kramer and Leder (1975)
Diels–Alder reaction	$Sc(OTf)_3$	Selectivity enhancement Solvent tunability	Clifford et al. (1998) Oakes et al. (1999)
Isobutane–isobutylene alkylation	HF, PPHF, H_2SO_4	Solvent replacement Increased reactivity	Olah et al. (1999)
Friedel–Crafts alkylation and acylation of aromatics	$AlCl_3$, $AlBr_3$		Buelow et al. (1998)
Hydrogenation			
α-Enamides to α-aminocarboxylates	(*R,R*)-Et-DuPHOS-Rh^+X^- [$X = B(Ar_F)_4$ or CF_3SO_3]	Selectivity enhancement Gas miscibility Solvent replacement	Burk et al. (1995)
Tiglic acid to α-methylcarboxylic acid	$Ru(OCOCH_3)_2(H_8$-BINAP)	Solvent replacement	Jessop et al. (1995c) Xiao et al. (1996)
Prochiral imines to chiral amines	Ir/Perfluoroalkyl modified-phosphinodihydrooxazoles/$B(Ar_F)_4$	Enhanced separations (CESS[a])	Kainz et al. (1999)
Chiral enamine reductions	(*R,R*)-Et-DuPHOS-Rh^+X^- [$X = B(_F)_4$ or CF_3SO_3]	Solvent replacement	Buelow et al. (1998)

(*continued*)

Table 2.1 (continued)

Reaction	Catalyst/Ligand	Advantage(s)	Reference(s)
Chiral reduction of *N*-acylhydrazones	(*R*,*R*)-Et-DuPHOS-RH^+X^- [X = $B(Ar_F)_4$ or CF_3SO_3]	Solvent replacement	Buelow et al. (1998)
Cyclopropene to cyclopropane	$MnH(CO)_5$	Gas miscibility	Jessop et al. (1995b)
Isoprene to 2-methylbutane	$Rh(hfacac)R_2PCH_2CH_2PR_2)$ [R = C_6H_4-*m*-$(CH_2CH_2C_6F_{13})$]	Solvent replacement	Kainz et al. (1998)
Oxidation Catalysis			
Olefinic alcohol epoxidation with *t*-BuOOH	$VO(O\mathit{i}Pr)_3$	Solvent replacement	Pesiri et al. (1998) Pesiri et al. (1999)
Olefinic alcohol epoxidation with *t*-BuOOH	V(IV)O(salen)	Solvent replacement	Haas and Kolis (1998b)
Chiral olefinic alcohol epoxidation with *t*-BuOOH	$Ti[OCH(CH_3)_2]_4$	Solvent replacement	Pesiri et al. (1998)
Olefin epoxidation with *t*-BuOOH/H_2O	$Mo(CO)_6$	Solvent replacement	Pesiri et al. (1998)
Alkene epoxidation	$Mo(CO)_6$	Selectivity enhancement	Haas and Kolis (1998a)
Phase transfer oxidation of alkenes	$RuO_4/NaIO_4$, OsO_4	Solvent replacement	Morgenstern et al. (1996) Buelow et al. (1998)
Tetrahdyrofuran oxidation with O_2	[(dcpe)Rh(hfacad)]	Solvent replacement	Loeker et al. (1998)
Cyclohexane oxidation	FeCl(tpfpp)	Solvent oxidative stability	Wu et al. (1997)
Cyclohexene epoxidation	FeCl(tfpp)	Selectivity enhancement Solvent oxidative stability	Birnbaum et al. (1999)

Fluoroether-functionalized anthraquinones for H_2O_2 synthesis	Pd/Al_2O_3	Enhanced mass transfer Gas miscibility Enhanced separations	Hâncu and Beckman (1999a, 1999b)
Acrylate esters to acetals	$PdCl_2/CuCl_2$, $PdCl_2/CuCl$	Solvent replacement	Jia et al. (1999)
Carbon–Carbon Bond Formation			
Carbonylation			
Propylene to butanals	$CO_2(CO)_8$	Solvent replacement Selectivity enhancement Solvent tunabiilty	Rathke et al. (1991) Rathke et al. (1992) Klingler and Rathe (1994) Guo and Akgerman (1997, 1999)
1-Octene to nonals	$HRh(CO)[P(p\text{-}CF_3C_6H_4)_3]_3$, *trans*-$RhCl(CO)[P(p\text{-}CF_3C_6H_4)_3]_2$	Increased activity	Palo and Erkey (1998) Palo and Erkey (1999a, 1999b)
1-Octene to nonals	Rh(COD)(hfacac) $P(C_6H_4\text{-}m\text{-}CH_2CH_2C_6F_{13})_3$, $P(C_6H_4\text{-}p\text{-}CH_2CH_2C_6F_{13})_3$, $P(OC_6H_4\text{-}p\text{-}CH_2CH_2C_6F_{13})_3$	Increased activity Gas miscibility Enhanced separations	Kainz et al. (1997) Koch and Leitner (1998)
Styrenes to arylpropanals	Rh(COD)(hfacac) (*R*, *S*)-3-H^2P^6-BINAPHOS	Selectivity enhancement	Francio and Leitner (1999)
Iodobenzyl alcohol to phthalide	$PdCl_2L_2$ [L = PMe_3, PPh_3, MeCN, $P(OEt)_3$, $P(OPh)_3$, $PPh(OMe)_2$, $PPh_2(OMe)$]	Increased activity	Kayaki et al. (1999)
1,6-Enynes to bicyclic ketones	$CO_2(CO)_8$	Increased activity	Jeong et al. (1997)

(*continued*)

Table 2.1 (continued)

Reaction	Catalyst/Ligand	Advantage(s)	Reference(s)
Palladium-Catalyzed Coupling Reactions			
Heck coupling of phenyl iodide with alkenes	$Pd(OCOCH_3)_2$/ $P(C_6H_5)_{3-n}[(CH_3CH_2C_6F_{13})n$ ($n = 1, 2$)	Solvent replacement	Carroll and Holmes (1998)
Heck coupling of phenyl iodide with alkenes	$Pd(OCOCH_3)_2$/tris[3,5-bis(trifluoromethyl) phenyl]phosphine	Solvent replacement	Morita et al. (1998) Cacchi et al. (1999)
Suzuki coupling of phenyl iodide with boronic acids	$PdCl_2L_2/P(C_6H_5)_{3-n}$ $(CH_3CH_2C_6F_{13})_n$ ($n = 1, 2$)	Solvent replacement	Carroll and Holmes (1998)
Stille coupling of phenyl iodide with vinyl tin reagents	$Pd_2(dba)_3$/tris[3,5-bis(trifluoromethyl)phenyl]-phosphine	Solvent replacement	Morita et al. (1998)
Sonogashira coupling or alkynes with phenyl iodide	Aryl halide and alkyne (Sonogashira)	Solvent replacement	Carroll and Holmes (1998)
Heck coupling of phenyl iodide with acrylates	$Pd(O_2CCF_3)$, $Pd[CF_3C(O)CHC(O)CF_3]_2$	Solvent replacement	Shezad et al. (1999)
Hydrovinylation	$(Allyl)Ni(II)B(Ar_F)_4)_4$/ 1-azaphospholene	Solvent replacement	Wegner and Leitner (1999)
Olefin Metathesis			
Ring-opening polymerization (ROMP) of norbornene	$Ru[(H_2O)_6](OTs)_2$	Solvent tunability	Mistele et al. (1996)
Ring-closing metathesis (RCM) of dienes	M-carbene	Solvent tunability	Kainz et al. (1998) Fürstner et al. (1997)

CESS, catalytic extraction using supercritical solvent.

and sulfuric acid were used as catalysts. An additional benefit was the substantial reduction in the amounts of liquid acids used for these reactions, thus alleviating the environmental and safety hazards associated with these toxic and highly corrosive materials.

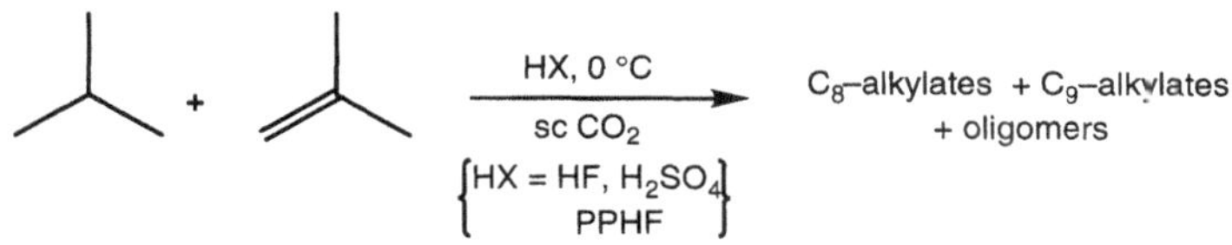

(2.2)

Clifford and Rayner have reported enhanced selectivities for Lewis acid catalyzed Diels–Alder cycloadditions in sc CO_2 relative to organic solvents. It is well known that Lewis acids catalyze Diels–Alder reactions and can improve their rate and selectivity (Dell, 1998; Santelli, 1996). The *endo*:*exo* product selectivity for the Sc(Otf)$_3$-catalyzed reaction between cyclopentadiene and alkyl acrylates (eq. 2.3) could be optimized to 24:1 by controlling the density of sc CO_2 (Clifford et al., 1997, 1998; Oakes et al., 1999), whereas in toluene the *endo*:*exo* ratio was 10:1.

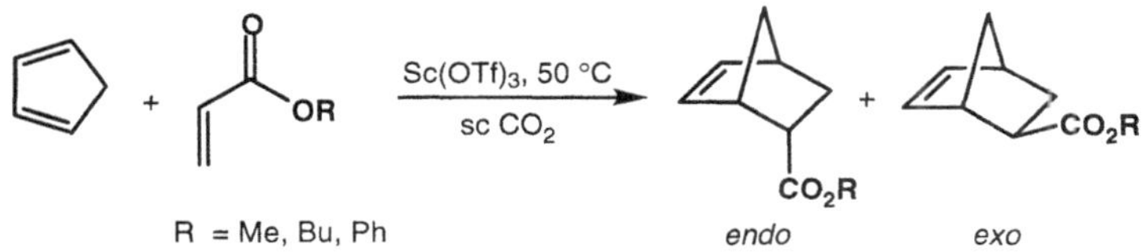

(2.3)

Catalytic Hydrogenation

Supercritical carbon dioxide offers several potential benefits for catalytic hydrogenation reactions, since the rate of many hydrogenation reactions run in organic solvents is proportional to hydrogen concentration, and is often limited by the rate of diffusion of H_2 from the gas to the liquid phase. Furthermore, the enantioselectivity of hydrogenation for a number of prochiral substrates using transition-metal complexes with chiral ligands is often sensitive to hydrogen concentration (Noyori, 1994; Sun et al., 1996). The complete miscibility of H_2 in CO_2 has been a main driver for several groups' investigations into catalytic hydrogenation. The report by Jessop et al. (1994b) of the catalytic hydrogenation of CO_2 in sc CO_2 using trimethylphosphine complexes of ruthenium—for example $RuH_2(PMe_3)_4$—clearly demonstrated the advantages of the miscibility of gaseous reagents. Additional details on the use of these catalysts are described in Chapter 3 of this book.

Despite the potential advantages of increased rates for catalytic hydrogenation reactions, there have been few reports on achiral homogeneously catalyzed reductions of unsaturated substrates. The first example of a hydrogenation reaction in sc CO_2 was the $MnH(CO)_5$-catalyzed hydrogenation of cyclopropene via a radical mechanism (Jessop et al., 1995a). Using $MnH(CO)_5$ or $CoH(CO)_4$, olefins were either hydrogenated or hydroformylated, depending on the olefin structure. The selectivity for hydrogenation of 3,3-dimethyl-1,2-diphenylcyclopropene in sc CO_2 was 66–70%, which is close to the selectivity in pentane, suggesting a solvent cage strength comparable to liquid alkanes. The sc CO_2-soluble ponytail complex $Rh(hfacac)(R_2PCH_2CH_2PR_2)$ [hfacac = bis(1,1,1,6,6,6-hexafluoropentane-2,4-dionato), R = C_6H_4-*m*-$(CH_2)_2(CF_2)_5CF_3$)] was shown by the Leitner group (Kainz et al., 1998) to be an active catalyst for the hydrogenation of isoprene. However, the rate of the reaction in sc CO_2 was considerably slower than the hydrogenation by the analogous dppp [1,3-bis(diphenylphosphino)propane] complex in organic solvents, due to either decomposition of the Rh catalyst or deactivation of the catalyst by forming a Rh formate complex.

Several groups have reported on highly enantioselective hydrogenation in dense-phase carbon dioxide using chiral bidentate phosphine or related ligands on Rh, Ru, or Ir (Scheme 2.1). Burk et al. (1995) found that asymmetric catalytic hydrogenation of α-enamides in sc CO_2 can lead to enantioselectives equal or superior to those observed in conventional organic solvents. A cationic Rh complex, **1**, containing the chiral 1,2-bis (*trans*-2,5-diethylphospholano)benzene (Et-DuPHOS) ligand (Burk et al., 1993), was rendered soluble in supercritical CO_2 through the use of the highly lipophilic tetrakis[3,5-bis(trifluoromethyl)-phenyl]borate ($B(Ar_F)_4^-$) counterion. Hydrogenation (50 °C, H_2 partial pressure 14 bar, total pressure

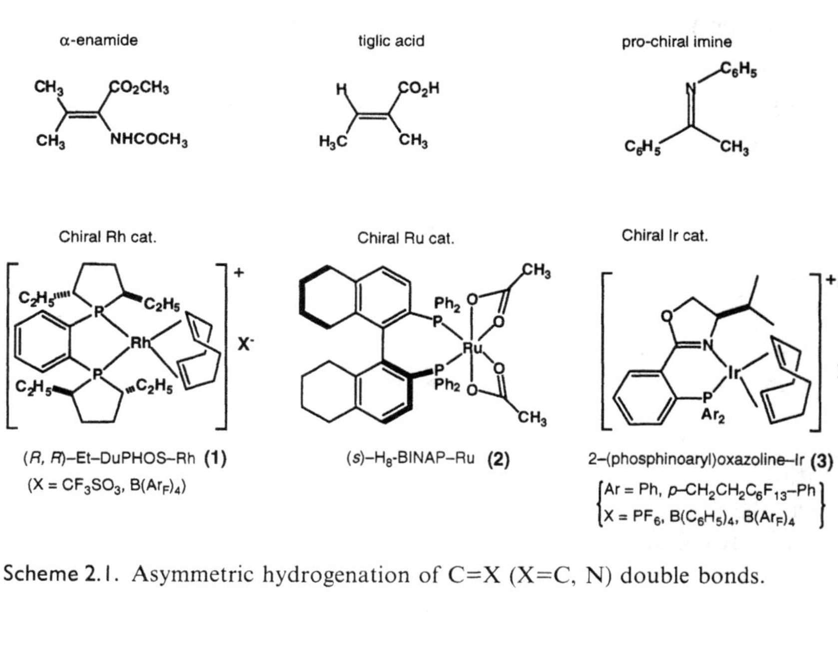

Scheme 2.1. Asymmetric hydrogenation of C=X (X=C, N) double bonds.

340 bar) proceeded smoothly to provide α-amino acid derivatives (eq. 2.4) with enantiomeric excesses (ee's) that were comparable to those achieved in conventional solvents.

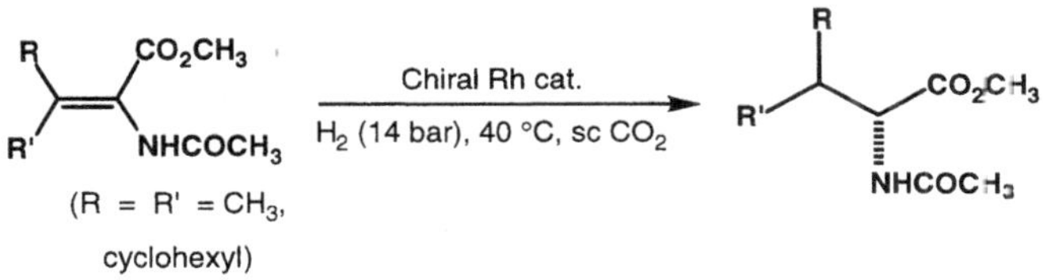

(2.4)

The hydrogenation reaction is quite general and over 30 substrates (enamide esters, enamides, hydrazones, enol acetates) have been reduced in sc CO_2 using the DuPHOS–Rh complexes, resulting in enantioselectivities that are equal or superior to conventional solvents (Buelow et al., 1998). For two β, β-substituted α-enamides, including a valine derivative, the ee's were found to be superior to those obtained in either methanol or hexane (Table 2.2). Preliminary kinetic studies reveal that hydrogenation reactions can be faster in supercritical CO_2 than in hexane, presumably due to the high miscibility of hydrogen gas in supercritical CO_2.

Concurrent with the initial report of Burk and Tumas (Burk et al., 1995), Ikariya and Noyori (Jessop et al., 1995c; Xiao et al., 1996) found that a CO_2-soluble complex, $Ru(OCOCH_3)_2(H_8\text{-binap})$, **2**, employing a partially hydrogenated derivative of BINAP in order to maximize solubility, was an active catalyst in sc CO_2 for the asymmetric hydrogenation of α,β-unsaturated carboxylic acids such as tiglic acid (Scheme 1). The ee of the products prepared in sc CO_2 (81%) was comparable to that in methanol (82%) and greater than that in hexane (73%) (all at 50 °C and 30 bar H_2). Lowering the H_2 partial pressure increased the ee in methanol but not in sc CO_2, most likely a manifestation of the already high miscibility of H_2. Leitner and Pfaltz (Kainz et al., 1999) have reported the use of Ir

Table 2.2. Asymmetric Hydrogenation of α-Enamides with (*R*, *R*)-Et-DuPHOS-RH Catalyst

Substrate	%ee in		
	MeOH	Hexane	sc CO_2
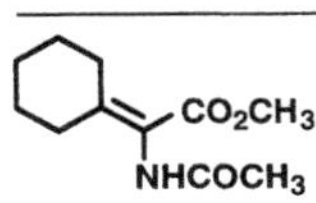	81.8	76.2	96.8
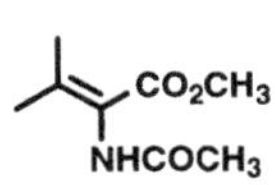	62.6	69.5	84.7

complexes with chiral phosphinodihydrooxazole ligands (**3** in Scheme 1) for the enantioselective hydrogenation of imines. Highly fluorinated alkyl chains were attached to the phosphine phenyl rings in the *para* position in order to increase solubility in sc CO_2, which resulted in no loss of activity or selectivity. Among the several Ir complexes tested in organic solvents, only the cationic complexes with the $B(Ar_F)_4^-$ counteranion gave a higher ee, up to 81%. Another important feature of this work was the selective extraction process by which the catalyst could be isolated from product and recycled several times without appreciable loss in activity and enantioselectivity (named the CESS process for Catalytic Extraction using Supercritical Solvent).

Transition-metal-catalyzed hydroboration, a reduction using B–H bonds, of styrene derivatives has been demonstrated to occur in supercritical carbon dioxide using tunable Rh(I) complexes as catalysts. In the case of vinyl anisole (eq. 2.5), significantly higher regioselectivities were reported for the reaction in this medium than in THF or perfluoro(methylcyclohexane) (Carter et al., 2000).

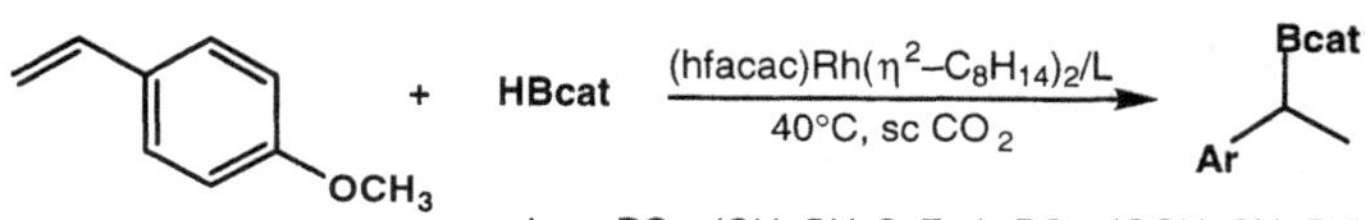

(2.5)

L = $PCy_2(CH_2CH_2C_6F_{13})$; $PCy_2(OCH_2CH_2C_6F_{13})$

Although not all of the factors that influence homogeneous hydrogenation and hydroboration in sc CO_2 are fully understood, it is clear that the use of sc CO_2 can lead to an increase in selectivity for some reactions. Additional work is needed to understand the opportunities for further selectivity enhancements and catalyst separation/recycle strategies. Even sc CO_2 systems that exhibit similar selectivities to those obtained in organic solvents could offer a practical, environmentally responsible method for the production of many important chiral building blocks.

Oxidation Catalysis

Epoxidation of olefins, including asymmetric epoxidations, mediated by early transition-metal catalysts and peroxidic oxidants have been studied in a wide range of organic solvents since the 1970s (Conte et al., 1997; Sharpless and Verhoeven, 1979). The well-documented rates, selectivities and mechanistic information for this important transformation have allowed researchers to benchmark activity and selectivity in dense-phase carbon dioxide for a reaction that does not involve gaseous reagents (Jessop et al. 1999). Several groups have reported the transition-metal-mediated catalyzed epoxidations of olefins using peroxides and molybdenum, vanadium, or titanium catalysts (Scheme 2.2).

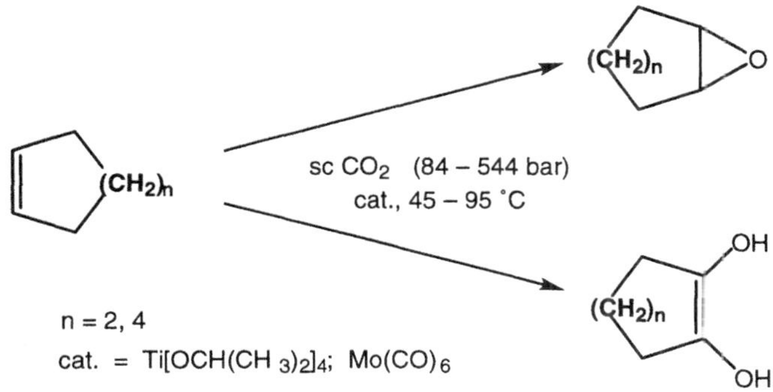

Scheme 2.2. Oxidation of cycloalkanes by *tert*-butyl hydroperoxide in sc CO_2.

Molybdenum hexacarbonyl is known to react with hydroperoxides, typically in refluxing aromatic solvents or neat olefin, to form an olefin oxidation catalyst. The epoxidation of 2,3-dimethylbutene by cumene hydroperoxide and $Mo(CO)_6$ was found to proceed with high conversions in sc CO_2, using small amounts of organic cosolvents (Jessop, 1998). Walther (Kreher et al., 1998) found that epoxidation of cyclooctene by $Mo(CO)_6$ and *t*-BuOOH occurs with high selectivity at lower temperatures (45 °C) in CO_2 than in organic solvents. Aqueous *t*-BuOOH leads to the oxidation of cyclohexene to the 1,2-diol, presumably via hydrolysis of the intermediate epoxide (Pesiri et al., 1998). Haas and Kolis (1998a) found that the selectivity for diols versus epoxides is strongly influenced by both the water content and the reaction temperature.

Pesiri et al. (1999) carried out a detailed kinetic study on the epoxidation of a number of allylic and homoallylic alcohols with $V(O\textit{i}Pr)_3$ and *t*-BuOOH (eq. 2.6). While complete conversions and good selectivities were observed, the rates of reaction in liquid CO_2 [$k = 9\ M^{-1}\ s^{-1}$ for (*Z*)-non-3-en-1-ol] were comparable to those measured for nonpolar organic solvents, such as hexane ($3\ M^{-1}\ s^{-1}$) or toluene ($17\ M^{-1}\ s^{-1}$). They also found that vanadyl ion immobilized in insoluble polymers can also catalyze the epoxidation of olefins. A diastereoselective epoxidation of allylic alcohols has been reported using a vanadyl-Salen catalyst precursor in sc CO_2 (Haas and Kolis, 1998b), and an enantioselective epoxidation of an allylic alcohol was reported using a chiral titanium catalyst in liquid CO_2 at 0 °C (eq. 2.7) (Pesiri et al., 1998). In both cases, selectivities similar to those reported in hydrocarbon solvents (Woodward et al., 1991) were obtained.

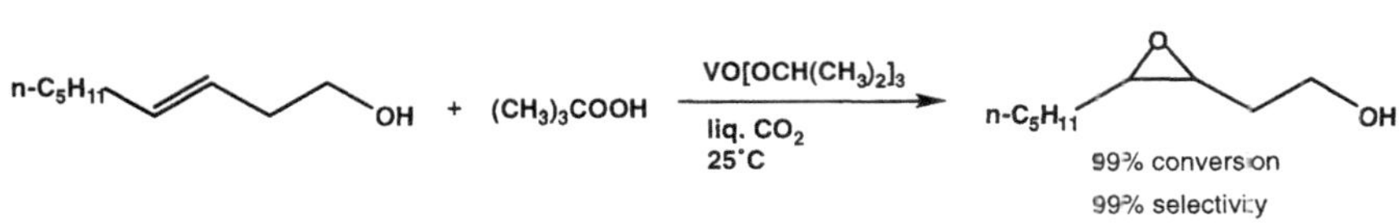

(2.6)

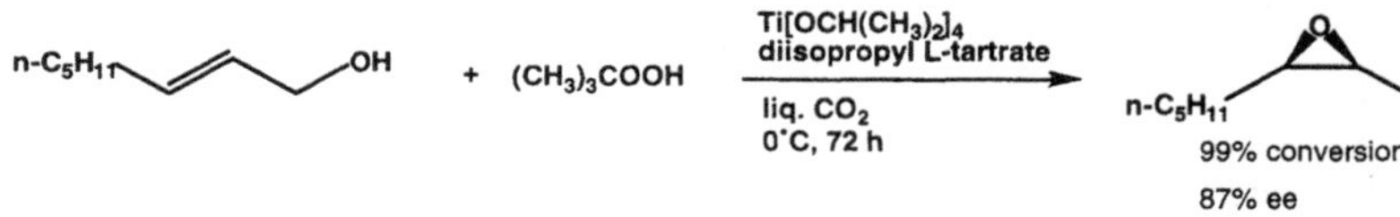

(2.7)

Selective oxidation of hydrocarbons using molecular oxygen (O_2) remains a key challenge in catalysis. While sc CO_2 should be ideally suited as a solvent for oxidation catalysis due to high O_2 solubility, low barriers to diffusion, oxidative stability, and the potential for separations, there have been few reports of homogeneous catalysis using O_2 as a selective oxidant. An approach using a traditional rhodium catalyst for the oxidation of tetrahydrofuran to γ-butyrolactone using O_2 in sc CO_2 was recently reported by Leitner (Jessop and Leitner, 1999). Turnover numbers of 135 per Rh were achieved in sc CO_2, but these results were not dissimilar to those obtained using gaseous O_2 as a reagent in organic solvents. Highly halogenated metalloporphyrins (Lyons et al., 1995) have been employed for the oxidation of cyclohexane and cyclohexene in sc CO_2 by Koda (Wu et al. 1997) and Birnbaum et al. (1999), respectively. Koda's group used FeCl[tpfpp) (tpfpp = 5,10,15,20-tetrakis(perfluorophenyl)porphyrin] and a sacrificial aldehyde to oxidize cyclohexane to predominantly cyclohexanone and cyclohexanol, with a 2.9% yield. The LANL group used tetrakis(perfluorophenyl)porphyrins with additional halogens on the porphyrin ring and found slightly higher selectivities for epoxide formation in sc CO_2 relative to benzene, methylene chloride, or acetonitrile. It is likely that the reactions in both studies involve radical intermediates, either bound or free from the porphyrin catalyst, which compete with other free-radical-forming autooxidation pathways. The oxidative stability of CO_2 toward radicals was shown to prevent competitive reactions via solvent oxidation.

In summary, a number of reports clearly demonstrate that olefin oxidations with peroxidic reagents and metal catalysts can proceed in dense-phase CO_2. It is not surprising that the rates and yields of these reactions in sc CO_2 were similar to the analogous reactions performed in organic solvents, given that the reactions involve relatively nonpolar transition states and no gaseous reagents. Except for incremental improvements, sc CO_2 will likely represent an environmentally benign alternative solvent only for these types of reactions. Catalytic reactions with oxygen in sc CO_2, on the other hand, represent a promising area of research that has not yet been fully developed.

Production of Hydrogen Peroxide: Hydrogenation and Oxidation

Hydrogen peroxide is synthesized commercially by the reduction of anthraquinone with hydrogen and the subsequent oxidation of the dihydroanthraquinone with molecular oxygen. This reaction is typically carried out in a batch multiphase process (organic, water, gas phase) in organic solvents, and can be fraught with separation difficulties as well as safety issues for the oxidation reaction.

Hâncu and Beckman (1999a, 1999b) have reported an elegant demonstration of the use of CO_2 as a solvent for the catalytic synthesis of hydrogen peroxide (Scheme 2.3) (1999a, 1999b) using anthraquinones modified with solubility-enhancing CO_2-philic moieties (either fluoroether or polysiloxane groups). The overall reactivity could be tuned by varying either the length of the CO_2-philic tail or the substitution pattern on the anthraquinone framework. The kinetics of the platinum-catalyzed hydrogenation step showed that the diffusion coefficients varied inversely with the molecular weight of the CO_2-philic tail, and that mass-transfer limitation became negligible for anthraquinones with shorter tails. More impressive, though, is the demonstration that using a homogeneous liquid CO_2 solution could prevent contamination of the aqueous product, eliminate the gas-phase and mass-transfer limitations of the multiphase process, and offer the possibility of running the reaction under continuous flow conditions.

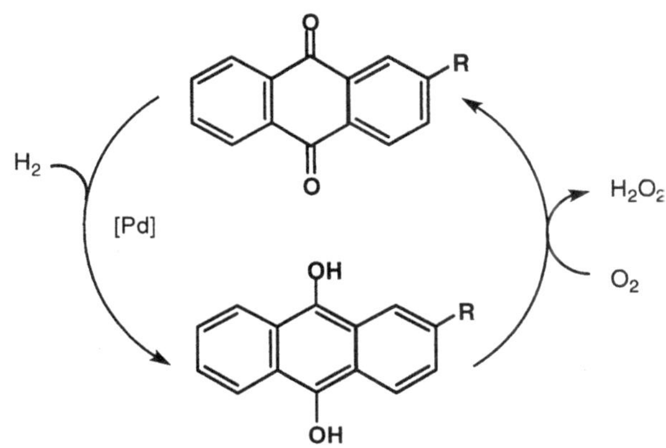

Scheme 2.3. Catalytic cycle for H_2O_2 synthesis using modified anthraquinones.

Carbon dioxide may also be an excellent solvent for the direct combination of H_2 and O_2 to form H_2O_2 (D. Hâncu and E. J. Beckman, unpublished results, 1999).

Carbon–Carbon Bond-Forming Reactions

Catalytic formation of carbon–carbon bonds is a powerful tool for construction of complex molecular architectures, and has been developed extensively for applications in organic synthesis. Three main classes of carbon–carbon bond forming reactions have been studied in sc CO_2: carbonylation (with particular attention paid to the hydroformylation of α-olefins), palladium-catalyzed coupling reactions involving aromatic halides, and olefin metathesis.

Carbonylation

Hydroformylation (the oxo process) involves the addition of H_2 and CO to an olefin to form aldehydes (eq. 2.8), which have a number of important industrial applications. Extensive mechanistic studies have shown that this reaction involves migratory insertion of a bound alkyl group (formed by insertion of an olefin into a metal hydride) into a bound CO, followed by reductive elimination of the aldehyde. The rate-limiting step for the hydroformylation in liquids is either the reaction of olefin and $HCo(CO)_4$ or the reaction of the acyl complex with H_2 to liberate the product aldehyde. The high miscibility of CO in sc CO_2 is therefore not necessarily a major factor in determining the rate of the hydroformylation. Typically, for α-olefins, linear aldehydes are preferred to branched products, and considerable effort has gone into controlling the selectivity of this reaction.

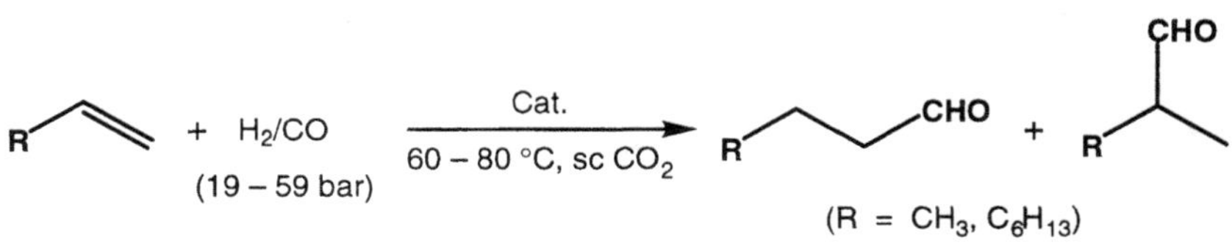

(2.8)

Rathke et al. (1991) reported the first example of homogeneous hydroformylation in sc CO_2, with propylene, H_2, CO, and a $Co_2(CO)_8$ precatalyst. The rate was found to be slightly lower in sc CO_2 compared with hydrocarbon solvents such as methylcyclohexane and heptane. The selectivity for the linear aldehyde butanal was slightly higher (88%) in sc CO_2 ($P = 56$ bar for H_2 and CO) than in benzene (83%) at slightly higher pressure ($P = 80$ bar for H_2 and CO). Although the effect of temperature on the linear: branched ratio in CO_2 is similar to that observed in organic solvents, this selectivity was found to increase from 73% to 81% by doubling the pressure at constant temperature.

Since Rathke and Klingler's pioneering work, several research groups have extended the range of metal complexes that can serve as hydroformylation catalysts. Leitner (Kainz et al., 1997; Koch and Leitner, 1998) found that a CO_2-soluble Rh complex with polyfluoroalkyl-substituted triarylphosphine or triarylphosphite ligands (Rh/ligand = 1/4–1/10) in sc CO_2 effected hydroformylation of 1-octene to give the linear aldehyde in a good yield and with 76–86% selectivity. Somewhat higher rates for 1-octene were measured in sc CO_2 relative to organic solvents or to liquid CO_2 when unmodified catalysts such as $Rh(COD)[CF_3C(O)CHC(O)CF_3]$ (COD = 1,5-cyclooctadiene) were employed, consistent with previous research indicating that modification of the rhodium center with CO_2-solubilizing ligands slows down the reaction but enhances the selectivity for linear aldehyde.

Utilization of metal complexes containing phosphines modified with fluorinated ponytails (e.g., $P[m\text{-}CH_2CH_2(CF_2)_6F\text{-}C_6H_4]$) led to higher regioselectivities than those found in conventional solvents (Koch and Leitner, 1998), and side reactions such as hydrogenation or olefin isomerization were suppressed in sc CO_2. Rhodium complexes such as $HRh(CO)[P(p\text{-}CF_3\text{-}C_6H_4)_3]_3$ and *trans*-$RhCl(CO)[P(p\text{-}CF_3\text{-}C_6H_4)_3]_2$ have also been reported to be active catalysts for the hydroformylation of 1-octene (Palo and Erkey, 1998, 1999a, 1999b). An additional feature of the Rh/fluorinated phosphine catalysts is their insolubility in common organic species, which permits not only facile separation of products from catalysts but also subsequent recycling of the precious metal catalysts without any noticeable changes in activity or selectivity (Koch and Leitner, 1998).

Recently, high enantioselectivity and regioselectivity were reported by Leitner in the rhodium-catalyzed asymmetric hydroformylation of *para*-substituted styrenes (Francio and Leitner, 1999) using a perfluoroalkyl-substituted ligand [(*R,S*)-BINAPHOS] in sc CO_2 (eq. 2.9). The regioselectivity to the branched isomer was found to be 88% in benzene (60 °C), which increased to 92–94% in sc CO_2, with a slight dependence on solvent density (pressure) and temperature. Enantioselectivities for the branched isomer ranged from 90% to 94% ee for both solvents.

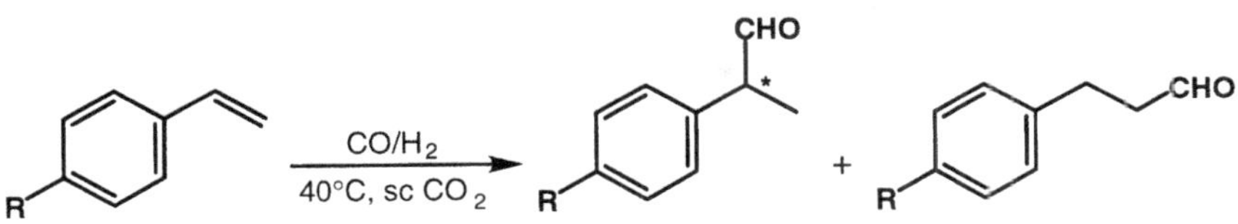

(2.9)

Ikariya and co-workers reported an efficient palladium-catalyzed carbonylation of aryl halides in sc CO_2 (eq. 2.10) (Kayaki et al., 1999). 2-Iodobenzyl alcohol was converted to the phthalide in the presence of $PdCl_2L_2$ [L = PMe_3, PPh_3, MeCN, $P(OEt)_3$, $P(OPh)_3$, $PPh(OMe)_2$, $PPh_2(OMe)$] with higher rates in sc CO_2 than in toluene.

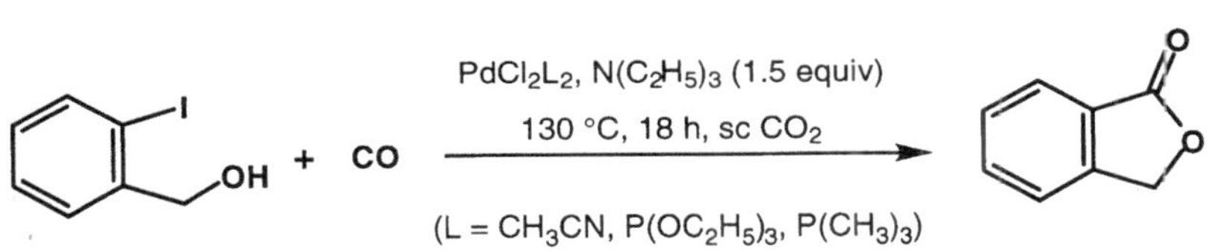

(2.10)

Intramolecular Pauson–Khand reactions to prepare [3.3.0]-bicyclooctenones from several 1,6-enynes (eq. 2.11) were reported to occur efficiently in sc CO_2 using a $Co_2(CO)_8$ catalyst (Jeong et al., 1997). The reactions proceed very well regardless of the substitution pattern of the acetylene moiety, although substitution on the olefin fragment yields a mixture of products.

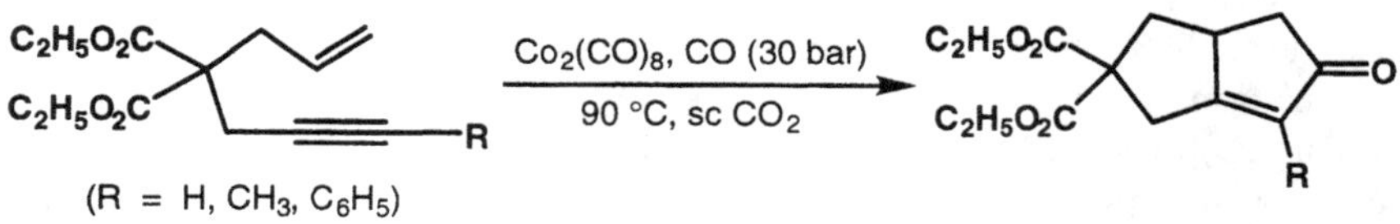

(2.11)

Palladium-Catalyzed Coupling Reactions

Results describing the use of sc CO_2 as a medium for palladium-catalyzed coupling reactions (Scheme 2.4) have shown that yields and selectivies are similar to those found for organic solvents. The insolubility of traditional phosphine-based catalysts led to the synthesis of the fluorinated analogues containing $P(C_6H_5)_{3-n}(CH_2CH_2C_6F_{13})_n$ ($n = 1$ or 2) ligands (Carroll and Holmes, 1998). Yields and turnover numbers for the Heck, Suzuki, and Sonogashira reactions were comparable to those in organic solvents.

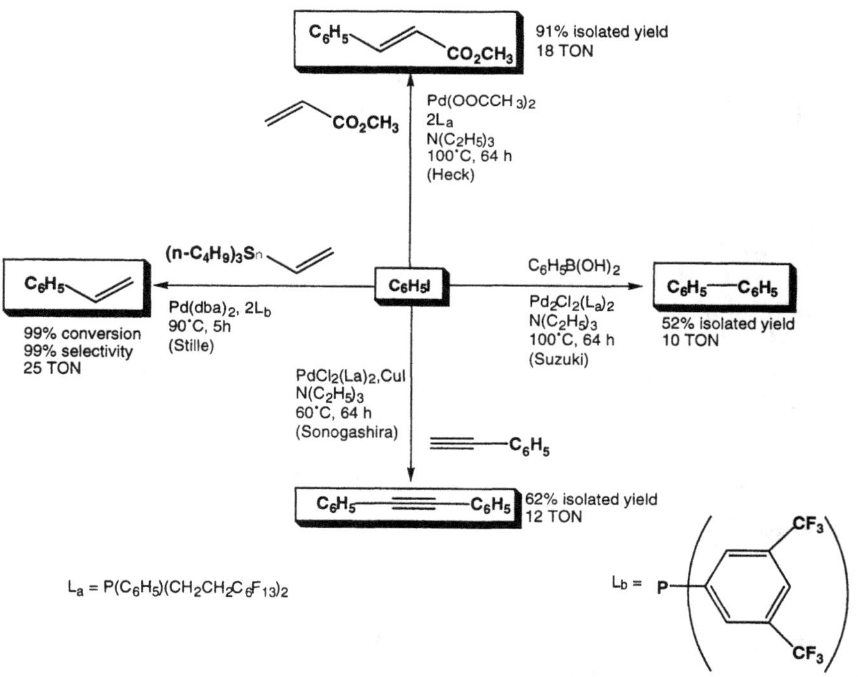

Scheme 2.4. Carbon–carbon bond-forming reactions in sc CO_2 catalyzed by CO_2-soluble Pd complexes. TON = mole product per mole catalyst.

Tumas and co-workers (Morita et al., 1998) also used fluorinated phosphine ligands for coupling reactions of phenyl iodide with alkenes or vinyl tin reagents. Complexes of a Pd(0) precursor with triarylphosphines containing bis-(trifluoromethyl) groups ($P[3,5\text{-}(CF_3)_2\text{-}C_6H_3]_3$) were more soluble than unfluorinated systems in sc CO_2 and led to higher conversions. Kinetics on the Stille coupling reactions with vinyl tributyl tin revealed that reaction rates were roughly one-half of those in toluene.

In general, the few studies performed so far suggest that C–C bond-forming reactions in sc CO_2 alone are not faster and do not offer greater yields compared with liquid solvents, although the use of water as a cosolvent for $Pd(OAc)_2$-catalyzed Heck reactions in sc CO_2 *has* led to enhanced reaction rates (Bhanage et al., 1999a). An as-yet untapped advantage to using sc CO_2 for these reactions likely lies in tunable local solute/solute clustering effects, which have proven beneficial in stoichiometric reactions.

Olefin Metathesis

The production of higher olefins and complex ring structures via olefin metathesis has been an important tool for synthetic chemists in both industry and academia. Ring-opening metathesis polymerization (ROMP) and ring-closing metathesis (RCM) have both been explored in sc CO_2. The ring-opening metathesis polymerization of norbornene in sc CO_2 using $[Ru(H_2O)_6](OTs)_2$ (Ts = *p*-toluenesulfonyl) to produce high-quality polymer in high yield was reported by DeSimone (Mistele et al., 1996). Catalyst solubility was increased by the addition of methanol as a cosolvent with a concomitant decrease in the *cis* selectivity of the polymer from 83% to 30%, indicating that the polymer microstructure may be controlled by changing the polarity of the reaction medium. Ruthenium– and molybdenum–carbene complexes were also shown to be quite active for olefin metathesis in sc CO_2. These catalysts have been used for ring-closing metathesis of dienes to cyclic olefins (eq. 2.12) (Fürstner et al., 1997; Kainz et al., 1998). Dienes with N–H functionality were converted to heterocycles in high yields without the need for the protecting groups necessary in conventional solvents. That the density of the sc CO_2 was found to influence the selectivity to the cyclized product is an interesting result that is not yet fully understood.

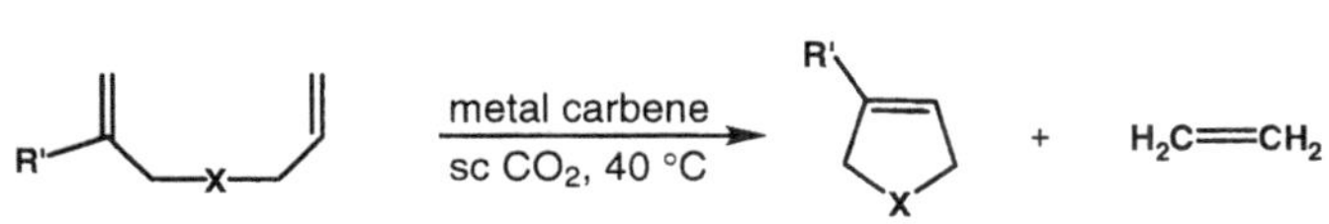

(2.12)

R' = H; **X** = N–Ts, $(CH_2)_7CO_2(CH_2)_3$, $(CH_2)_3CO_2(CH_2)_2CO_2(CH_2)_3$
R' = CH_3; **X** = $CH_2C(O)C(CH_3)_2$

Heterogeneous Catalysis

While homogeneous catalysts can be well characterized in solution, difficulties in separation and recovery have limited their applications to large-scale chemical synthesis, except for a few cases where the catalysts are highly active (e.g., acetic acid synthesis, hydroformylation, and hydrocyanation of butadiene to adiponitrile). A large number of industrial processes (Thomas and Thomas, 1997) are carried out with heterogeneous catalysts in the liquid or gas phase, particularly for large-scale and commodity chemical synthesis. Both gas-phase and liquid-phase reactions may be amenable to improvements in activity, selectivity, or catalyst life when run in sc CO_2. Heterogeneously catalyzed hydrogenations have received considerable attention (several pilot demonstrations for foodstuffs or specialty chemical synthesis are in the works in Europe), although other reactions remain largely unexplored (summarized in Table 2.3).

Acid Catalysis

Subramaniam and co-workers have investigated the use of industrial Pd/γ-Al_2O_3 catalysts for 1-hexene isomerization (Clark and Subramaniam, 1996; McCoy and Subramaniam, 1995). Supercritical CO_2 is thought to readily solubilize hexene oligomerization/polymerization products ("coking precursors", which result from peroxide impurities in the olefin feed; Ginosar and Subramaniam, 1995), thus precluding the deposition of high-molecular-weight alkene oligomers in the catalyst pores. This inhibition of coke deposition leads to both improved catalyst activity and extended catalyst lifetime compared with subcritical reaction pressures.

The enhanced diffusivity of polynuclear compounds in sc CO_2 has been utilized to enhance catalyst lifetimes in both 1-butene/isoparaffin alkylations (Clark and Subramaniam, 1998; Gao et al., 1996). The former may be catalyzed using a number of solid acid catalysts (zeolites, sulfated zeolites, etc.), and the use of sc CO_2 as a solvent/diluent permits the alkylations to be carried out at relatively mild temperatures, leading to the increased production of valuable trimethylpentanes (which are used as high-octane gasoline blending components). The enhancement of product *selectivity* in the latter process is believed to result from rapid diffusion of ethylbenzene product away from the Y-type zeolite catalysts, thus preventing product isomerization to xylenes.

Friedel–Crafts alkylation has been studied by Poliakoff in a continuous-flow reactor (Hitzler et al., 1998a). The reaction of mesitylene and anisole with propene or 2-propanol over a solid acid catalyst (based on a Deloxan support) in sc CO_2 provided exclusive formation of the monoalkylated products at 50% conversion. Use of the continuous-flow reactor prevents catalyst deactivation, and permits use of comparatively small reactors. The

Table 2.3. Examples of Heterogeneous Catalysis in Dense-Phase Carbon Dioxide

Reaction	Catalyst/Ligand	Advantage(s)	Reference(s)
Acid Catalysis			
Alkylation of 1-butene with isobutane	H-USY zeolite sulfated zirconia	Increased activity Selectivity enhancement Longer catalyst lifetime	Clark and Subramaniam (1998) Subramaniam and Clark (1999)
Ethylation of benzene with ethylene	Y-type zeolites	Selectivity enhancement Longer catalyst lifetime	Gao et al. (1996) Gao et al. (1997)
Friedel–Crafts alkylation of mesitylene with 2-propanol	Polysiloxane-supported solid acid	Solvent tunability Selectivity enhancement	Hitzler et al. (1998a, 1998b)
Isomerization of 1-hexene to internal isomers	$Pt/\gamma\text{-}Al_2O_3$	Increased activity Selectivity enhancement Longer catalyst lifetime	Subramaniam and McHugh (1986) Saim and Subramaniam (1990) Baptist-Nguyen and Subramaniam (1992) McCoy and Subramaniam (1995) Ginosar and Subramaniam (1995) Clark and Subramaniam (1996)
Diels-Alder reaction	Amorphous silica	Selectivity enhancement	Weinstein et al. (1999)
Dehydration of alcohols to ethers, acetals, and ketals	Deloxan ASP, Amberlyst	Selectivity enhancement	Gray et al. (1999)
Esterification			
Esterification of oleic acid with methanol	Sulfonic macroporous ion exchange resin	Increased activity	Vieville et al. (1993, 1994)

(*continued*)

Table 2.3 (continued)

Reaction	Catalyst/Ligand	Advantage(s)	Reference(s)
Hydrogenation			
CO_2 (+ Me_2NH)	Rh, Ir Pd, Pt, Ru / Sol-gel derived silica matrix	Increased activity Enhanced mass transfer Gas miscibility	Krocher et al. (1996a, 1996b, 1998, 1999)
Acetophenone hydrogenation	5% Pd/APII Deloxan	Solvent tunability Selectivity enhancement Enhanced mass transfer Gas miscibility	Hitzler and Poliakoff (1997) Hitzler et al. (1998a, 1998b)
m-Cresol hydrogenation	5% Pd/Deloxan	Solvent tunability Selectivity enhancement Enhanced mass transfer Gas miscibility	Hitzler et al. (1998a, 1998b)
Benzaldehyde hydrogenation	5% Pd/Deloxan	Solvent tunability Selectivity enhancement Enhanced mass transfer Gas miscibility	Hitzler et al. (1998a, 1998b)
Furan hydrogenation	5% Pd/Deloxan	Enhanced mass transfer Gas miscibility	Hitzler et al. (1998a, 1998b)
Isophorone hydrogenation	5% Pd/Deloxan	Solvent tunability Selectivity enhancement Enhanced mass transfer Gas miscibility	Hitzler et al. (1998a, 1998b)
Cyclohexanone hydrogenation	3% Ru/Deloxan	Enhanced mass transfer Gas miscibility	Hitzler et al (1998a, 1998b)

Hydrogenation of fats, oils, etc.	Pd/C	Increased activity Selectivity enhancement Longer catalyst lifetime	Tacke (1995) Tacke et al. (1996)
Cyclohexene hydrogenation	5% Pd/Deloxan	Increased activity Enhanced mass transfer Gas miscibility	Hitzler and Poliakoff (1997)
Hydrogenation of 1,2-(methylenedioxy)-4-nitrobenzene	1% Pd/APII Deloxan	Enhanced mass transfer Gas miscibility	Hitzler and Poliakoff (1997)
Hydrogenation of α, β-unsaturated aldehydes	Pt/Al_2O_3 (unmodified)	Selectivity enhancement	Bhanage et al. (1999c)
Oxidation Catalysis			
Toluene oxidation with O_2	CoO/Al_2O_3 CoO-MoO_3/Al_2O_3 MoO_3/Al_2O_3	Increased activity Selectivity enhancement	Dooley and Knopf (1987)
Propene oxidation	CuI/CuO	Selectivity enhancement	Gaffney and Sofranko (1992) Gaffney and Sofranko (1993)
Toluene to benzaldehyde	Co/Al_2O_3	Increased activity Selectivity enhancement Enhanced mass transfer Gas miscibility	Dooley and Knopf 1987
Propene to propylene glycol	$CuI/Cu_2O/MnO_2$ on MgO or γ-Al_2O_3	Selectivity enhancement Enhanced mass transfer Gas miscibility	Gaffney and Sofranko (1992) Gaffney and Sofranko (1993)
Ethanol to acetaldehyde	Pt/TiO_2	Enhanced mass transfer Gas miscibility	Zhou and Akgerman (1995)
Aromatics and acetaldehyde to CO_2 and H_2O	Pt/Al_2O_3, Pt/TiO_2	Enhanced mass transfer Gas miscibility	Zhou and Akgerman (1995)

high diffusivity of supercritical fluids allows the reduction of mass transport restrictions at the surface of the solid catalyst compared with liquid-phase reactions. Compared with gas-phase reactions, sc CO_2 appears to reduce the coking of catalysts, thereby preventing premature deactivation of the catalytic active site.

Hydrogenation

Supercritical carbon dioxide is particularly well suited to the heterogeneous hydrogenation of unsaturated organic molecules. This is in part due to increased solubility of hydrogen in sc CO_2 relative to that in traditional reaction media and to the ease of product/catalyst separation, but it is also a result of the potential elimination of mass transport resistance between hydrogen and the substrate under sc CO_2 conditions. Accordingly, transition-metal catalysts on organic or inorganic supports have been successfully employed to catalyze the hydrogenation of a wide variety of substrates (Hitzler et al., 1998), including alkenes (Bertucco et al., 1997; Devetta et al., 1999; Hitzler and Poliakoff, 1997; Minder et al., 1995), alkynes, aldehydes, ketones (Devetta et al., 1999; Hitzler et al., 1998a), vegetable fats and oils, and free fatty acids and acid esters (Tacke, 1995; Tacke et al., 1996).

Poliakoff et al. showed that use of heterogeneous catalysts in a continuous-flow reactor allows for hydrogenation of large volumes of material with a relatively small reaction vessel (Hitzler et al., 1998b), a potentially very useful feature for a scaled-up process. They also showed that even a small flow of sc CO_2 can maintain the temperature of the highly exothermic hydrogenation because supercritical fluids, with their favorable heat-transfer properties, can effectively remove heat generated by the reaction. Hence, the outcome of the reaction in terms of activity and selectivity can be manipulated by using sc CO_2 as the carrier medium. For example, using this flow reaction system, isophorone, a 2-cyclohexenone derivative, was selectively hydrogenated at the olefinic bond at flow rates up to 2.0 mL/min (118 bar CO_2, 200 °C) over Deloxan-supported Pd to give 3,3,5-trimethylcyclohexanone with high selectivity and quantitative conversion (eq. 2.13). The versatility of this experimental setup and catalyst system is exemplified by the fact that hydrogenation products **I, II, III**, or **IV** may be obtained selectively from acetophenone and benzophenone (eq. 2.14) through careful tuning of reaction conditions.

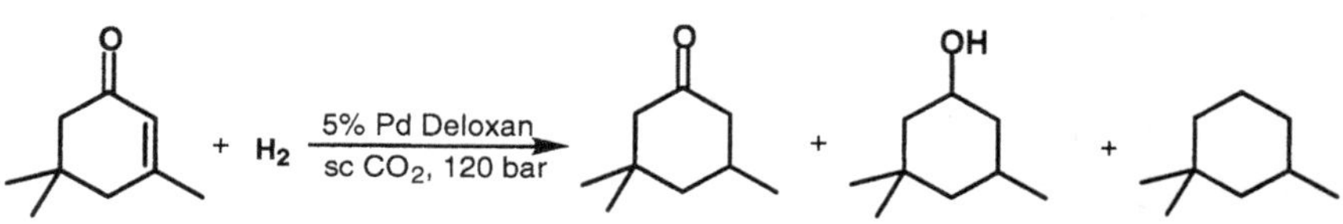

(2.13)

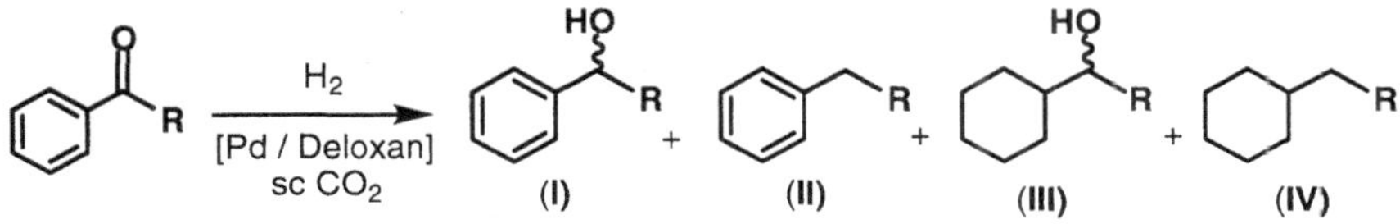

(2.14)

Oxidation Catalysis

Dooley and Knopf (1987) have reported the partial oxidation of toluene with air in sc CO_2 using heterogeneous catalysts. The Al_2O_3-supported catalysts used in this work included 5% CoO, 5% CoO/10% MoO_3, and 10.5% MoO_3. Although all of these solid catalysts were known to be suitable for oxidizing organic compounds using O_2, the cobalt oxides were observed to have the greatest activity. At low rates and conversions, the observed products included benzaldehyde, benzyl alcohol, and cresol isomers, in addition to minor amounts of condensation products and carbon oxides. By comparison to the low-pressure vapor-phase process, the high-pressure sc CO_2 process afforded not only increased selectivity for partial oxidation products but also improved reaction rates.

Gaffney and Sofranko (1993) investigated the selective oxidation of propene to propylene glycol in sc CO_2 using a heterogeneous catalyst comprising of copper iodide, copper oxide, and a reducible metal. Higher propylene glycol selectivity was achieved in sc CO_2 (90%) than in subcritical aqueous solvent (50%), with no measurable loss of iodide from the catalyst after 50 h under reaction conditions.

There have also been several reports of the total oxidation of organics to CO_2 and H_2O in sc CO_2 using heterogeneous catalysts (Mukhopadhyay and Srinivas, 1996; Srinivas and Mukhopadhyay, 1997). The high gas miscibility of CO_2 appears to offer several advantages in these reactions.

Biphasic Catalysis

Despite the remarkable activity and selectivity attainable through sophisticated ligand design in homogeneous catalysis, broad commercial application has been hampered by the need to recover and/or recycle expensive catalysts and ligands. A number of approaches, termed "'biphasic catalysis", have been advanced where a soluble catalyst is immobilized in one liquid phase (often aqueous) and the substrates and products are isolated in a separate immiscible phase (Baker and Tumas, 1999; Cornils and Herrmann, 1998). Several recent reports have focused on systems using only water and CO_2, where water-soluble catalysts and CO_2-soluble substrates and products can be isolated in separate phases (Table 2.4). Hydrogenation of cinnamal-

Table 2.4. Examples of Biphasic Catalysis in H_2O/sc CO_2 Mixtures

Reaction	Catalyst/Ligand	Advantage(s)	Reference(s)
Hydrogenation of cinnamaldehyde	Ru/TPPTS	Gas miscibility Enhanced separations	Bhanage et al. (1999a)
Phase transfer oxidative cleavage of cycloalkenes	RuO_4/$NaIO_4$ in H_2O/CO_2		Morgenstern et al. (1996)
Phase transfer dihydroxylation of cycloalkenes	OsO_4/$NaIO_4$ in H_2O/CO_2		Buelow et al. (1998)
Heck vinylation	$Pd(OAc)_2$/ TPPTS	Enhanced activity	Bhanage et al. (1999b)
Hydrogenation of styrene and 1-alkenes in CO_2/H_2O emulsions	Rh/TPPTS/ Surfactants	Gas miscibility Enhanced separations Increased activity	Jacobson et al. (1999)

dehyde has been reported using a biphasic water/sc CO_2 system and a water-soluble Ru catalyst containing sulfonated phosphine ligands (Bhanage et al., 1999b). The conversions were found to be ca. four times higher than those in toluene/water mixtures (38% vs. 11%). Jacobson et al. (1999) have observed similar rate enhancements for styrene hydrogenation using a water-soluble Rh complex with dppts ligands [dppts = $C_6H_5P(C_6H_4SO_3Na)_2$]. Turnover frequencies at 50% conversion were measured to be 4 h^{-1} in toluene/water versus 26 h^{-1} in sc CO_2/water. The main advantages of these systems are the possibility for facile separation of the catalyst-containing aqueous phase from the product-containing sc CO_2 phase and the subsequent ability to recycle the metal catalyst. However, the pH of the aqueous phase in water/CO_2 mixtures has been measured to be ca. 3 (Holmes et al., 1999), which can be either an advantage or a potential problem for certain catalytic reactions.

The use of sc CO_2 instead of toluene as a solvent leads to some rate enhancement in these two systems, although it is clear that this activity is still not practical for most nonpolar, nonvolatile substrates. Significant improvements to the biphasic water/supercritical CO_2 system were accomplished by forming H_2O/CO_2 emulsions using newly developed surfactants (Jacobson et al., 1999). Three different surfactants were used that form water in CO_2 (w/c) or CO_2 in water (c/w) emulsions: (1) anionic surfactant perfluoropolyether ammonium carboxylate, (2) cationic Lodyne 106A, and (3) nonionic poly(butylene oxide)-*b*-poly(ethylene oxide). The low interfacial tension, γ, between water and CO_2 (17 mNm^{-1} at pressures above 70 bar), which is significantly lower than water/alkane systems (30–60 mNm^{-1}), together with the increased surface areas (up to 10^5 m^2 L^{-1} emulsion) allowed for enhanced reaction rates across the interphase. The w/c and c/w emulsions also offered a distinct advantage compared with the more commonly used water/oil emulsions in that they can be easily broken by simply decreasing the pressure, thereby allowing the catalyst-containing aqueous phase to be reused. Hydrogenations of several 1-alkenes using the water-soluble catalyst $RhCl(tppds)_3$, [tppds = tris(3,5-disulfonatophenyl)phosphine] were performed in these emulsion systems. For styrene, the turnover frequencies increased from 26 h^{-1} in CO_2/water to 150–300 h^{-1} for the emulsions formed from the three different surfactants. The turnover frequencies (TOFs) at 50% conversion for 1-octene, 1-decene, and 1-eicosene were measured to be 140 h^{-1}, 110 h^{-1}, and 30 h^{-1}, respectively. Catalyst recycling was also demonstrated.

Conclusions

Catalysis is the cornerstone of the chemical, fuel processing, pharmaceutical, and agrochemical industries, and advances in this area will be key to the development of "greener" processes for chemical synthesis. While catalysis in sc CO_2 is still an emerging area of research, much work has been done that has clearly demonstrated that a wide range of catalyzed reactions may be conducted under homogeneous and heterogeneous conditions in this medium. Researchers have taken advantage of many of the properties of CO_2 as a solvent, in some cases leading to improvements in reaction selectivity, overall activity, product recovery, and catalyst recycling. These advances have been especially pronounced for reactions involving gaseous reagents, such as hydrogenation and hydroformylation, classes of reactions that are carried out industrially (using traditional reaction media) on very large scales.

While dense-phase CO_2 will not provide a panacea for green chemistry, future research in homogeneous and heterogeneous catalysis is warranted in a number of areas. A number of regio- or stereoselective reactions, including asymmetric catalysis, stand to potentially benefit from tunable solvent properties. The use of molecular oxygen for the selective oxidation of hydrocarbons, particularly through nonradical pathways, the use of multiple catalysts for tandem or cascading reactions, and the use of CO_2 as both a solvent *and as a reagent* all represent fertile grounds for future research in CO_2.

Multiphase approaches appear to be very promising. The ability of well-defined organic species of moderate molecular weight either to transport reagents or catalysts across immiscible phases or to facilitate the convergence of substrates and catalysts in a small reaction volume are two means by which surfactants (Chapter 8 in this book; McClain et al., 1996) may be exploited for use in CO_2. The abilities of diblock or multiblock surfactants to form micelles or reverse micelles, and then for these metastable structures to reversibly disassemble by varying the operating pressure, represent powerful and underutilized tools for enhancing and optimizing not only reactivity and selectivity but also product isolation and catalyst recycling. Immobilizing homogeneous catalysts in distinct phases, such as CO_2-soluble or swellable polymers, polymeric or inorganic membranes, or ionic liquids, could bridge the gap between homogeneous and heterogeneous catalysis. The area of catalysis in dense-phase CO_2 is still young, and we anticipate exciting advances in the near future.

References

Baiker, A. *Chem. Rev.* **1999**, *99*, 453–473.

Baker, R. T.; Tumas, W. *Science* **1999**, *284*, 1477–1479.

Baptist-Nguyen, S.; Subramaniam, B. *AIChE J.* **1992**, *38*, 1027–1037.

Bertucco, A.; Canu, P.; Devetta, L.; Zwahlen, A. G. *Ind. Eng. Chem. Res.* **1997**, *36*, 2626–2633.

Bhanage, B. M.; Ikushima, Y.; Shirai, M.; Arai, M. *Chem. Comm.* **1999a**, 1277–1278.

Bhanage, B. M.; Ikushima, Y.; Shirai, M.; Arai, M. *Tetrahedron Lett.* **1999b**, *40*, 6427–6430.

Bhanage, B. M.; Ikushima, Y.; Shirai, M.; Arai, M. *Catal. Lett.* **1999c**, *62*, 175–177.

Birnbaum, E. R.; LeLacheur, R. M.; Horton, A. C.; Tumas, W. *J. Mol. Cat. A*, **1999**, *139*, 11–24.

Buelow, S.; Dell'Orco, P.; Morita, D. K.; Pesiri, D. R.; Birnbaum, E.; Borkowsky, S. L.; Brown, G. H.; Feng, S.; Luan, L.; Morgenstern, D. A.; Tumas, W. In *Frontiers in Benign Chemical Synthesis and Processing*; Anastas, P. T., Williamson, T. C., Eds.; Oxford University Press: 1998; pp. 264–285.

Burk, M. J.; Feaster, J. E.; Nugent, W. A.; Harlow, R. L. *J. Am. Chem. Soc.* **1993**, *115*, 10125–10138.

Burk, M. J.; Feng, S. G.; Gross, M. F.; Tumas, W. *J. Am. Chem. Soc.* **1995**, *117*, 8277–8278.

Cacchi, S.; Fabrizi, G.; Gasparrini, F.; Villani, C. *Synlett* **1999**, 345–347.

Carroll, M. A.; Holmes, A. B. *Chem. Commun.* **1998**, 1395–1396.

Carter, C. A. G.; Nolan, S.; Baker, R. T.; Tumas, W. *Chem. Commun.* **2000**, 347.

Chandler, K.; Culp, C. W.; Lamb, D. R.; Liotta, C. L. Eckert, C. A. *Ind. Eng. Chem. Res.* **1998**, *37*, 3252–3259.

Clark, M. C.; Subramaniam, B. *Chem. Eng. Sci.* **1996**, *51*, 2369–2377.

Clark, M. C.; Subramaniam, B. *Ind. Eng. Chem. Res.* **1998**, *37*, 1243–1250.

Clifford, A. A.; Pople, K.; Gaskill, W. J.; Bartle, K. D.; Rayner, C. M. *Chem. Commun.* **1997**, 595–596.

Clifford, A. A.; Pople, K.; Gaskill, W. J.; Bartle, K. D.; Rayner, C. M. *J. Chem. Soc., Faraday Trans.* **1998**, *94*, 1451–1456.

Collman, J. P.; Hegedus, L. S.; Norton, J. R.; Finke, R. G. *Principles and Applications of Organotransition Metal Chemistry*; University Science Books: Mill Valley, CA, 1987.

Conte, V.; DiFuria, F.; Licini, G. *Appl. Catal., A* **1997**, *157*, 335–361.

Cornils, B.; Herrmann, W. A. *Applied Homogeneous Catalysis with Organometallic Compounds*; VCH: New York, 1996.

Cornils, B.; Herrmann, W. A. *Aqueous-Phase Organometallic Catalysis*; Wiley-VCH: Weinheim, 1998.

Dell, C. P. J. Chem. Soc., *Perkin Trans. 1* **1998**, *22*, 3873–3905.

Devetta, L.; Giovanzana, A.; Canu, P.; Bertucco, A.; Minder, B. J. *Catal. Today* **1999**, *48*, 337–345.

Dooley, K. M.; Knopf, F. C. *Ind. Eng Chem. Res.* **1987**, *26*, 1910–1916.

Francio, G.; Leitner, W. *Chem. Commun.* **1999**, 1663–1664.

Fürstner, A.; Koch, D.; Langemann, K.; Leitner, W.; Six, C. *Angew. Chem., Int. Ed. Eng.* **1997**, *36*, 2466–2469.

Gaffney, A. M.; Sofranko, J. A. Symposium on Catalytic Selective Oxidation, American Chemical Society, Washington, DC, 1992, 1273–1279.

Gaffney, A. M.; Sofranko, J. A. U.S. Patent 5,210,336, 1993.

Gao, Y.; Shi, Y. F.; Zhu, Z. N.; Yuan, W. K. *3rd International Symposium on High-Pressure Chemical Engineering*, Zurich, Switzerland, 1996; pp. 151–156

Gao, Y.; Liu, H. Z.; Shi, Y. F.; Yuan, W. K. *4th International Symposium on Supercritical Fluids*; Sendai, Japan, 1997; pp. 531–534.

Ginosar, D. M.; Subramaniam, B. *J. Catal.* **1995**, *152*, 31–41.

Guo, Y.; Akgerman, A. *Ind. Eng. Chem. Res.* **1997**, *36*, 4581–4585.

Guo, Y.; Akgerman, A. *J. Supercrit. Fluids* **1999**, *15*, 63–71.

Gray, W. K.; Smail, F. R.; Hitzler, M. G.; Ross, S. K.; Poliakoff, M. *J. Am. Chem. Soc.* **1999**, *121*, 10711–10718.

Haas, G. R.; Kolis, J. W. *Organometallics* **1998a**, *17*, 4454–4460.

Haas, G. R.; Kolis, J. W. *Tetrahedron Lett.* **1998b**, *39*, 5923–5926.

Hâncu, D.; Beckman, E. J. *Ind. Eng. Chem. Res.* **1999a**, *38*, 2824–2832.

Hâncu, D.; Beckman, E. J. *Ind. Eng. Chem. Res.* **1999b,** *38*, 2833–2841.

Hitzler, M. G.; Poliakoff, M. *Chem. Commun.* **1997**, 1667–1668.

Hitzler, M. G.; Smail, F. R.; Ross, S. K.; Poliakoff, M. *Chem. Commun.* **1998a**, 359–360.

Hitzler, M. G.; Smail, F. R.; Ross, S. K.; Poliakoff, M. *Org. Process Res. & Dev.* **1998b**, *2*, 137–146.

Holmes, J. D.; Ziegler, K. J.; Audriani, M.; Lee, C. T., Jr.; Bhargava, P. A.; Steytler, D. C.; Johnston, K. P. *J. Phys. Chem. B* **1999**, *103*, 5703–5711.

Jacobson, G. B.; Lee, T.; Johnston, K. P.; Tumas, W. *J. Am. Chem. Soc.* **1999**. *121*, 11902–11903.

Jeong, N.; Hwang, S. H.; Lee, Y. W.; Lim, J. S. *J. Am. Chem. Soc.* **1997**, *119*, 10549–10550.

Jessop, P. G. *Top. Catalysis* **1998**, *5*, 95–103.

Jessop, P. G.; Leitner, W., Eds. *Chemical Synthesis Using Supercritical Fluids*; Weinheim: Wiley-VCH, 1999.

Jessop, P. G.; Ikariya, T.; Noyori, R. *Nature* **1994**, *368*, 231–233.

Jessop, P. G.; Hsiao, Y.; Ikariya, T.; Noyori, R. *Chem. Commun.* **1995**, 707–708.

Jessop, P. G.; Ikariya, T.; Noyori, R. *Organometallics* **1995b**, *14*, 1510–1513.

Jessop, P. G.; Ikariya, T.; Noyori, R. *Science* **1995c**, *269*, 1065–1069.

Jessop, P. G.; Hsiao, Y.; Ikariya, T.; Noyori, R. *J. Am. Chem. Soc.* **1996**, *118*, 344–355.

Jessop, P. G.; Ikariya, T.; Noyori, R. *Chem. Rev.* **1999**, *99*, 475–493.

Jia, L.; Jiang, H.; Li, J. *Chem. Comun.* **1999**, 985–986.

Kainz, S.; Koch, D.; Baumann, W.; Leitner, W. *Angew. Chem., Int. Ed. Engl.* **1997**, *36*, 1628–1630.

Kainz, S.; Koch, D.; Leitner, W. In *Selective Reactions of Metal Activated Molecules*; Werner, H., Schreier, W., Eds. Vieweg: Wiesbaden, 1998; p. 151.

Kainz, S.; Brinkmann, A.; Leitner, W.; Pfaltz, A. *J. Am. Chem. Soc.* **1999**, *121*, 6421–6429.

Kayaki, Y.; Noguchi, Y.; Iwasa, S.; Ikariya, T.; Noyori, R. *Chem. Commun.* **1999**, 1235–1236.

King, M. B.; Bott, T. R., Eds. *Extraction of Natural Products Using Near-Critical Solvents*; Blackie Academic & Professional: Bishopbriggs, U.K., 1993.

Klingler, R. J.; Rathke, J. W. *J. Am. Chem. Soc.* **1994**, *116*, 4772–4785.

Koch, D.; Leitner, W. *J. Am. Chem. Soc.* **1998**, *120*, 13398–13404.

Kramer, G. M.; Leder, F. U.S. Patent 3,880,945, 1975.

Kreher, U.; Schebesta, S.; Walther, D. *Z. Anorg. All. Chem.* **1998**, *624*, 602–612.

Krocher, O.; Koppel, R. A.; Baiker, A. 3rd International Symposium on High-Pressure Chemical Engineering, Zurich, Switzerland, 1996a, 91–96.

Krocher, O.; Koppel, R. A.; Baiker, A. *Chem. Commun.* **1996b**, 1497–1498.

Krocher, O.; Koppel, R. A.; Froba, M.; Baiker, A. *J. Catal.* **1998**, *178*, 284–298.

Krocher, O.; Koppel, R. A.; Baiker, A. *J. Mol. Catal. A* **1999**, *140*, 185–193.

Loeker, F.; Koch, W.; Leitner, W. In *Selective Oxidations in Petrochemistry*; Emig, C., Kohlpainter, C., Eds. DGMK: Hamburg, 1998; p. 209.

Lyons, J. E.; Ellis, P. E.; Myers, H. K. *J. Catal.* **1995**, *155*, 59–73.

McClain, J. B.; Betts, D. E.; Canelas, D. B.; Samulski, E. T.; DeSimone, J. M.; Londono, J. D.; Cochran, H. D.; Wignall, G. D.; Chillura-Martino, D.; Triolo, R. *Science* **1996**, *274*, 2049–2052.

McCoy, B. J.; Subramaniam, B. *AIChE J.* **1995**, *41*, 317–323.

McHugh, M.; Krukonis, V. *Supercritical Fluid Extraction*; Butterworth-Heinemann: Boston, MA, 1994.

Minder, B.; Mallat, T.; Pickel, K. H.; Steiner, K.; Baiker, A. *Catal. Lett.* **1995**, *34*, 1–9.

Mistele, C. D.; Thorp, H. H.; DeSimone, J. M. *J. Macromol. Sci. Pure Appl. Chem., A33* **1996**, 953–960.

Morgenstern, D. A.; LeLacheur, R. M.; Morita, D. K.; Borkowsky, S.; Feng, S.; Brown, G. H.; Luan, L.; Gross, M. F.; Burk, M. J.; Tumas, W. In *Green Chemistry: Designing Chemistry for the Environment*; Anastas, P. T., Williamson, T. C., Eds.; American Chemical Society: Washington, DC, 1996; pp. 132–151.

Morita, D. K.; Pesiri, D. R.; David, S. A.; Glaze, W. H.; Tumas, W. *Chem. Commun.* **1998**, 1397–1398.

Mukhopadhyay, M.; Srinivas, P. *Ind. Eng. Chem. Res.* **1996**, *35*, 4713–4717.

Noyori, N. *Asymmetric Catalysis in Organic Synthesis*; Wiley: New York, 1994.

Oakes, R. S.; Heppenstall, T. J.; Shezad, N.; Clifford, A. A.; Rayner, C. M. *Chemical Comunications* **1999**, 1459–1460.

Olah, G. A.; Marinez, E.; Torok, B.; Prakash, G. K. S. *Catal. Lett.* **1999**, *61*, 105–110.

Palo, D. R.; Erkey, C. *Ind. Eng. Chem. Res.* **1998**, *37*, 4203–4206.

Palo, D. R.; Erkey, C. *Ind. Eng. Chem. Res.* **1999a**, *38*, 2163–2165.

Palo, D. R.; Erkey, C. *Ind. Eng. Chem. Res.* **1999b**, *38*, 3786–3792.

Parshall, G. W.; Ittel, S. D. *Homogeneous Catalysis*; John Wiley and Sons New York, 1992.

Pesiri, D. R.; Morita, D. K.; Glaze, W.; Tumas, W. *Chem. Comun.* **1998**, 1015–1016.

Pesiri, D. R.; Morita, D. K.; Walker, T.; Tumas, W. *Organometallics* **1999**, *18*, 4916–4924.

Rathke, J. W.; Klingler, R. J.; Krause, T. R. *Organometallics* **1991**, *10*, 1350–1355.

Rathke, J. W.; Klingler, R. J.; Krause, T. R. *Organometallics* **1992**, *11*, 585–588.

Saim, S.; Subramaniam, B. *J. Supercrit. Fluids* **1990**, *3*, 214–221.

Santelli, M. In *Lewis Acids and Selectivity in Organic Synthesis*; CRC Press Boca Raton, 1996.

Sharpless, K. B.; Verhoeven, T. R. *Aldrichimica Acta* **1979** *12*, 63.

Shezad, N.; Oakes, R.; Clifford, A. A.; Rayner, C. M. *Tetrahedron Lett.* **1999**, *40*, 2221–2224.

Srinivas, P.; Mukhopadhyay, M. *Ind. Eng. Chem. Res.* **1997**, *36*, 2066–2074.

Subramaniam, B.; Clark, M. C. U.S. Patent 5,907,075, 1999.

Subramaniam, B.; McCoy, B. J. *Ind. Eng. Chem. Res.* **1994**, *33*, 504–508.

Subramaniam, B.; McHugh, M. A. *Ind. Eng. Chem. Process Des. Dev.* **1986**, *25*, 1–12.

Sun, Y. K.; Landau, R. N.; Wang, J.; Leblond, C.; Blackmond, D. G. *J. Am. Chem. Soc.* **1996**, *118*, 1348–1353.

Tacke, T. *Chem.-Anlagen Verfahren* **1995**, *11*, 19.

Tacke, T.; Wieland, S.; Panster, P. 3rd International Symposium on High-Pressure Chemical Engineering, Zurich, Switzerland, 1996, pp. 17–21.

Thomas, J. M.; Thomas, W. J. *Principles and Practice of Heterogeneous Catalysis*; VCH: Weinheim, 1997.

Vieville, C.; Mouloungui, Z.; Gaset, A. *Ind. Eng. Chem. Res.* **1993**, *32*, 2065–2068.

Vieville, C.; Mouloungui, Z.; Gaset, A. 3rd International Symposium on Supercritical Fluids, Strasbourg, France, 1994, pp. 19–24.

Wegner, A.; Leitner, W. *Chem. Commun.* **1999**, 1583–1584.

Weinstein, R. D.; Renslo, A. R.; Danheiser, R. L.; Tester, J.W. *J. Phys. Chem. B* **1999**, *103*, 2878–2887.

Woodard, S. S.; Finn, M. G.; Sharpless, K. B. *J. Am. Chem. Soc.* **1991**, *113*, 106–113.

Wu, X. W.; Oshima, Y.; Koda, S. *Chem. Lett.* **1997**, 1045–1046.

Xiao, J. L.; Nefkens, S. C. A.; Jessop, P. G.; Ikariya, T.; Noyori, R. *Tetrahedron Lett.* **1996**, *37*, 2813–2816.

Zhou, L. B.; Akgerman, A. *Ind. Eng. Chem. Res.* **1995**, *34*, 1588–1595.

Zhou, L. B.; Erkey, C.; Akgerman, A. *AIChE J.* **1995**, *41*, 2122–2130.

3

Carbon Dioxide as a Reactant and Solvent in Catalysis

TAKAO IKARIYA
RYOJI NOYORI

An increased awareness of global atmospheric carbon levels and heightened efforts to recover industrial emissions prior to their release into the environment has led to the availability of an unprecedented amount of carbon dioxide for industrial utilization. Unfortunately, chemical utilization of carbon dioxide as an industrial feedstock is limited by thermodynamic and kinetic constraints. Toxic carbon monoxide, the main competitor in many processes, is used in industry instead because CO_2 is perceived to be less reactive and its efficient catalytic conversion has remained elusive. The major commercial uses of CO_2 today are in beverages, fire extinguishers, and refrigerants, where inert physical properties such as oxidative and thermodynamic stability are advantageous. It is this stability that has limited the use of CO_2 to only a very few synthetic chemical processes (urea, aspirin, carbonates) despite the enormous availability of this resource. The conversion of CO_2 into useful organic compounds will likely rely on the use of metal catalysts to lower energy inputs. Increasingly, the use of supercritical carbon dioxide appears to offer significant advantages in the catalytic activation of CO_2 to yield useful products.

Liquid or supercritical CO_2 (sc CO_2) can be used as a reaction medium and can potentially replace conventional organic solvents to serve as an environmentally benign reaction medium (Ikariya and Noyori, 1999; Jessop and Leitner, 1999; Jessop et al., 1995b; Noyori, 1999). A supercritical fluid (SCF) is any substance that has a temperature and pressure higher than their critical values and which has a density close to or higher than its critical density (Jessop and Leitner, 1999; Jessop et al., 1995b). Carbon dioxide has a critical temperature of 31.0 °C and a critical pressure of 71.8 bar. The supercritical region of the phase diagram is the one at temperatures higher than the T_c and pressures higher than the P_c at which the liquid and gas phases become indistinguishable (Figure 3.1). Below T_c, liquid CO_2 can be maintained under relatively modest pressures. Subcritical liquid CO_2 behaves like any other nonpolar liquid solvent. Properties such as density

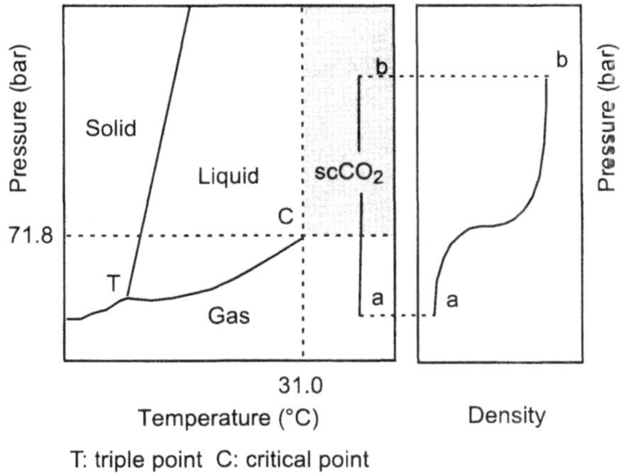

Figure 3.1. Phase diagram of CO_2 and its density as a function of pressure.

(Figure 3.1) are continuous above the T_c and discontinuous below it. Thus, intermediate densities can be obtained in the supercritical region. Supercritical fluids exist as a single phase, combining some of the advantageous gas-phase properties (such as high miscibility with other gases, high diffusivity, and relatively weak molecular association) with liquid-phase properties (such as the ability to dissolve and transport organic compounds) (Darr and Poliakoff, 1999; Ikariya and Noyori, 1999; Jessop and Leitner, 1999; Jessop et al., 1995b, 1999; Noyori, 1999).

Among the chemical advantages, the most pertinent to homogeneous catalysis in sc CO_2 is the miscibility of gases, such as H_2, with sc CO_2. The concentration of H_2 in a supercritical mixture of H_2 (83.7 bar) and CO_2 (118.2 bar) at 50 °C is 3.2 M, while the concentration of H_2 in tetrahydrofuran (THF) under the same pressure is merely 0.4 M (Jessop et al., 1996). This property of sc CO_2 must allow significant rate enhancement of reactions for which the rate is greater than zeroth order in H_2 concentration. The usefulness of this property has been noticed previously. The high concentration of CO_2 in sc CO_2 could also be advantageous in any reaction that incorporates CO_2. It is possible therefore that the use of sc CO_2 as a solvent will allow the more widespread use of CO_2 as an organic carbon source—a concept presented in a number of reports, patents, and applications.

Carbon dioxide is a linear molecule with equivalent C–O distances of 1.16 Å (Vol'pin and Kolomnikov, 1973). The bond strength in CO_2 is measured to be $D = 127.1$ kcal mol^{-1}, relatively weak compared with the CO bond in carbon monoxide ($D = 258.2$ kcal mol^{-1}) (Bard et al., 1985; Latimer, 1952; Weast, 1978). Resonance structures of the CO_2 molecule as illustrated in Figure 3.2 show that its chemical reactivity is associated either with the presence of carbon–oxygen double bonds and lone-paired electrons on the oxygen atoms or with the electrophilic carbon atom. The quantitative mole-

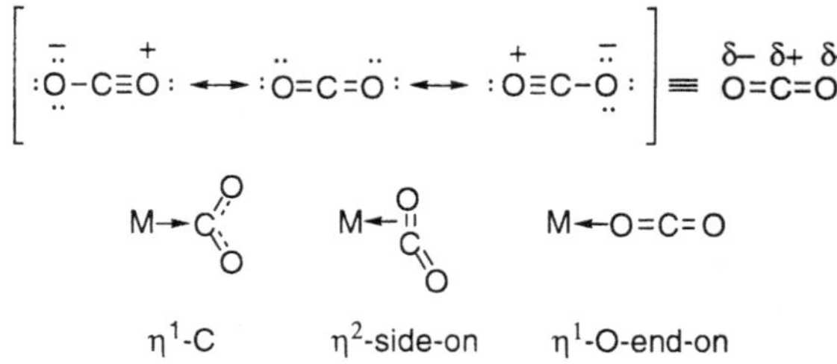

Figure 3.2. Resonance structure of the CO_2 molecule and three binding modes of CO_2 to a mononuclear metal center.

cular orbital (MO) energy-level diagram of linear CO_2 molecules shows that the first ionization potential is estimated to be as high as 13.7 eV and the lowest unnocupied molecular orbital (LUMO) level is about 3.8 eV, indicating that CO_2 is a relatively weaker donor of electrons but has a high electron affinity with the central carbon atom (Vol'pin and Kolomnikov, 1973). The CO_2 molecule is not necessarily inert but interacts, for example, with transition metals in several manners that can lead to its activation. Figure 3.2 exhibits three binding modes of CO_2 to metallic centers, although such interaction may not be necessary for chemical reactions. Many of the aspects of the coordination chemistry of CO_2 have been extensively reviewed (Behr, 1988b; Jessop et al., 1995c; Vol'pin and Kolomnikov, 1973; Yin and Moss, 1999). Thus, CO_2 has been used as chemical raw material for a long time. A few industrial processes based on CO_2 are used, mainly syntheses of bulk chemicals such as inorganic carbonates (Group 1 and 2 elements, the Solvay process, used as pigments and in the glass and ceramic industry), urea (fertilizers, animal feeding additives, resins), organic carbonates, pharmaceuticals (Kolbe–Schmitt reaction, aspirin), and methanol synthesis via the water gas shift reaction (WGSR) (Aresta, 1990; Behr, 1988b). The total amount used in synthetic chemistry is around 100 Mt per year. These industrial applications have been reviewed extensively (Aresta, 1990; Behr 1988b; Eliasson, 1994).

Carbon dioxide (CO_2) is a remarkable C_1 feedstock because of the vast amounts of carbon that exist in this form or in carbonate rocks and because of its low cost and low toxicity. Along with water, CO_2 is an end product of metabolic systems and of most combustion processes and is thermodynamically very stable ($\Delta G = -94.3$ kcal mol^{-1}, cf. -32.8 kcal mol^{-1} for CO) (Behr, 1988b; Aresta, 1990). Therefore, its utilization requires reactions with certain high-energy substances or electroreductive processes. In nature, a large proportion of CO_2 in the air is being fixed via photosynthesis in which the chloroplasts of the plant promote the reaction of CO_2 with water to yield carbohydrates and oxygen. Sunlight is crucial to convert the energy-poor CO_2 to energy-rich molecules. Thus, glucose is efficiently produced on the surface of leaves and its volume is estimated to reach about

2×10^{11} per year (Behr, 1988b). Because photosynthesis is an ideal chemical reaction in terms of efficiency and environmental concerns, chemists aim to establish such a perfect reaction—an artificial photosynthesis that would provide an efficient transformation technology.

In order to utilize CO_2 as a reactant or reaction medium, a large amount of concentrated CO_2 with a high purity is required. Although recovery and disposal of diluted CO_2 from sources such as the atmosphere (0.03% in volume) or exhaust gases generated by transportation and heating systems are either not feasible or economically unfavorable because of concentrations that are too low, concentrated CO_2 is available from sources such as flue gases from power plants (about 20% of CO_2) and industrial processes. In addition, a considerable amount of CO_2 is potentially available after extraction from flue gases in natural gas wells if a practical way of using large volumes of CO_2 is developed (Aresta, 1990). The focus is especially on the utilization of CO_2 under supercritical conditions for both homogeneous catalysis by metal-containing complexes and some heterogeneously catalyzed reactions for illustrative purposes. The reader is referred to other chapters in this book for discussion of phase-transfer catalysis, polymerization, uncatalyzed reactions, and radical reactions in sc CO_2.

The Reactivity of CO_2

Liquid CO_2 and sc CO_2 are not always inert reaction media. Carbon dioxide will react under at least some conditions and with some reagents. The CO_2 molecule inserts readily into M–H, M–R, M–OR, or M–NR_2 bonds in transition-metal complexes (Aresta 1990; Behr, 1988b; Jessop et al., 1995c; Yin and Moss, 1999) and reacts with secondary or primary amines to form carbamate salts (Takeshita and Kitamoto, 1988; Wright and Moore, 1948). As mentioned earlier, catalytic reactions of CO_2 are well known in industrial processes (Behr, 1988b). There have been many reports of catalytic reactions of CO_2 with various kinds of substrates such as strained hetereocycles and hydrocarbons including aromatics, alkenes, alkynes, and 1,3-dienes (Behr, 1988b; Palmer and Van Eldik, 1983; Yin and Moss, 1999). These reactions have already been reviewed by several authors (Behr, 1988a; Braunstein et al., 1988; Darensbourg, 1990; Darensbourg et al., 1988; Denise and Sneeden, 1982; Eisenberg and Hendricksen, 1979; Inoue and Yamazaki, 1982; Sneeden, 1982; Walther, 1987; Yin and Moss, 1999). They are of great industrial interest, but are still in the research phase. Based on the stoichiometric reactions and the many examples of the fixation of CO_2, the reactivity of CO_2 will continue to be researched in order to attain CO_2 fixation.

Catalytic hydrogenation of CO_2 is one of the most promising approaches to CO_2 fixation. The hydrogenation of CO_2 to CO, hydrocarbons, and alcohols is thermodynamically favorable because of the concomitant production of water. However, hydrogenation to formic acid is not thermodynamically favorable. In order to reduce CO_2 beyond the level of formic acid, an oxygen sink is required. If reasonable yields are to be obtained, the formic acid product must be stabilized with a reagent such as a base (giving formate salts), an alcohol (giving formate esters), or an epoxide (giving diol formates). Formate esters and formamides are often produced via formic acid, but it is a mistake to assume that the production of these products by CO_2 hydrogenation necessarily requires formic acid as an intermediate. The products of CO_2 hydrogenation depend on the conditions and catalyst (Jessop et al., 1995c). High-temperature hydrogenation conditions, usually along with heterogeneous catalysis, produce CO, CH_3OH, or CH_4. Mild conditions usually give formic acid or its derivatives such as alkyl formates or formamides (Jessop et al., 1995c).

Hydrogenation of CO_2

Hydrogenation of CO_2 to CH_3OH could result in a new, large-scale utilization of CO_2 because CH_3OH can be used as a storage and transport medium for hydrogen gas in addition to being an important source of various chemicals and energy (Behr, 1988b; Eliasson, 1994). The CH_3OH is made commercially from synthesis gas, a mixture of CO and H_2 in a 1:2 molar ratio. When gas with a composition of $CO{:}H_2 = 1{:}3$, generated from steam reforming of natural gases, is used instead, CO_2 is needed to counterbalance the surplus H_2. The reaction leads to CH_3OH with the formation of H_2O. Thus, a considerable amount of CO_2 is being transformed into an organic bulk chemical, at about 16×10^6 t per year (data from 1986) (Behr, 1988b; Eliasson, 1994) via this indirect method of producing CH_3OH from CO_2 through CO. On the other hand, many heterogeneous catalyst systems such as Cu/ZrO_2, Cu/ZnO, or Cu/Al_2O_3 have been investigated for direct methanol synthesis from CO_2 and H_2 (Paul and Pradier, 1994). The catalyst efficiency is strongly dependent on the catalyst component and reaction conditions. The maximum efficiency can be attained at around 250 °C for thermodynamic reasons. The selectivity to CH_3OH is of the order of 50–60%. Hydrocarbons and CO are the major by-products (Paul and Pradier, 1994).

Homogeneous hydrogenation using transition-metal complexes as catalysts under mild conditions gives formic acid and its derivatives with high selectivities (Jessop et al., 1995c). The first homogeneously catalyzed hydrogenation of CO_2 to formic acid was reported in 1976 by Inoue and Hashimoto who used transition metal complexes of Groups 9 and 10 as catalysts (Inoue et al., 1976). Subsequently, a variety of homogeneous catalysts of the second- and third-row metals of Groups 8 through 10, usually with halides or hydride as anionic ligands and tertiary phosphines as neutral ligands, have been found to be effective for this reaction (Jessop et al., 1995c). The most active catalyst, found by Leitner, is the bidentate phosphine–Rh complex, Rh(hfacac)[$Cy_2P(CH_2)_3PCy_2$] (hfacac = hexafluoroacetylacetonato; Cy = cyclohexyl) (Fornika et al., 1995; Hutschka et al., 1997; Leitner, 1995). Carbon dioxide hydrogenation with this Rh catalyst in a DMSO/$N(C_2H_5)_3$ mixture occurred rapidly with a maximum turnover frequency (TOF) of 1335 h^{-1} at 39.4 bar and 25 °C (Figure 3.3) (Fornika et al., 1995). Addition of a base improves the enthalpy and bases used include inorganic bases and trialkylamines. Dialkylamines are also usable—especially for aqueous systems, because of higher solubility. The use of aqueous medium is an attractive approach to fixation of CO_2 because water is the usual solvent for the recovery and concentration of CO_2 from processed flue gases. Leitner has cleverly used a water-soluble rhodium complex, $Rh(Cl)(tppts)_3$ [tppts = tris(3-sulfonatophenyl)phosphine], as the catalyst for hydrogenation of CO_2, which gave formate with initial TOFs of 7260h^{-1} at 81 °C and 1335 h^{-1} at 23 °C in the presence of dimethylamine under 39.4 bar (CO_2:H_2 = 1:1) (Figure 3.4) (Leitner et al., 1998). On the other hand, the Ru complex with tppts ligand, $RuCl_2(tppts)_3$, gave a low TOF of 6 h^{-1} at 23 °C. Interestingly, water-soluble Ru and Rh complexes with 3-sulfonatophenyldiphenylphosphine (tppms), $RuCl_2(tppms)_2$ and $RhCl(tppms)_3$, have proved to be efficient catalysts for hydrogenation of hydrogen carbonate to formate in water under mild conditions, 4.9–19.7 bar of CO_2 and 34.5–59.1 bar of H_2 (Joó et al., 1999). The reaction rate was highly dependent on the pH of the solution tested and, at around pH = 8.3, the maximum TOF exceeded 260 h^{-1} with the Rh complex. An increase of the H_2 pressure resulted in an increase in the rate of the reaction with both catalysts. Carbon dioxide in the gas phase was detrimental with the Ru catalyst but was essential for the high rate of the reaction with the Rh complex. Further mechanistic investigation is necessary.

Performing hydrogenation of CO_2 in sc CO_2 allows one to take advantage of the miscibility of H_2 in sc CO_2. We showed that dissolving $N(C_2H_5)_3$, 0.0006 equivalents of $RuH_2[P(CH_3)_3]_4$, and 83.7 bar of H_2 in 50 mL of sc CO_2 (total pressure 206.8 bar) led to rapid production of formic acid,

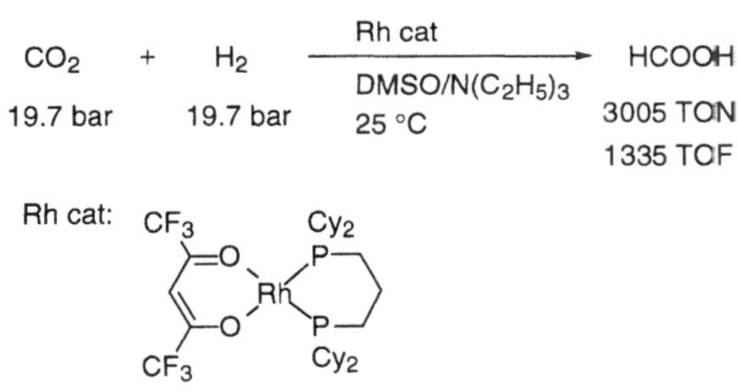

Figure 3.3. Carbon dioxide hydrogenation catalyzed by a Rh complex in a mixture of DMSO and $N(C_2H_5)_3$.

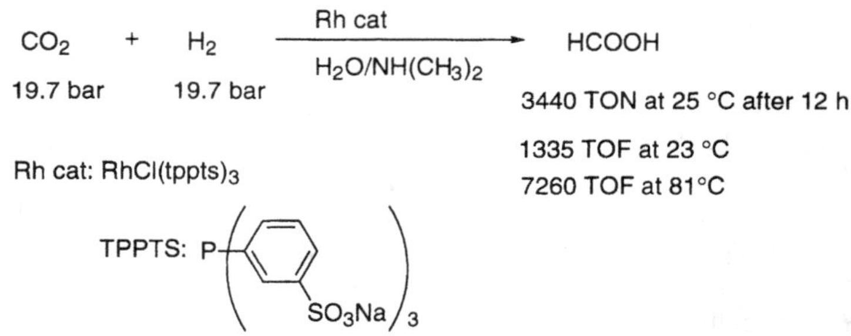

Figure 3.4. Carbon dioxide hydrogenation catalyzed by a water-soluble Rh complex in H_2O. The initial rate was determined from the data obtained within the first 10 min of the reaction. At 81 °C, the catalyst was deactivated after 10 min.

as the 2:1 adduct with the amine, with a TOF of 680 h^{-1} at 50 °C (Figure 3.5) (Ikariya et al., 1993, 1995; Jessop et al., 1994a, 1994b, 1995, 1996, 1999). A comparison of different catalyst precursors showed that $RuH_2[P(C_6H_5)_3]_4$, which has moderate activity in conventional solvents, was less active than $RuH_2[P(CH_3)_3]_4$ because of the low solubility in sc CO_2 and that $RuCl_2[P(CH_3)_3]_4$ had an induction period but was very active thereafter.

The rate of the reaction in sc CO_2 is highly dependent on the temperature, the pressure, and the reagent concentrations. The dependence of rate on reagent concentration seems to be more a function of phase behavior than a function of the stoichiometry of the rate-determining step. Although the reaction system was homogeneous at the start, the product precipitated as the reaction proceeded. Addition of 0.02 equivalent of water doubled the rate of the reaction to 1400 h^{-1} (Figure 3.5). Adding a small quantity of

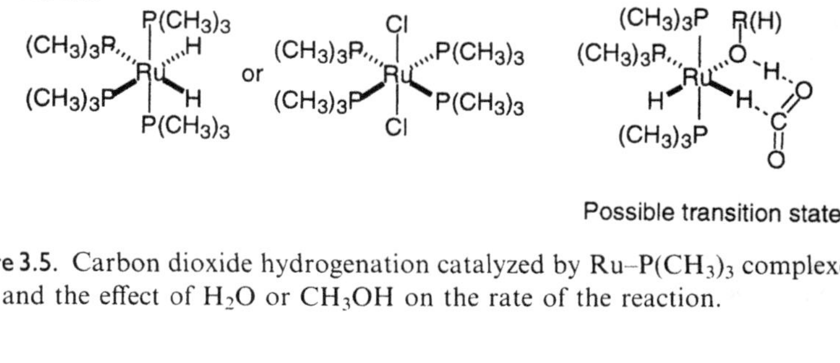

Figure 3.5. Carbon dioxide hydrogenation catalyzed by Ru–$P(CH_3)_3$ complexes in sc CO_2 and the effect of H_2O or CH_3OH on the rate of the reaction.

methanol dissolved into the supercritical solution was even more effective at accelerating the reaction (over 4000 h^{-1}). The great enhancement of the reaction rate is tentatively attributed to the hydrogen-bond stabilization of the transition state as shown in Figure 3.5. Adding too much water or methanol caused the rate to drop dramatically because of the formation of a liquid water or methanol phase (Table 3.1). The rate was also dependent on the amount of $N(C_2H_5)_3$ in the vessel; the rate increased with increasing amounts of $N(C_2H_5)_3$ up to a limit, beyond which the mixture was no longer homogeneous and the rate dropped significantly. Addition of DMSO accelerated the reaction so much that we were unable to determine whether the single-phase (dissolved DMSO) or the biphasic (liquid DMSO) conditions led to the higher rate.

The use of methanol as an additive at 80 °C led to the synthesis of methyl formate in addition to formic acid (Figure 3.6) (Ikariya et al., 1994a; Jessop et al., 1995a, 1996), and the use of biphasic conditions (liquid methanol) caused an increase, not a decrease, in the rate of hydrogenation of sc CO_2 (Jessop et al., 1996).

With primary or secondary amines as the base for sc CO_2 hydrogenation at 100 °C, mono- or dialkylformamides were synthesized in TON up to 420,000, far higher than any previously reported (Figure 3.6) (Ikariya et al., 1994a, 1995; Jessop et al., 1994a, 1994b, 1996). This system was complicated by the fact that the secondary or primary amine formed a liquid

Table 3.1. Hydrogenation of CO_2 in sc CO_2 and Other Solvents and the Effects of Water and Other Additives on the Initial Rate of Reaction[a]

Catalyst (μmol)	Additive (mmol)	Water (mmol)	Phase[b]	Time (h)	TOF (h^{-1})
2.9	—	560	2	1	34
2.0	THF, 190[c]	0.1	2	1	84
3.2	CH_3CN, 290[c]	0.1	2	1	330
3.4	CH_3OH, 250	0	2	0.5	1500
2.2	—	0	1	1	680
2.2	—	0.1	1	1	1400
2.5	CH_3OH, 13	0	1	0.5	>4000[d]
2.9	DMSO, 1.6	0	1	0.5	>3300[d]
2.5	DMSO, 140[e]	0	2	0.5	>4000[d]

[a]Reaction conditions: cat $RuH_2[P(CH_3)_3]_4$ 2.2–3.3 μmol, 5.0 mmol $N(C_2H_5)_3$, H_2 78.8–83.7 bar, total pressure 197.0–206.8 bar, 50 °C, 50-mL reaction vessel.

[b]Number of phases visible at the start of the reaction.

[c]15 mL.

[d]The initial rate must be higher than the rate shown.

[e]10 mL.

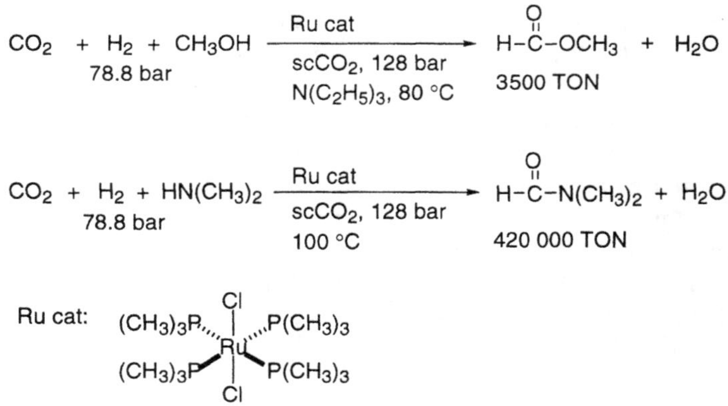

Figure 3.6. Methyl formate and DMF synthesis via CO_2 hydrogenation catalyzed by a Ru–$P(CH_3)_3$ complex in sc CO_2.

carbamate salt on contact with CO_2. Thus, two phases were in the vessel right from the start of each reaction, the sc CO_2 phase and a liquid salt phase. Although it was not possible to determine if the catalyst remained in the sc CO_2 phase throughout the reaction, the $RuCl_2[P(CH_3)_3]_4$ complex was shown to be soluble in sc CO_2 and not very soluble in the liquid carbamate. A standard sc CO_2 reaction procedure for the hydrogenation of CO_2 under otherwise identical conditions gave dimethylformamide (DMF) in a good yield. Several factors probably contributed to the high activity, including the favorable electronic properties of the catalyst and the ability of the SCF to dissolve both H_2 and the product DMF, as well as favorable phase control.

Subsequent studies by the group of Baiker (Kröcher et al., 1996) demonstrated that a heterogenized analogue of $RuCl_2[P(CH_3)_3]_4$ supported by a sol–gel-derived silica matrix was also very active for the synthesis of DMF, giving a TON of up to 110,800. The catalyst was prepared by the co-condensation of $RuCl_2[P(CH_3)_2(CH_2)_2Si(OC_2H_5)_3]_3$ with $Si(OC_2H_5)_4$. Supported Ir, Pd, Pt, and Rh catalysts were not as effective; the Rh complex was reduced to metal as had been observed in the homogeneous reaction (Jessop et al., 1996). The complex $RuCl_2(DPPE)_2$ [DPPE = 1,2-bis(diphenylphosphino)ethane], which is not soluble in sc CO_2, was a practical catalyst, showing high reaction rates in a 500-mL reaction vessel (up to 740,000 TON at the initial 18% conversion after 2 h) (Ikariya et al., 1994a). Baiker also showed that lower but still reasonable rates of DMF formation could be obtained at subcritical pressures with $RuCl_2(DPPE)_2$. The length of the carbon backbone in the diphosphine ligand affected the catalytic activity, with Ru complexes containing bis(diphenylphosphino)methane (dppm) and especially DPPP being inferior to the complex containing dppe. These results confirmed that the properties of the ligands have very strong effects on catalytic activity as observed in solution reaction with Rh complexes (Fornika et al., 1995).

Our group found that $RuCl_2(DPPE)_2$ was soluble in liquid dimethylammonium dimethylcarbamate where the hydrogenation of CO_2 occurred (Suzuki et al., 2001). Since dimethylamine is capable of strong interaction with the metal complexes, a significant rate enhancement observed in the DMF formation might be attributed to ligand acceleration by dimethylamine via a similar transition state as illustrated in Figure 3.5. As shown in Figure 3.7, the time course of DMF formation with both the $P(CH_3)_3$ and DPPE catalysts under identical supercritical conditions clearly showed that the catalyst performance of both complexes was almost identical: initial rapid DMF formation followed by a lowering of the rate at the end of the reaction probably due to inhibition by the large amount of water formed during the reaction. Favorable phase control might be important to maintain the catalyst for a long time.

Copolymerization of Epoxides and CO_2

Metal-catalyzed reactions of CO_2 and epoxides that give polycarbonates and/or carbonates have been extensively investigated as a potentially effective CO_2 fixation (Beckman, 1999; Inoue, 1987). The possible reaction mechanism is illustrated in Figure 3.8 (Darensbourg et al., 1999). The repetition of the reaction sequence in which CO_2 inserts into a metal–alkoxide bond, followed by ring-opening of the epoxide with the metal carbonate forms the alternating copolymer. In 1969, this copolymerization was first reported by Inoue and Tsuruta who used a Zn catalyst derived from

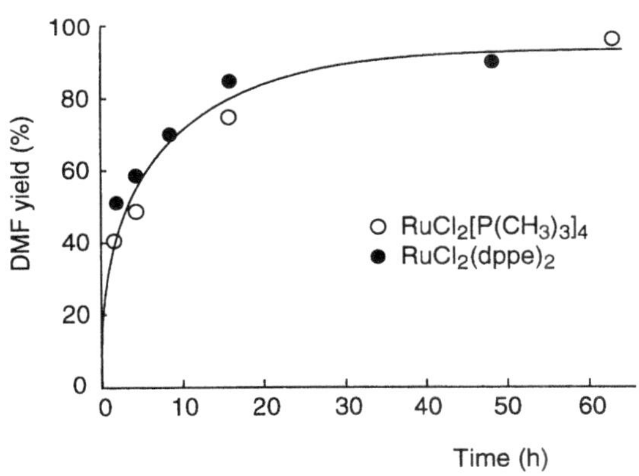

Figure 3.7. Time versus DMF yield plots for DMF formation via CO_2 hydrogenation catalyzed by $RuCl_2[P(CH_3)_3]_4$ and $RuCl_2(dppe)_2$. Reaction conditions: Ru cat 3.1–3.2 μmol, S/C = 10,000, P_{H2} = 82.7-84.7 bar, P_{CO2} = 124.1–128.0 bar, 100 °C

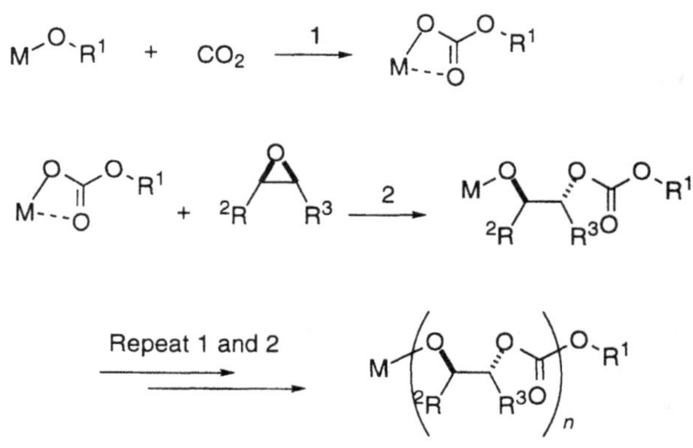

Figure 3.8. Mechanism of copolymerization of CO_2 and epoxides.

$Zn(C_2H_5)_2$ and H_2O (Inoue, 1976; Inoue et al., 1969). Since then, there have been a number of reports of polycarbonate formation from CO_2 and epoxides with Zn-based catalysts in organic solvents (DePasqueale 1973; Inoue et al., 1971a, 1971b; Kobayashi et al., 1973, Kuran et al., 1976; Rokicki and Kuran, 1981; Tsuda et al., 1976). A significant improvement in the activity of catalysts has been made by several research groups who used liquid CO_2 or sc CO_2 as a reactant and solvent for the copolymerization (Beckman, 1999). Darensbourg (Darensbourg et al., 1995) found Zn(II) glutarate as a heterogeneous catalyst effected the precipitation copolymerization of CO_2 and propene oxide in sc CO_2, leading to the formation of polycarbonate with an M_w of about 10^4 in 10–20% yields at 60 °C or 85 °C at 20.7–80.8 bar. Above the critical pressure of CO_2, the polymer contained a relatively high percentage of carbonate linkage to the ether linkage—over 90% compared with less than 75% under low-pressure conditions.

When a CO_2-soluble monomeric bulky phenoxy Zn(II) catalyst, Zn(2,6-diphenylphenoxide)$_2$(THF or ether)$_2$, was used as an initiator, the copolymerization of CO_2 and cyclohexene oxide at 80 °C and 50.2 bar produced over 400 g of polymer per gram of Zn (Figure 3.9) (Darensbourg and Holtcamp, 1995). The productivity attained was an order of magnitude higher than with previous catalysts (on the order of 50 g of polymer per gram of Zn). The catalytic activity was strongly influenced by the structure of the catalyst and the reaction conditions. As the substituents on the phenolate ligands of the catalysts varied from $CH(CH_3)_2$, to C_6H_5, to $C(CH_3)_3$, to CH_3 groups, the yield of the polymer increased from 477 g to 1441 g of polymer per gram of Zn over a reaction period of 69 h (Darensbourg et al., 1999). The productivity of the reaction was not improved by conducting the reaction under supercritical conditions, at 139.8 bar and 80 °C, and gave the product at 357 g per gram of Zn with 90% carbonate linkages. The tacticity

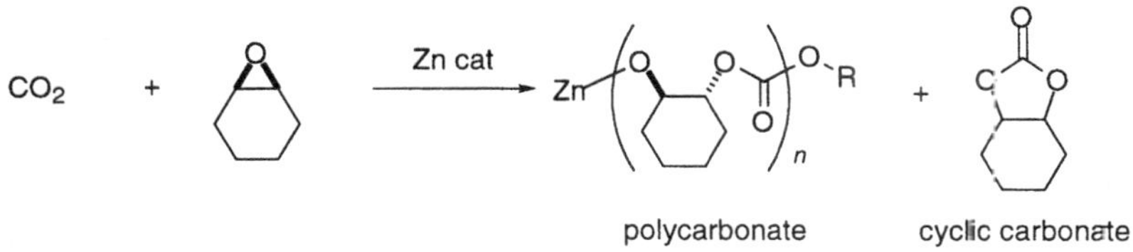

Zn cat		Medium (bar)	Temp (°C)	Time (h)	Yield (g polymer/g Zn)
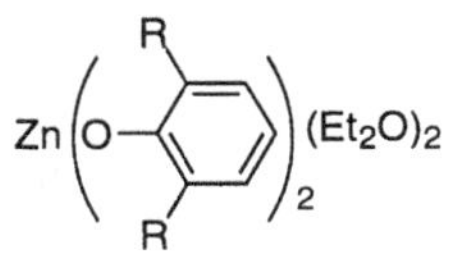	R = CH_3	CO_2, 50.2	80	69	1441
	$CH(CH_3)_2$	CO_2, 50.2	80	69	477
	$C(CH_3)_3$	CO_2, 50.2	80	69	677
	C_6H_5	CO_2, 50.2	80	69	602
	C_6H_5	$scCO_2$, 435.3	80	69	357
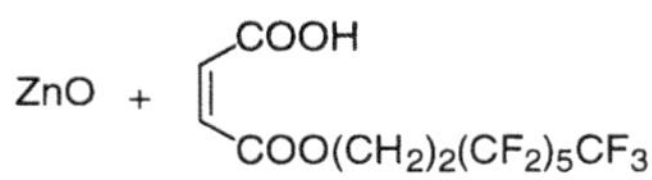		$scCO_2$, 136	110	24	457

Figure 3.9. Copolymerization of CO_2 and cyclohexene oxide catalyzed by Zn complexes in liquid CO_2 and sc CO_2.

of the copolymer chain was predominantly syndiotactic and the molecular weights of the methanol-insoluble products were in the range from $M_w = 45 \times 10^3$ to 173×10^3. Beckman (Super and Beckman, 1997; Super et al., 1998) found that a CO_2-soluble Zn catalyst prepared from ZnO and mono(3,3,4,4,5,5,6,6,7,7,8,8,8-tridecafluorooctyl)malate has proved to efficiently effect the copolymerization of CO_2 and cyclohexene oxide in sc CO_2 at 90–110 °C over a wide range of pressures (67.0–334.8 bar) (Figure 3.9). The maximum polymer yield exceeded 450 g per gram of Zn at 110 °C and 133.9 bar. A very careful study of phase behavior during the reaction revealed that the cyclohexene oxide–CO_2 binary system is the best choice for the effective copolymerization because of favorable mass transfer and high reactant concentration in the same phase containing the catalyst (Super and Beckman, 1997; Super et al., 1998).

Coates made a great improvement in the catalytic activity for copolymerization of CO_2 and epoxides by using a well-defined Zn catalyst with bulky β-diketiminate ligands, Zn(bdi)(OAc) or Zn(bdi)OCH_3 [bdi = 2-(2,6-dialkylphenylamino)-4-(2,6-dialkylphenylimino)-2-pentene] (Figure 3.10) (Cheng et al., 1998). Although the bdi–Zn complexes were inactive for the homopolymerization of epoxide, the addition of low-pressure CO_2 to the neat solution of epoxide containing the Zn catalyst resulted in the rapid production of the polycarbonate with an $M_w = 31.0 \times 10^3$ and a narrow polydispersity, $M_w/M_n = 1.11$, at 50 °C and 6.7 bar. The initial rate of the

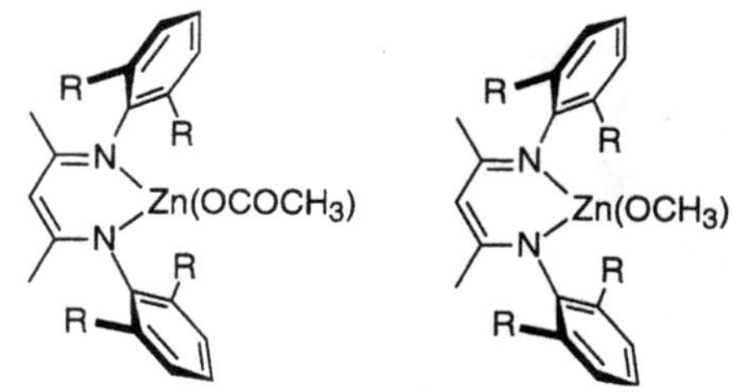

Figure 3.10. Highly active Zn(II)-based catalysts for the alternating copolymerization of CO_2 and epoxides. R = $CH(CH_3)_2$.

reaction was one or two orders of magnitude higher than the previously reported results. The reaction proceeded in a living manner to form syndiotactic polymers with predominantly carbonate linkages. In these copolymerizations, CO_2 acts as both a reactant and a medium and no solvent is required.

Reactions of CO_2 with Unsaturated Hydrocarbons

Although the reactions of CO_2 with unsaturated hydrocarbons such as alkenes, alkynes, and 1,3-dienes have been extensively investigated, their catalytic efficiencies are still unsatisfactory. Readers are referred to many outstanding review articles on this subject (Behr, 1988b). The supercritical conditions of CO_2 have been tested for CO_2 fixation into useful chemicals, in which the reaction efficiencies were improved in some cases. The coupling of alkynes with CO_2 catalyzed by $Ni(cod)_2/P(CH_3)_3$ (cod = 1,5-cyclooctadiene) proceeded well in sc CO_2 to give 2-pyrone derivatives with over 90% selectivity and a TON of 3 for 50 h at 40 °C (Dinjus and Fornika, 1996; Reetz et al., 1993). This selectivity, but not the rate, was greater than that obtainable in a THF/acetonitrile solution. The rate in sc CO_2 was improved by increasing the temperature; at 95 °C, a TON of 9 was obtained after about 9 h. Fourier transform infrared (FTIR) measurements (Kreher et al., 1998) showed that some catalyst deactivation was occurring by the formation of nickel carbonyl complexes. The source of the CO was believed to be from the nickel-catalyzed reduction of the CO_2 by $P(CH_3)_3$.

The reaction of propargylic alcohols and sc CO_2 in the presence of a trialkylphosphine as a catalyst gave cyclic carbonates in an excellent yield (Ikariya and Noyori, 1999). Dixneuf reported that the reaction proceeded without solvent, but not in nonpolar solvents such as toluene. The reaction efficiency in sc CO_2 was superior to that in solution phase (Fournier et al., 1989; Journier et al., 1991). The TON reached 1200 and the TOF exceeded 400. The sufficient concentration of CO_2, as well as the high reactivity of the ion-pair intermediate in sc CO_2, is responsible for such high efficiency.

Conclusions

Carbon dioxide has been used as a raw material in the syntheses of bulk chemicals such as urea or organic and inorganic carbonates, and there are also many examples in the literature where the catalytic reactions of CO_2-producing chemicals have been extensively developed. However, these reactions cannot make any substantial contribution to the problem of the control of CO_2 concentration in the air, even if considerable amounts of CO_2 are used, because most of these organic chemicals have a relatively short lifetime compared with that of inorganic carbonates and are back-converted into CO_2 in a short time. One should develop new synthetic procedures for CO_2 fixation to obtain such chemicals, which are produced by toxic or more expensive materials in conventional technology. The catalytic hydrogenation of CO_2 is among the most straightforward, attractive methods for this purpose. Research described in this review has shown that high catalytic efficiency, yields, and reaction rates can be obtained from CO_2 with optimum conditions and catalysts. It is obvious that the thermodynamic stabilities of the CO_2 and H_2O are predominant in the consideration of the reactions of CO_2 with H_2. In the hydrogenation of CO_2, water acts as an oxygen sink, but the hydrogen, which is thus consumed, is prepared industrially by the water gas shift reaction in which CO_2 itself is the ultimate oxygen sink. In this way, CO_2 acts as its own oxygen sink. This irony is a problem for the supposedly beneficial environmental impact of the use of CO_2 as a carbon source; its reduction—for example, to methanol—using H_2 from the WGSR does not result in the net consumption of CO_2, but rather in the net production of CO_2. If, in the future, solar or hydrothermally powered electrolysis were to become a major method for the production of H_2, then CO_2 hydrogenation would result in the overall consumption of CO_2. Despite the cost of H_2, in areas of the world where transportation costs increase the price of methanol or formic acid, then production from CO_2 and H_2 may be economically feasible. Industrial plants using heterogeneous catalysts for this process have been used or are being prepared by several companies.

References

Aresta, M. In *Enzymatic and Model Carboxylation and Reduction Reactions for Carbon Dioxide Utilization*; Aresta, M., Scloss, J. V., Eds.; NATO ASI Series C, Vol. 314; Kluwer Academic: Dordrecht, 1990; pp. 1–18.

Bard, A. J.; Paesons, R.; Jordan, J., Eds. *Standard Potentials in Aqueous Solution*; IUPAC, Physical and Analytical Chemistry Division, Marcel-Dekker: New York, 1985.

Beckman, E. J. *Science* **1999**, *283*, 946–947.

Behr, A. *Angew. Chem. Int. Ed. Engl.* **1988a**, *27*, 661–678.

Behr, A., Ed. *Carbon Dioxide Activation by Metal Complexes*; VCH: Weinheim, 1988b.

Braunstein, P.; Matt, D.; Nobel, D. *Chem. Rev.* **1988**, *88*, 747–764.

Cheng, M.; Lobkovsky, E. B.; Coates, G. W. *J. Am. Chem. Soc.* **1998**, *120*, 11018–11019.

Darensbourg, D. J. In *Enzymatic and Model Carboxylation and Reduction Reactions of Carbon Dioxide Utilization*; Aresta, M., Schloss, J. V., Eds; NATO ASI Series C, Vol. 314; Kluwer Academic: Dordrecht, 1990; pp. 43–64.

Darensbourg, D. J.; Holtcamp, M. W. *Macromolecules* **1995**, *28*, 7577–7579.

Darensbourg, D. J.; Bauch, C. G.; Ovalles, C. In *Catalytic Activation of Carbon Dioxide*; Ayers, W. M., Ed.; ACS Symposium Series 363; American Chemical Society: Washington, DC, 1988; pp. 26–41.

Darensbourg, D. J.; Stafford, N. W.; Katsurao, T. *J. Mol. Catal.* **1995**, *104*, L1–L4.

Darensbourg, D. J.; Holtcamp, M. W.; Struck, G. E.; Zimmer, M. S.; Niezgoda, S. A.; Rainey, P.; Robertson, J. B.; Draper, J. D.; Reibenspies, J. H. *J. Am. Chem. Soc.* **1999**, *121*, 107–116.

Darr, J. A.; Poliakoff, M. *Chem. Rev.* **1999**, *99*, 495–541.

Denise, B.; Sneeden, R. P. A. *CHEMTECH* **1982**, 108–112.

DePasqueale, R. J. *J. Chem. Soc., Chem. Commun.* **1973**, 157–158.

Dinjus, E.; Fornika, R. In *Applied Homogeneous Catalysis with Organometallic Compounds*; Cornils, B., Hermann, W. A., Eds; VCH: Weinheim, 1996; Vol. 2, pp. 1048–1072.

Eisenberg, R.; Hendricksen, D. E. *Adv. Catal.* **1979**, *28*, 79–172.

Eliasson, B. In *Carbon Dioxide Chemistry: Environmental Issues*; Paul, J., Pradier, C.-M., Eds.; Royal Society of Chemistry: London, 1994; pp. 5–15.

Fornika, R.; Görls, H.; Seemann, B.; Leitner, W. *J. Chem. Soc., Chem. Commun.* **1995**, 1479–1481.

Fournier, J.; Bruneau, C.; Dixneuf, P. H. *Tetrahedron Lett.* **1989**, *30*, 3981–3982.

Hutschka, F.; Dedieu, A.; Eichberger, M.; Fornika, R.; Leitner, W. *J. Am. Chem. Soc.* **1997**, *119*, 4431–4443.

Ikariya, T.; Noyori, R. In *Transition Metal Catalyzed Reactions*; Murahashi, S.-I, Davies, S. G., Eds.; IUPAC, Blackwell Science; New York, 1999; pp. 1–28.

Ikariya, T.; Jessop, P. G.; Noyori, R. Japan Tokkai Patent 5-274721, 1993.

Ikariya, T.; Jessop, P. G.; Hsiao, Y.; Noyori, R. Japan Tokkai Patent 6-125,402, 1994a.

Ikariya, T.; Jessop, P. G.; Noyori, R. Japan Tokkai Patent 6-125,401, 1994b.

Ikariya, T.; Hsiao, Y.; Jessop, P. G.; Noyori, R. Eur. Patent Appl. Patent 0 652 202 A1, 1995.

Inoue, S. *CHEMTECH* **1976**, September, 588–594.

Inoue, S. In *Carbon Dioxide as a Source of Carbon*; Aresta, M., Forti, G., Eds.; Reidel Publishing Co.: Dordrecht, 1987; pp. 331–337.

Inoue, S.; Yamazaki, N., Eds. *Organic and Bio-Organic Chemistry of Carbon Dioxide*; Kodansha: Tokyo, 1982.

Inoue, S.; Koinuma, H.; Tsuruta, T. *J. Polym. Sci., Polym. Lett. B* **1969**, *7*, 287–292.

Inoue, S.; Koinuma, H.; Tsuruta, T. *Polym. J.* **1971a**, *2*, 220–224.

Inoue, S.; Koinuma, H.; Yokoo, Y.; Tsuruta, T. *Macromol. Chem.* **1971b**, *143*, 97–104.

Inoue, Y.; Izumida, H.; Sasaki, Y.; Hashimoto, H. *Chem. Lett.* **1976**, 863–864.

Jessop, P.G.; Leitner, W., Eds. *Chemical Synthesis Using Supercritical Fluids*; Wiley-VCH: Weinheim, 1999.

Jessop, P. G.; Hsiao, Y.; Ikariya, T.; Noyori, R. *J. Am. Chem. Soc.* **1994a**, *116*, 8851–8852.

Jessop, P. G.; Ikariya, T.; Noyori, R. *Nature* **1994b**, *368*, 231–233.

Jessop, P. G.; Hsiao, Y.; Ikariya, T.; Noyori, R. *J. Chem. Soc., Chem. Commun.* **1995a**, 707–708.

Jessop, P. G.; Ikariya, T.; Noyori, R. *Science*, **1995b**, *269*, 1065–1069.

Jessop, P. G.; Ikariya, T.; Noyori, R. *Chem. Rev.* **1995c**, *95*, 259–272; and references therein.

Jessop, P. G.; Hsiaro, Y.; Ikariya, T.; Noyori, R. *J. Am. Chem. Soc.* **1996**, *118*, 344–355.

Jessop, P. G.; Ikariya, T.; Noyori, R. *Chem. Rev.* **1999**, *99*, 475–493.

Joó, F.; Laurenczy, G.; Nádasdi, L.; Elek, J. *Chem. Commun.* **1999**, 971–972.

Journier, J. M.; Fournier, J.; Bruneau, C.; Dixneuf, P. H. *J. Chem. Soc., Perkin Trans.* **1991**, *1*, 3271–3274.

Kobayashi, M.; Inoue, S.; Tsuruta, T. *J. Polym. Sci., Polym. Chem. Ed.* **1973**, *11*, 2383–2385.

Kreher, U.; Schebesta, S.; Walther, D. Z. *Anorg. Allg. Chem.* **1998**, *624, 602–612.*

Kröcher, O.; Köppel, R. A.; Baiker, A. *Chem. Commun.* **1996**, 1497–1498.

Kröcher, O; Köppel, R. A.; Baiker, A. *Chem. Commun.* **1997**, 453–454.

Kuran, W.; Pasynkiewicz, S.; Skupinska, J.; Rokicki, A. *Makromol. Chem.* **1976**, *177*, 11–20.

Latimer, W. L., Ed. *The Oxidation States of the Elements and Their Potentials In Aqueous Solutions*, 2nd ed.; Prentice-Hall: New York, 1952.

Leitner, W. *Angew. Chem., Int. Ed. Engl.* **1995**, *34*, 2207–2221.

Leitner, W.; Dinjus, E.; Gaßner, F. In *Aqueous-Phase Organometallic Catalysis*; Cornils, B., Hermann, W. A., Eds.; Wiley-VCH, Weinheim, 1998; pp. 486–498.

Noyori, R., Ed. Supercritical Fluids, Special Issue. *Chem. Rev.* **1999**, *99*.

Palmer, D. A.; Van Eldik, R. *Chem. Rev.* **1983**, *83*, 651–731.

Paul, J.; Pradier, C.-M.; Eds. *Carbon Dioxide Chemistry: Environmental Issues*; Royal Society of Chemistry: London, 1994.

Reetze, M. T.; Könen, W.; Strack, T. *Chimia* **1993**, *47*, 493.

Rokicki, A.; Kuran, W. *J. Macromol. Sci., Rev. Macromol. Chem.* **1981**, *C21*, 135–136.

Sneeden, R. P. A. In *Comprehensive Organometallic Chemistry*; Wilkinson, G., Stone, F. G. A., Abel, E. W., Eds.; Pergamon Press: Oxford, 1982: Vol. 8, pp. 225–283.

Super, M.; Beckman, E. J. *Macromol. Symp.* **1998**, *127*, 89–108.

Super, M.; Berluche, E.; Costello, C.; Beckman, E. *Macromolecules* **1997**, *30*, 368–372.

Suzuki, T.; Kayaki, Y.; Ikariya, T. *Chem. Lett.* **2001**, 1016–1017.

Takeshita, K; Kitamoto, A. *J. Chem. Eng. Jpn.* **1988**, *21*, 411–417.

Tsuda, T.; Chujo, Y.; Saegusa, T. *Chem. Commun.* **1976**, 415–416.

Vol'pin, M. E.; Kolomnikov, I. S. *Pure Appl. Chem.* **1973**, *33*, 567–581.

Walther, D. *Coord. Chem. Rev.* **1987**, *79*, 135–174.

Weast, R. C., Ed. *Handbook of Chemistry and Physics*; CRC: Cleveland, 1978; Vol. 58.

Wright, H. B.; Moore, M. B. *J. Am. Chem. Soc.* **1948**. *70*, 3865–3866.

Yin, X.; Moss, J. R. *Coord. Chem. Rev.* **1999**, *181*, 27–59.

4

Free-Radical Chemistry in Supercritical Carbon Dioxide

J. M. TANKO

During the 1990s, the chemical industry has focused on ways to reduce and prevent pollution caused by chemical synthesis and manufacturing. The goal of this approach is to modify existing reaction conditions and/or to develop new chemistries that do not require the use of toxic reagents or solvents, or that do not produce toxic by-products. The terms "environmentally benign synthesis and processing" and "green chemistry" have been coined to describe this approach where the environmental impact of a process is as important an issue as reaction yield, efficiency, or cost.

Most chemical reactions require the use of a solvent that may serve several functions in a reaction: for example, ensuring homogeneity of the reactants, facilitating heat transfer, extraction of a product (or by-product), or product purification via chromatography. However, because the solvent is only indirectly involved in a reaction (i.e., it is not consumed), its disposal becomes an important issue. Thus, one obvious approach to "green chemistry" is to identify alternative solvents that are nontoxic and/or environmentally benign. Supercritical carbon dioxide (sc CO_2) has been identified as a solvent that may be a viable alternative to solvents such as CCl_4, benzene, and chloroflurocarbons (CFCs), which are either toxic or damaging to the environment.

The critical state is achieved when a substance is taken above its critical temperature and pressure (T_c, P_c). Above this point on a phase diagram, the gas and liquid phases become indistinguishable. The physical properties of the supercritical state (e.g., density, viscosity, solubility parameter, etc.) are intermediate between those of a gas and a liquid, and vary considerably as a function of temperature and pressure.

The interest in sc CO_2 specifically is related to the fact that CO_2 is nontoxic and naturally occurring. The critical parameters of CO_2 are moderate ($T_c = 31\ ^\circ C$, $P_c = 74$ bar), which means that the supercritical state can be achieved without a disproportionate expenditure of energy. For these two reasons, there is a great deal of interest in sc CO_2 as a solvent for chemical reactions. This chapter reviews the literature pertaining to free-radical reactions in sc CO_2 solvent.

The characteristic feature of a neutral free-radical is that it possesses a single unpaired electron. Being one electron short of a full octet, a radical is electron-deficient. Also, because of the unpaired electron, a radical is also paramagnetic. Radicals are generally very reactive and short-lived, and thus are produced as intermediates in organic reactions. Unlike other organic reactive intermediates (carbocations, carbanions, singlet carbenes) that are diamagnetic, radicals can be studied by electron paramagnetic resonance (EPR) spectroscopy, which can provide detailed information about their structure and environment.

Most free-radical reactions of synthetic value are chain reactions, the key steps of which are illustrated in Scheme 4.1. In the initiation step, a reactive radical is generated from a nonradical precursor (initiator). In many cases, this can be accomplished thermally. For instance, peroxides possess a weak oxygen–oxygen bond and, consequently, undergo homolytic dissociation upon heating: ROOR $\rightarrow$ 2RO$\cdot$. Free radicals can also be generated photochemically, radiolytically, or by electron transfer from appropriate precursors.

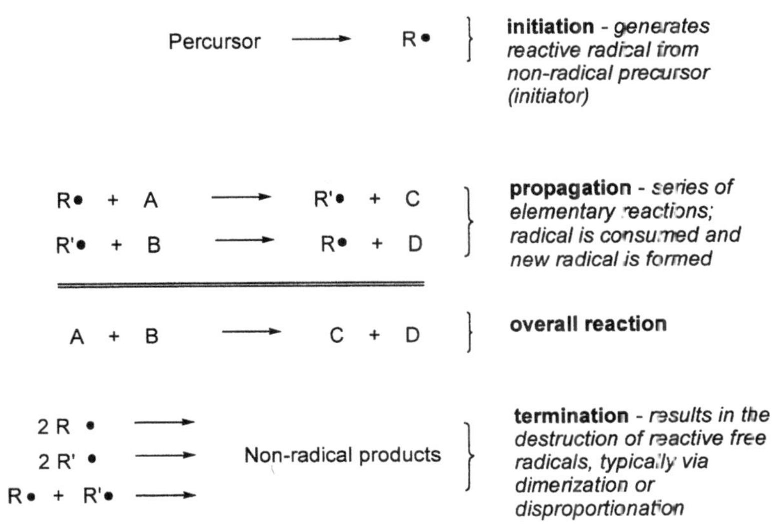

Scheme 4.1

The propagation steps (Scheme 4.1) are characterized by a series of elementary reactions in which a radical either reacts with a closed shell (diamagnetic) molecule or undergoes unimolecular rearrangement. In either case, a new free-radical is formed that reacts in the subsequent propagation step. The final propagation step generates the same radical that is used in the first step, thereby completing the chain and allowing the sequence to be repeated. For most synthetically useful free-radical chain processes, the chain length—defined approximately as the number of times the propagation steps are repeated for each radical produced in the initiation step—is very high. Consequently, the propagation steps account for most of the reactants consumed and products formed in the process, and the mathematical sum of the propagation steps is equal to the overall reaction stoichiometry.

Finally, the termination step(s) (Scheme 4.1) result in the destruction of free-radicals, thereby breaking (terminating) the chain. Typically, these steps involve two radicals that react via dimerization or disproportionation. The rate constants for these steps are very high (typically at the diffusion-controlled limit). However, because the steady-state concentrations of radicals are very small, these steps do not become important until later stages of the reaction (i.e., when the rates of the propagation steps slow down because the reactants have been consumed). Generally, for reactions characterized by high chain lengths, the yield of products generated by the termination steps is negligible.

Radicals react with closed-shell (diamagnetic) molecules via two pathways: addition to a multiple bond (eq. 4.1) and atom abstraction (eq. 4.2). Thus, a suitable solvent for a radical reaction must not possess either abstractable atoms or reactive multiple bonds. Unfortunately, most of the solvents suitable for free-radical reactions (e.g., benzene, CCl_4) are damaging to the environment or carcinogenic.

$$R\bullet + A{=}B \rightarrow R\text{-}A\text{-}B\bullet \tag{4.1}$$

$$R\bullet + A\text{-}B \rightarrow R\text{-}A + B\bullet \tag{4.2}$$

Much of the research discussed in this chapter has demonstrated that sc CO_2 is a viable alternative solvent for free-radical reactions. Moreover, several of these studies demonstrate that the unique properties of a supercritical fluid, specifically the ability to change important solvent properties such as viscosity by varying pressure (and temperature), can be exploited to manipulate reaction yield and selectivity. Solvent viscosity is particularly important for reactions that are diffusion-controlled or those for which "cage-effects" are important.

The rates of bimolecular reaction between two extremely reactive species in solution may be limited by the rate at which the two species diffuse together; such reactions are referred to as diffusion-controlled. Solvent will affect the rate of these reactions by hindering the mobility of the reactants and, consequently, the rate constant is a function of solvent viscosity. Typically, radical–radical (termination) reactions are diffusion-controlled and thus their rates are governed by solvent viscosity.

Another way solvent viscosity can affect the outcome of a chemical process is when cage effects are important. Consider two very reactive species generated in close proximity to each other (e.g., A and B, produced simultaneously from a common precursor as a "solvent caged-pair," Scheme 4.2, where the brackets denote solvent molecules surrounding the various species in solution). Because of their high reactivity, A and B may react with each other before they have a chance to separate, yielding a different product than results from reaction of free A or free B in solution. Increased solvent viscosity will favor production of in-cage product P because the diffusion of A and B away from each other is hindered; lower viscosity will favor separation of A and B. Because of the high reactivity of many free-radicals, cage effects are relatively common in free-radical reactions, especially in cases where two radicals are generated simultaneously and in close proximity (e.g., from a free-radical initiator).

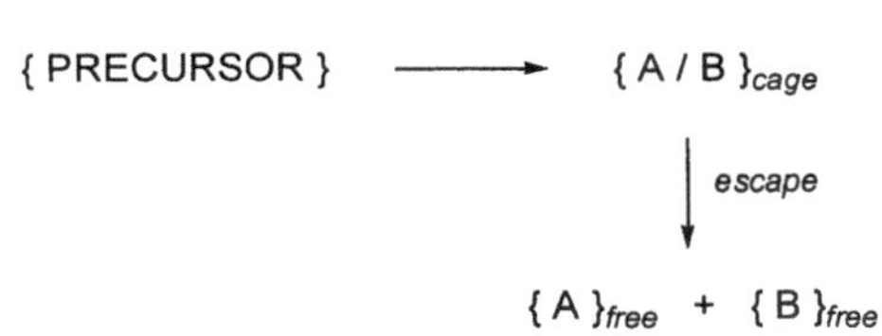

Scheme 4.2

There is one more unique feature of supercritical fluid solvents that will be a recurring theme in this chapter. Several studies have demonstrated that near the critical point, the density of the solvent about a solute is enhanced relative to the bulk density (solvent/solute clustering). As such, the mobility of the solute may be impeded to an extent greater than expected on the basis of the bulk viscosity. This phenomenon may also affect reactivity for reactions that are diffusion-controlled or for which cage effects are important, particularly near the critical point (*vide infra*).

Generation of Free Radicals

Thermolysis

Diazo compounds (diazenes) are an important class of compounds typically used as initiators of free-radical reactions. Upon heating, diazenes undergo decomposition as illustrated in Scheme 4.3. The concerted pathway is important for symmetrical diazenes ($R = R'$), while unsymmetrical diazenes ($R \neq R'$) decompose via the stepwise pathway, especially when $R\cdot$ is more stable than $R'\cdot$.

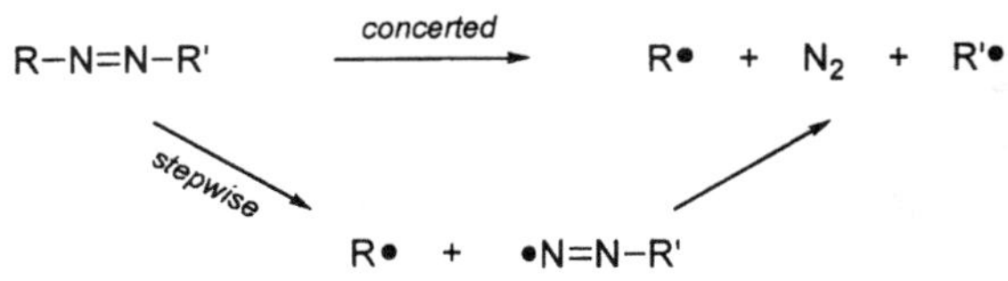

Scheme 4.3

Sigman and Leffler (1987) studied the decomposition of phenylazotriphenylmethane (PAT) in sc CO_2. The major products arising from this reaction are summarized in Scheme 4.4. Initially, upon decomposition of PAT, the phenyl diazenyl and triphenylmethyl radicals are produced as a caged-pair, which can either recombine in-cage (k_{-1}) or diffuse apart (cage-escape, k_{diff}) yielding free-radicals (Scheme 4.5). Because the rate constant for cage-escape depends on the viscosity of the solvent, it was anticipated that the observed rate constant for decomposition (k_{obs}) might show a dependence on solvent viscosity. By varying the sc CO_2 pressure, these workers were able to vary the viscosity from 0.25 to 1.02 mP, and found that k_{obs} *did* increase with decreasing viscosity (i.e., less cage-escape at higher viscosities). However, because the rate constants were erratic from run to run, the authors were reluctant to attach any significance to the observed viscosity effect (Sigman and Leffler, 1987).

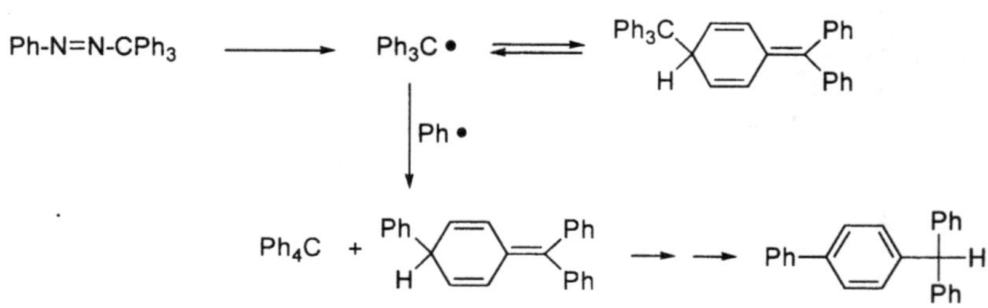

Scheme 4.4

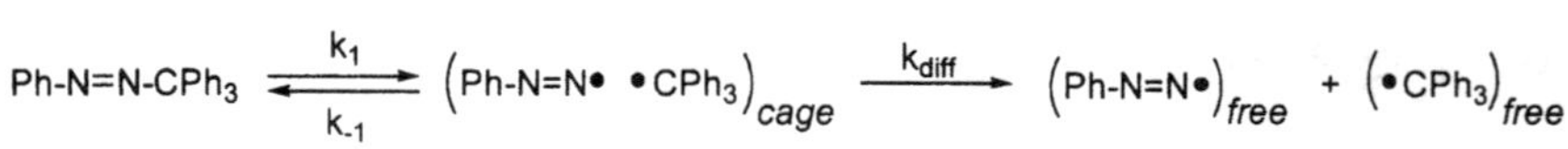

Scheme 4.5

Several years later, DeSimone and co-workers (Guan et al., 1993) examined the decomposition of the free-radical initiator 2,2′azobis(isobutyronitrile) (AIBN, Scheme 4.6) in sc CO_2 as a function of pressure. The rate constant was found to increase with increasing pressure (reaching a maximum at approximately 250 bar). At higher pressures, the rate constant decreased with increasing pressure. These observations were rationalized on the basis of two competing effects. At pressures between 100 and 250 bar, the dielectric constant of sc CO_2 increases significantly with pressure. Decompostion of diazenes such as AIBN are sensitive to solvent polarity because the decomposition is preceded by a *trans* → *cis* isomerization of the diazene, and the *cis* form is more polar. At higher pressures, the dielectric constant does not change significantly with pressure; the rate constant for decomposition decreases because the direct effect of pressure predominates (decomposition of AIBN has a volume of activation of $+24\ cm^3/mol$).

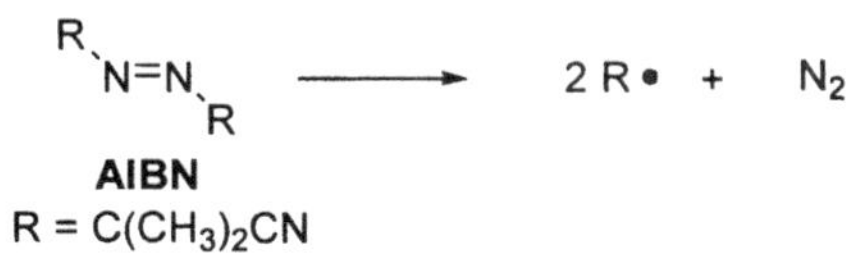

Scheme 4.6

Diacyl peroxides are another important source of free-radicals and, consequently, are also commonly used as initiators of free-radical reactions. There is a vast amount of data pertaining to the kinetics and mechanism of decomposition of these compounds in conventional solvents; there are a number of side reactions, both radical and ionic in nature, that complicate the kinetics of their decomposition. Generally, these compounds decompose by initial O–O bond cleavage that generates carboxyl radicals ($RCO_2\cdot$), which subsequently decarboxylate yielding $R\cdot$ (Scheme 4.7)

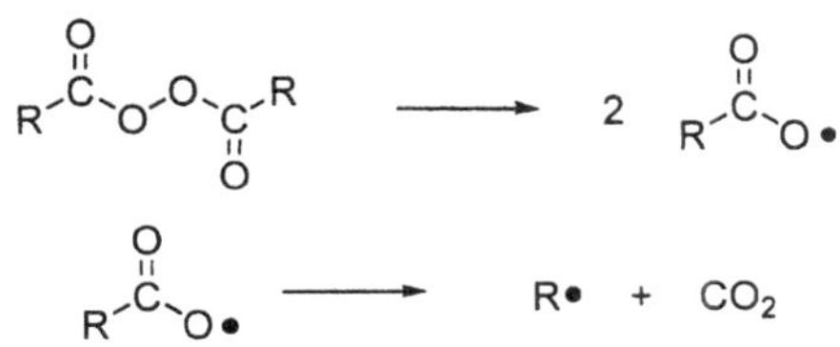

Scheme 4.7

As is the case in conventional solvents, competing radical and ionic processes (Scheme 4.8) were needed to explain the observed products and reaction kinetics in the decomposition of diacyl peroxides in sc CO_2 (Sigman et al., 1987). In addition to $ArCO_2R$ and other products attributable to free-radicals, carboxy inversion products were observed, which result from ionization of the diacyl peroxide (Scheme 4.8).

The kinetics of thermolysis processes can also be used to obtain important thermodynamic information regarding free-radicals in solution (e.g., bond dissociation energies, etc.) and, as is the case with conventional solvents, sc CO_2 can be used in this regard. For example, Roth et al. (1996) examined

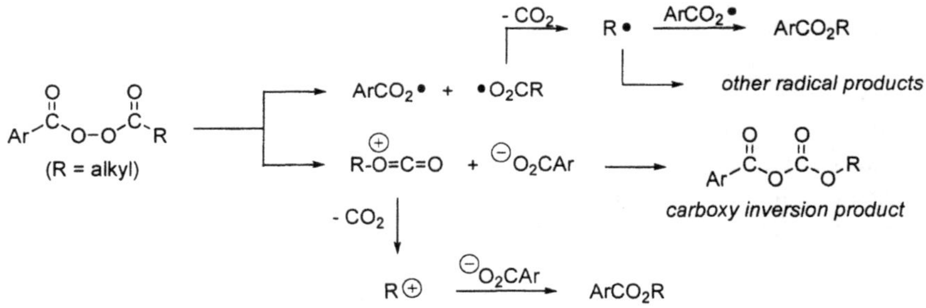

Scheme 4.8

the thermal decomposition of bitropenyl in sc CO_2 ($P = 350$ bar; $T = 120 - 170\,^{\circ}C$, eq. 4.3). A detailed analysis of the kinetics and activation parameters for this process led to a determination of the heat of formation of the tropenyl radical.

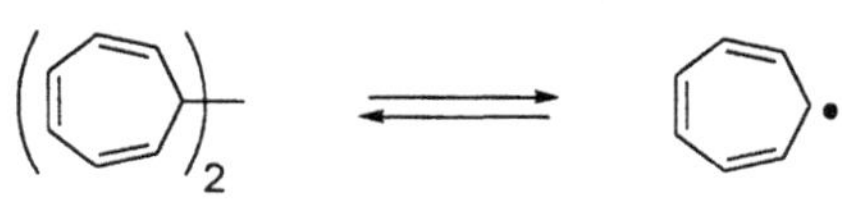 (4.3)

Photolysis: Norrish Type I Photoreaction

O'Shea et al. (1991) examined the photolysis of an unsymmetrical dibenzylketone in sc CO_2 (Scheme 4.9). The yield of bibenzyls R_1–R_1, R_1–R_2, and

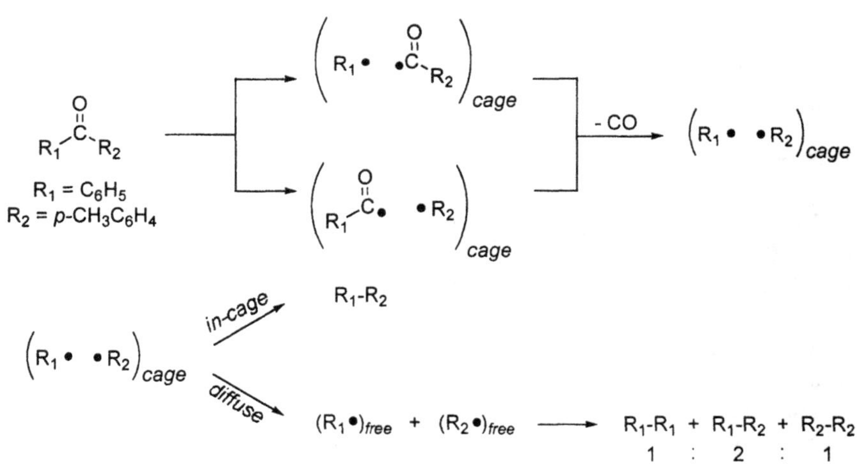

Scheme 4.9

R_2–R_2 was 25%, 50%, and 25%, respectively, and did not vary with pressure. The fact that a *statistical* 1:2:1 ratio of bibenzyls is produced, at all pressures examined including near the critical pressure, suggests that these products are produced by diffusive encounters of $(R_1.)_{free}$ and $(R_2.)_{free}$ rather than by in-cage coupling (which would yield solely R_1–R_2). Thus, these authors found no evidence for an *enhanced* cage-effect that might be attributable to solute solvent clustering.

Cage lifetimes are typically on the picosecond time regime, whereas acyl radicals (C=O) · have lifetimes on the order of nanoseconds (Roberts et al., 1993a, 1993b). As a consequence, one explanation for the failure to observe a cage-effect is that R_1. and R_2. are never formed as a geminate caged-pair (i.e., the cage disintegrates before decarbonylation occurs). To address this possibility, Andrew et al. (1995) examined the photo-Fries rearrangement of napthyl acetate (**1**) in sc CO_2. Photolysis of **1** (Scheme 4.10) generates a short-lived caged-pair (**2**) that either undergoes in-cage combination leading to Fries product **3** or undergoes cage-escape leading to α-naphthol (**4**). At high pressures, a 4:1 ratio of **3**:**4** was observed, which was invariant with pressure. This result is reasonable because over the relatively narrow range of pressures examined, CO_2 viscosity does not vary significantly. However, near the critical pressure the ratio spiked to >12:1, an effect that was attributed to solvent/solute clustering.

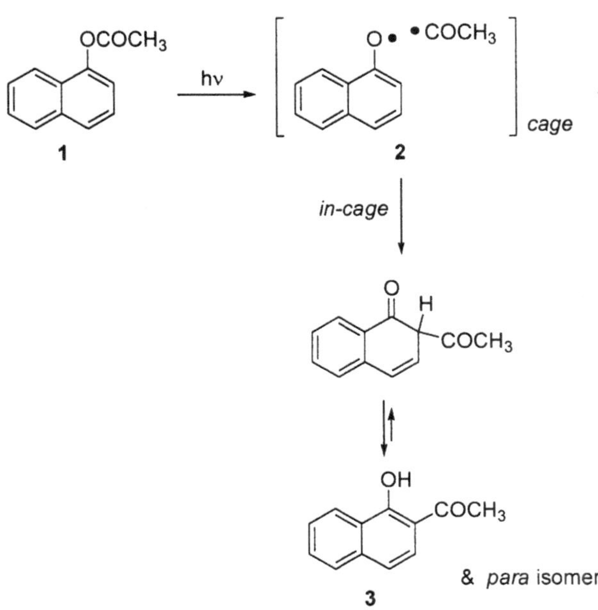

Scheme 4.10

Free-Radical Chain Reactions

This section discusses free-radical chain reactions that have been successfully conducted in sc CO_2. Notably missing from this section are free-radical polymerization reactions in sc CO_2, which are discussed in detail in other chapters of this book.

Oxidation of Hydrocarbons

The first example of a free-radical chain reaction successfully conducted in sc CO_2, which demonstrated the potential of this solvent for preparative scale chemistry, was a report from the McHugh group (Suppes et al., 1989) dealing with the oxidation of cumene (eq. 4.4). The propagation steps for this reaction are depicted in Scheme 4.11. Pressure (and thus viscosity) had little effect on the initiation, propagation, or termination rate constants. No unusual kinetic behavior was observed near the critical point.

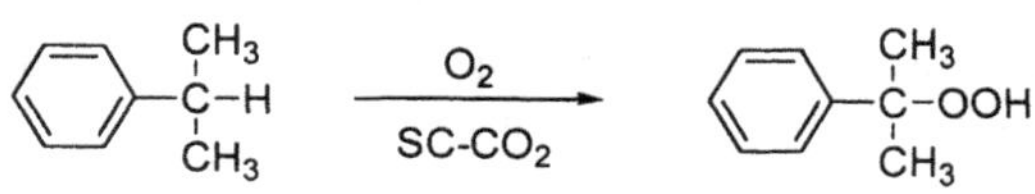

(4.4)

The oxidation of cyclohexane to cyclohexanone and cyclohexanol is an important industrial procedure used in the synthesis of adipic acid. Srinivas and Mukhopadhyay (1994) reported the oxidation of cyclohexane in sc CO_2, yielding cyclohexanone and cyclohexanol as the major reaction products. At the high temperatures employed in this study ($>137\,°C$), cyclohexyl hydroperoxide (c-$C_6H_{11}OOH$), which is produced by the mechanism outlined in Scheme 4.11, decomposes to cyclohexanone and cyclohexanol.

$$R\bullet + O_2 \longrightarrow RO_2\bullet$$

$$RO_2\bullet + RH \longrightarrow ROOH + R\bullet$$

Scheme 4.11

Free-Radical Halogenation

Alkyl halides are important starting materials in the synthesis and manufacture of a variety of organic compounds. Free-radical halogenation (Scheme 4.12) is an important methodology for the synthesis of alkyl halides from alkanes. Brominations (X = Br) tend to be very selective because of the

$$\left.\begin{array}{l} X\bullet + RH \longrightarrow HX + R\bullet \\ R\bullet + X_2 \longrightarrow RX + X\bullet \end{array}\right\} \textit{propagation}$$

$$\left. RH + X_2 \longrightarrow RX + HX \right\} \textit{overall}$$

Scheme 4.12

low reactivity of the bromine atom (i.e., the weakest C−H bond undergoes substitution). In contrast, chlorinations (X = Cl) exhibit little selectivity because the chlorine atom is highly reactive; all C−H bonds are susceptible to substitution and, generally, all possible monochlorides are produced (Russell, 1973).

In 1994, the free-radical bromination of toluene (and other alkylaromatics) in sc CO_2 was reported (Tanko and Blackert, 1994; Tanko et al., 1994). Product yields were similar to those obtained with conventional solvents. The relative reactivity of the "secondary" hydrogens of ethylbenzene versus the "primary" hydrogens of toluene on a per hydrogen basis, $r(2°/1°)$, were assessed via competition experiments and (a) did not vary with pressure, (b) were nearly identical to what is observed in conventional solvents.

The Ziegler reaction is an important method for the synthesis of benzylic or allylic bromides. *N*-bromosuccinimide (NBS) is the reagent used in this reaction, which proceeds via the mechanism outlined in Scheme 4.13. The role of NBS is to provide a low, steady-state concentration of Br_2 in solution so as to avoid competing electrophilic processes. The choice of solvent is critical for this reaction. It is important to use a solvent in which NBS is insoluble in order to prevent Br abstraction from NBS, which results in formation of the succinimidyl radical. Previously, CCl_4 was the solvent of choice. However, this study found that the reaction could be conducted in sc CO_2; reaction yields and selectivities were identical to those found in CCl_4.

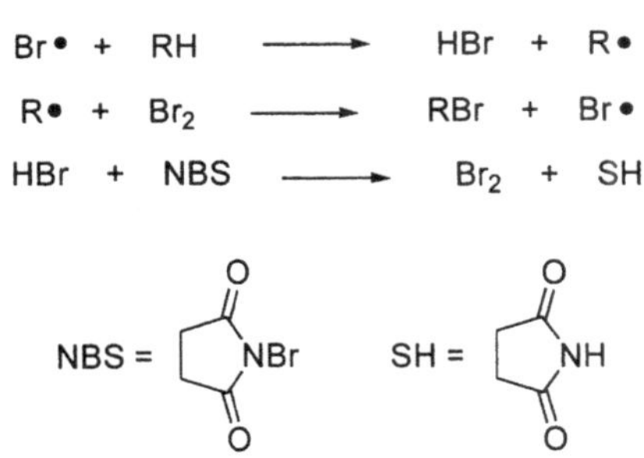

Scheme 4.13

The free-radical chlorination of alkanes can also be performed in sc CO_2 (Fletcher et al., 1998; Tanko et al., 1996). The free-radical chlorination of 2,3-dimethylbutane (23DMB, eq. 4.5) produces both 1° and 3° alkyl chlorides. In sc CO_2, the selectivity of the reaction (3° versus 1°) was intermediate between what is found in the gas phase and conventional solvents. At 40 °C, chlorine atom selectivity (per hydrogen) for the 3° versus 1° hydrogens of 23DMB [$S(3°/1°)$] is 3.97 gas phase. In solution (neat 23DMB), $S(3°/1°) = 3.08$. In sc CO_2, $S(3°/1°)$ varies with pressure, from ca. 3.8 (1200 psi) to 2.9 (9000 psi).

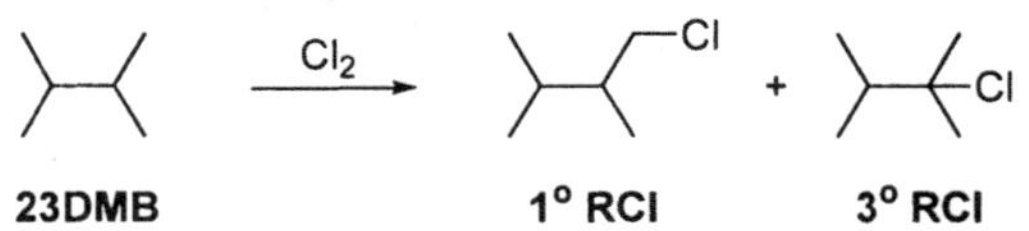

(4.5)

Free-radical chlorination also provides insight into the importance of cage-effects in sc CO_2 via the chlorine atom cage-effect (Raner et al., 1988; Skell and Baxter, 1985; Tanko and Anderson, 1988; Tanner et al., 1991). For the chlorine atom abstraction step in the free-radical chlorination of an alkane (depicted as RH_2 to indicate that more than one hydrogen can be abstracted), the geminate $RHCl/Cl\cdot$ caged-pair is partitioned among three pathways (Scheme 4.14): diffusion apart (k_{diff}), abstraction of hydrogen from RH_2 that makes up the cage walls (k_{RH_2}), and a second in-cage abstraction of hydrogen from the alkyl chloride (k_{RHCl}). While the k_{diff} and k_{RH_2} steps result in the formation of monochloride (RHCl, **M**), the k_{RHCl} step results in the formation of polychlorides (RCl_2, **P**) via $RCl\cdot + Cl_2 \rightarrow RCl_2 + Cl\cdot$.

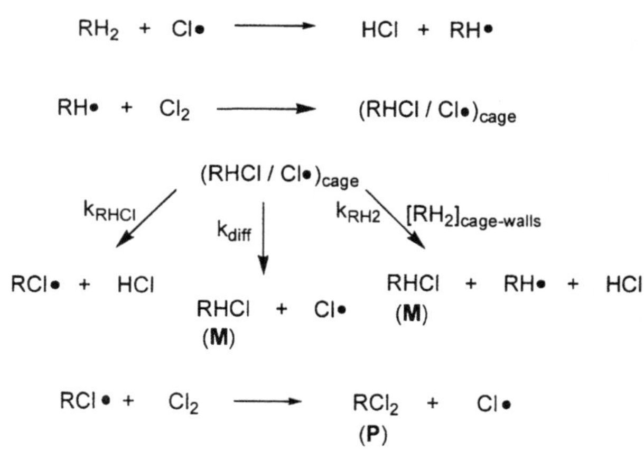

Scheme 4.14

The monochloride:polychloride ratio is a direct measure of the amount of cage-escape relative to in-cage reaction. Utilizing 2,3-dimethylbutane, neopentane, and cyclohexane as substrates, and both conventional and sc CO_2 solvents, the monochloride polychloride ratio was examined over a range of viscosities spanning 1.7 orders of magnitude (from conventional solvents such as $CFCl_3$, $CF_2ClCFCl_2$, CCl_4, etc., to sc CO_2). There was no indication of an enhanced cage-effect near the critical point in sc CO_2 solvent, and the magnitude of the cage-effect observed in sc CO_2 at all pressures examined was within the limits of what is anticipated based upon extrapolations from conventional solvents (Fletcher et al., 1998; Tanko et al., 1995).

Reductions Using R_3SnH

Since the 1980s, the use of free-radicals in organic synthesis has increased dramatically, primarily because of the exploitation of chemistry based upon tin-centered radicals. In this regard, one of the most important classes of reactions involves reductions using trialkyl stannanes (R_3SnH), which are useful in the reductive cleavage of C−X bonds (X = halogen, O, S, etc.) This chemistry is especially useful when the carbon-centered radical generated in the reaction is capable of undergoing unimolecular rearrangement, and it provides a powerful tool for the synthesis of five-membered rings (*vide infra*).

The propagation steps for this reaction are outlined in Scheme 4.15. The tin-centered radical abstracts X from R′–X, generating a carbon-centered radical, R′•, which subsequently abstracts hydrogen from the tin hydride, yielding the organic product, and completing the chain by regenerating the tin radical.

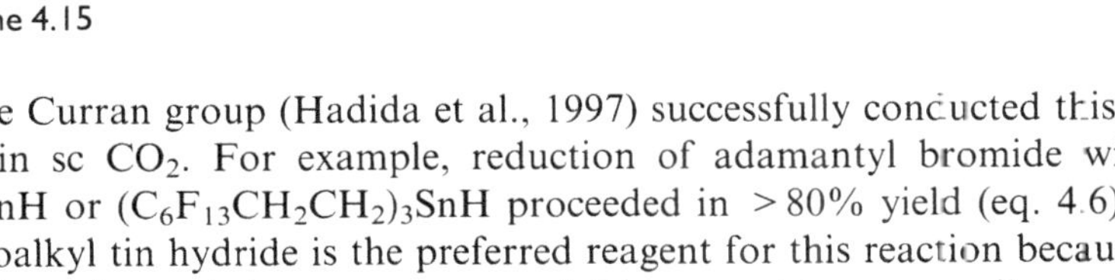

Scheme 4.15

The Curran group (Hadida et al., 1997) successfully conducted this reaction in sc CO_2. For example, reduction of adamantyl bromide with *n*-Bu_3SnH or $(C_6F_{13}CH_2CH_2)_3SnH$ proceeded in >80% yield (eq. 4.6). The fluoroalkyl tin hydride is the preferred reagent for this reaction because the reaction mixture is homogeneous, and this reagent is more easily separated from the reaction products than *n*-Bu_3SnH.

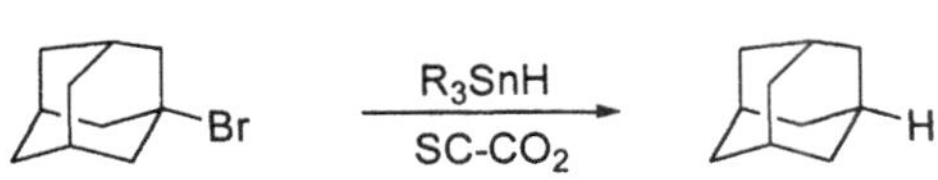

(4.6)

The use of this chemistry in a reductive cyclization (eq. 4.7) also proved successful. The key step in this reaction is the cyclization of the intermediate Δ^5-hexenyl radical **5** (eq. 4.8).

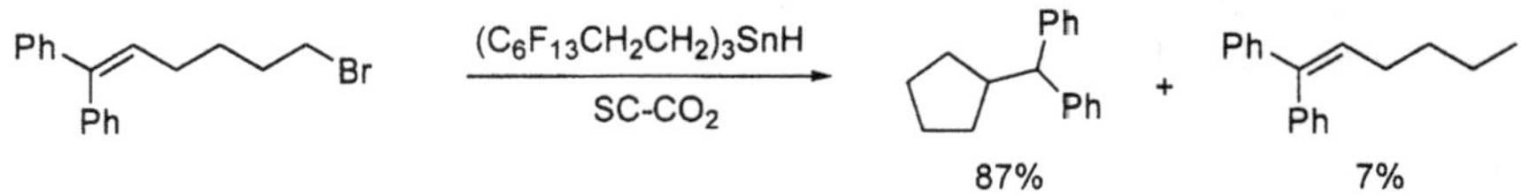

(4.7)

5 **6**

(4.8)

An intriguing observation reported by these workers was that the reduction of 9-iodoanthracene produced anthracene (71%) and 9-anthracenecarboxylic acid (10%), the latter of which may arise from addition of the 9-anthryl radical to one of the C=O bonds of CO_2 (eq. 4.9). If this hypothesis is correct, this would be the first example of a carbon-centered radical adding to CO_2 under supercritical conditions.

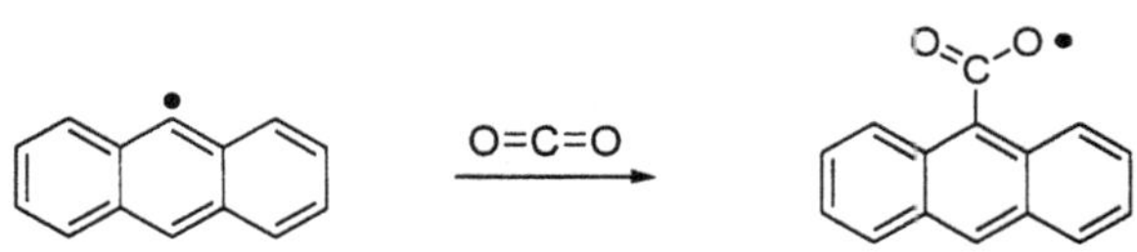

(4.9)

A plausible alternative, however, is that the mechanism involves homolytic aromatic substitution involving a carbon-centered radical generated by the addition of $R_3Sn\cdot$ to one of the O=C bonds of CO_2 (Scheme 4.16). [Addition of $R_3Sn\cdot$ to CO_2 was demonstrated to occur by Curran (Hadida et al., 1997); Leffler reported that no CO_2 addition products were formed in the thermolysis of PAT, which generates the highly reactive phenyl radical (Sigman et al., 1987)]. Regardless of the exact mechanism, this reaction provides a unique example of a free-radical-based method for carboxylation.

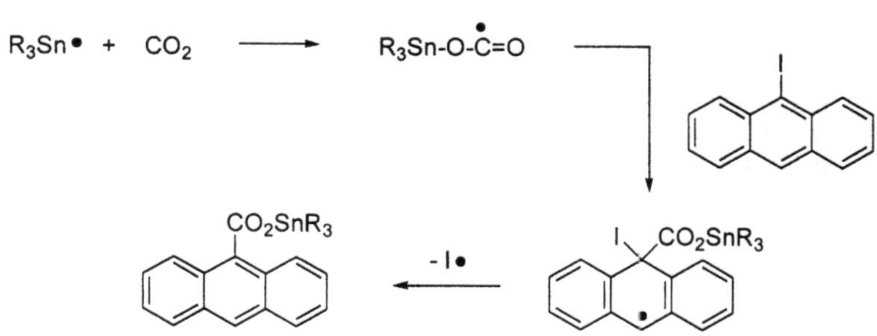

Scheme 4.16

Termination Reactions and Inhibition

Radical–radical reactions (dimerization and disproportionation, eqs. 4.10 and 4.11, respectively) are typically diffusion-controlled, and lead to the termination of free-radical chain processes (i.e., they react to form non-radical products, thereby terminating the chain). In its ground state, molecular oxygen is a diradical, and thus reacts very efficiently with organic radicals (eq. 4.12). Many chain reactions are inhibited by molecular oxygen because the radical produced (ROO·) is ineffective at propagating the chain. Consequently, great care is taken to exclude molecular oxygen when conducting a free-radical reaction.

$$2\,\mathrm{R}\bullet \longrightarrow \mathrm{R\text{-}R} \tag{4.10}$$

$$2\,\mathrm{R}\bullet \longrightarrow \mathrm{R_{-H}} + \mathrm{R_{+H}} \tag{4.11}$$

$$\mathrm{R}\bullet + \bullet\mathrm{O\text{-}O}\bullet \longrightarrow \mathrm{R\text{-}O\text{-}O}\bullet \tag{4.12}$$

In a series of papers, Chateauneuf and Brennecke (Roberts et al., 1993b, 1995) reported absolute rate constants for the dimerization of benzyl radical determined using laser flash photolysis (LFP). The termination rate constant ($2\ k_T$) was diffusion-controlled and, thus, decreased slightly with increasing pressure (e.g., $2\ k_T = 4 \times 10^{10}$ and $2 \times 10^{10}\ M^{-1}\ s^{-1}$ at 80 and 100 bar, respectively, at 35 °C) because of the increase in CO_2 viscosity. No evidence was found for any distortion in reaction rates that might arise from clustering phenomena.

The rate constant for reaction of benzyl radical with molecular oxygen was also measured via LFP in sc CO_2 (Roberts et al., 1995). The rate constant was found to increase slightly with increasing pressure (from $1.1 \times 10^9\ M^{-1}\ s^{-1}$ to $1.5 \times 10^9\ M^{-1}\ s^{-1}$ with an increase in pressure from 80 to 300 bar). This reaction is under activation control, and the increase in rate with increasing pressure is attributable to the direct effect of pressure (i.e., the activation volume is negative for this addition process).

EPR Studies of Radicals in Supercritical Carbon Dioxide

As noted at the beginning of this chapter, radicals are a unique class of organic intermediates because they can be studied using magnetic resonance techniques such as electron paramagnetic resonance spectroscopy (EPR). The EPR technique has proven valuable for ascertaining the local structure about a radical in a supercritical fluid solvent.

Nitroxyl radicals are an especially stable class of free-radicals whose properties are explicable, invoking the two resonance forms depicted in Scheme 4.17. Because the contribution of dipolar resonance form **8** to the

7 8

Scheme 4.17

resonance hybrid increases with increasing solvent polarity, the magnitude of the hyperfine coupling constant between the unpaired electron and N (a_N) provides a direct measure of solvent polarity (Tanko and Suleman, 1996).

Carlier and Randolph (1993), Ganapathy et al. (1996), and Batchelor (1998) used a_N to determine local solvent densities in several supercritical fluid solvents including CO_2. These studies helped establish the importance of solute/solvent clustering near the critical point of a supercritical fluid. Through an analysis of line shapes, these workers were also able to assess the rate constant for Heisenberg spin exchange between nitroxyl radicals. Later results suggested that these rate constants are not perturbed by solvent/solute clustering near the critical point (Batchelor, 1998).

Closing Remarks

An analysis of the literature discussed in this chapter reveals that sc CO_2 is a viable alternative solvent for free-radical reactions. Generally, CO_2 is unreactive toward organic radicals, though some tin-centered radicals undergo addition to the carbon–oxygen double bond of CO_2. For the most part, however, it is reasonable to conclude that most common free-radical reactions can be successfully conducted in sc CO_2 solvent. In some cases, sc CO_2 is actually a superior solvent with respect to issues pertaining to chemical reactivity (i.e., the potential to manipulate reaction rate and selectivity by varying pressure). In most cases, reaction yields are not compromised in sc CO_2, demonstrating that reaction efficiency and pollution prevention are not mutually exclusive.

Acknowledgments Financial support from the National Science Foundation (CHE-9524986) and the U.S. Environmental Protection Agency is gratefully acknowledged.

References

Andrew, D.; Des Islet, B. T.; Margaritis, A.; Weedon, A. C. Photo-Fries Rearrangement of Naphthyl Acetate in Supercritical Carbon Dioxide: Chemical Evidence for Solvent–Solute Clustering. *J. Am. Chem. Soc.* **1995**, *117*, 6132–6133.

Batchelor, S. M. Free Radical Motion in Super Critical Fluids Probed by EPR Spectroscopy. *J. Phys. Chem. B* **1988**, *102*, 615–619.

Carlier, C.; Randolph, T. W. Dense-Gas Solvent–Solute Clusters at Near Infinite Dilution: EPR Spectroscopic Evidence. *AIChE J.* **1993**, *39*, 876–884.

Fletcher, B.; Suleman, N. K.; Tanko, J. M. Free Radical Chlorination of Alkanes in Supercritical Carbon Dioxide: The Chlorine Atom Cage Effect as a Probe for Enhanced Cage Effects in Supercritical Fluid Solvents. *J. Am. Chem. Soc.* **1998**, *120*, 11839–11844.

Ganapathy, S.; Carlier, C.; Randolph, T. W.; O'Brien, J. A. Influence of Local Structural Correlations on Free-Radical Reactions in Supercritical Fluids: A Hierarchical Approach, *Ind. Eng. Chem. Res.* **1996**, *35*, 19–27.

Guan, Z.; Combes, J. R.; Menceloglu, Y. Z.; DeSimone, J. M. Homogeneous Free Radical Polymerizations in Supercritical Carbon Dioxide: 2. Thermal Decomposition of 2,2′-Azobis(isobutyronitrile), *Macromolecules* **1993**, *26*, 2663–2669.

Hadida, S.; Super, M. S.; Beckman, E. J.; Curran, D. P. Radical Reactions with Alkyl and Fluoroalkyl (Fluorous) Tin Hydride Reagents in Supercritical CO_2. *J. Am. Chem. Soc.* **1997**, *119*, 7406–7407.

O'Shea, K. E.; Combes, J. R.; Fox, M. A.; Johnston, K. P. Photolysis of Dibenzylketones in Supercritical Ethane and Carbon Dioxide *Photochem. Photobio.* **1991**, *54*, 571–576.

Raner, K. D.; Lusztyk, J.; Ingold, K. U. Kinetic Analysis of Alkane Polychlorination with Molecular Chlorine. Chlorine Atom/Monochloride Geminate Pairs and the Effect of Reactive "Cage Walls" on the Competition Between Monochloride Rotation and Chlorine Atom Escape. *J. Am. Chem. Soc.* **1988**, *110*, 3519–3524.

Roberts, C. B.; Zhang, J.; Brennecke, J. F.; Chateauneuf, J. E. Laser Flash Photolysis Investigations of Diffusion-Controlled Reactions in Supercritical Fluids. *J. Phys. Chem.* **1993a**, *97*, 5618–5623.

Roberts, C. B.; Zhang, J.; Chateauneuf, J. E.; Brennecke, J. F. Diffusion-Controlled Reactions in Supercritical CHF_3 and CO_2/Acetonitrile Mixtures, *J. Am. Chem. Soc.* **1993b**, *115*, 9576–9582.

Roberts, C. B.; Zhang, J.; Chateauneuf, J. E.; Brennecke, J. F. Laser Flash Photolysis and Integral Equation Theory to Investigate Reactions of Dilute Solutes with Oxygen in Supercritical Fluids. *J. Am. Chem. Soc.* **1995**, *117*, 6553–6560.

Roth, W. R.; Hunold, F.; Neumann, M.; Bauer, F. Sauerstoff-abfang von radikalen in überkritischem kohlendioxid; Bildungsenthalpie und rekombinationsbarriere des cyclohepatrienyl-radikals. *Liebigs Ann. Chem.* **1996**, 1679–1690.

Russell, G. A. Reactivity, Selectivity and Polar Effects in Hydrogen Atom Transfer Reactions. In *Free Radicals*; Koch, J. K., Ed.; Wiley: New York, 1973; Vol. I, pp. 275–331.

Sigman, M. E.; Leffler, J. E. Supercritical Carbon Dioxide. 3. The Decomposition of Phenylazotriphenylmethane in Supercritical Carbon Dioxide, *J. Org. Chem.* **1987**, *52*, 1165–1167.

Sigman, M. E.; Barbas, J. T.; Leffler, J. E. Supercritical Carbon Dioxide. 5. Carboxyinversion Reactions of Diacyl Peroxides. Alkyl Group Rearrangements and CO_2 Exchange. *J. Org. Chem.* **1987**, *52*, 1754–1757.

Skell, P. S.; Baxter, H. N. Multiple Substitutions in Radical-Chain Chlorinations. A New Cage Effect. *J. Am. Chem. Soc.* **1985**, *107*, 2823–2824.

Srinivas, P.; Mukhopadhyay, M. Oxidation of Cyclohexane in Supercritical Carbon Dioxide Medium. *Ind. Eng. Chem. Res.* **1994**, *33*, 3118–3124.

Suppes, G. J.; Occhiogrosso, R. N.; McHugh, M. A. Oxidation of Cumene in Supercritical Reaction Media. *Ind. Eng. Chem. Res.* **1989**, *28*, 1152–1156.

Tanko, J. M.; Anderson, F. E. Competititve Cage Kinetics. Relative Rates of Complexation of Chlorine Atom by Various Arenes. *J. Am. Chem. Soc.* **1988**, *110*, 3525–3530.

Tanko, J. M.; Blackert, J. F. Free-Radical Side-Chain Bromination of Alkylaromatics in Supercritical Carbon Dioxide, **1994**, *263*, 203–205.

Tanko, J. M.; Suleman, N. K. Solvent Effects in the Reactions of Neutral Free Radicals. In *Energetics of Organic Free Radicals*; Simões, J. A. M., Greenberg, A., Liebman, J. F., Eds.; SEARCH Series, Vol. 4; Blackie: Glasgow, 1996; pp. 224–293.

Tanko, J. M.; Blackert, J. F.; Sadeghipour, M. Supercritical Carbon Dioxide as a Medium for Conducting Free-Radical Reactions. In *Benign by Design: Alternative Synthetic Design for Pollution Prevention*; Anastas, P. T., Farris, C. A., Eds.; ACS Symposium Series 577; American Chemical Society: Washington, DC, 1994; pp. 98–113.

Tanko, J. M.; Suleman, N. K.; Fletcher, B. Viscosity-Dependent Behavior of Geminate Caged-Pairs in Supercritical Fluid Solvent. *J. Am. Chem. Soc.* **1996**, *118*, 11958–11959.

Tanner, D. D.; Oumar-Mahamat, H.; Meintzer, C. P.; Tsai, E. C.; Lu, T. T.; Yang, D. Viscosity-Dependent Cage Reactions. Multiple Substitutions in Radical Chain Chlorinations, *J. Am. Chem. Soc.* **1991**, *113*, 5397–5402.

5

Fluorous Phases and Compressed Carbon Dioxide as Alternative Solvents for Chemical Synthesis: A Comparison

WALTER LEITNER

The principal goal of basic research in chemical synthesis is the development of efficient tools for functional group transformations and for the assembly of building blocks during the construction of molecules with increasing complexity. Traditionally, new approaches in this area have focused on the quest for new reaction pathways, reagents, or catalysts. Comparably less effort has been devoted to utilize the reaction medium as a strategic parameter, although the use of solvents is often crucial in synthetically useful transformations.

The first choice for a solvent during the development of a synthetic procedure is usually an organic liquid, which is selected on the basis of its protic or aprotic nature, its polarity, and the temperature range in which the reaction is expected to proceed. Once the desired transformation is achieved, yield and selectivity are further optimized in the given medium by variation of temperature, concentration, and related process parameters. At the end of the reaction, the solvent must be removed quantitatively from the product using conventional workup techniques like aqueous extraction, distillation, or chromatography. If the synthetic procedure becomes part of a large-scale application, the solvent can sometimes be recycled, but at least parts of it will ultimately end up in the waste stream of the process.

Increasing efforts to develop chemical processes with minimized ecological impact and to reduce the emission of potentially hazardous or toxic organic chemicals have stimulated a rapidly growing interest to provide alternatives to this classical approach of synthesis in solution. At the same time, researchers have started to realize that the design and utilization of multifunctional reaction media can add a new dimension to the development of synthetic chemistry. In particular, efficient protocols for phase separations and recovery of reagents and catalysts are urgently required to provide innovative flow schemes for environmentally benign processes or for high-throughput screening procedures. Fluorous liquid phases and supercritical carbon dioxide (sc CO_2) have received particular attention among the various reaction media that are discussed as alternatives to classical organic solvents.

The aim of this chapter is to compare these two media directly and to critically evaluate their potential for synthetic organic chemistry. Firstly, the solvent properties related to solution chemistry will be discussed briefly, putting both media side-by-side. Then, different reaction types that were investigated in both media will be described and selected examples will be analyzed in detail. Finally, the chapter will conclude with some guidelines and prospects for the future development of the two approaches. For more details on the historical development and chemcial applications of the individual media, the interested reader is referred to other chapters of this book and to review articles on fluorinated liquids (Barthel-Rosa and Gladysz, 1999; Betzemeier and Knochel, 1999; Curran, 1998; Horváth, 1998) and sc CO_2 (Dinjus et al., 1996; Jessop and Leitner, 1999; Jessop et al., 1995a; Leitner, 1999; Morgenstern et al., 1996; Noyori, 1999; Savage et al., 1995).

Basic Properties of Fluorous Phases and Carbon Dioxide as Reaction Media

Physical Properties as Related to Chemical Synthesis

The term "fluorous" has been coined in analogy to "aqueous" (Horváth and Rábai, 1994a) and is now widely accepted to designate liquid solvents that are characterized by a high degree of fluorination, which makes them sufficiently distinct from conventional organic solvents. Some frequently used fluorinated solvents, their trade names or acronyms, and some of their properties are compiled in Table 5.1. Typically, boiling points of fluorous phases range from 60 °C to 110 °C, although up to 250 °C can be reached with some perfluorinated amines. Several perfluorinated alkanes remain liquid even at temperatures below −60 °C. The densities of perfluorinated compounds are considerably larger than unity and they constitute the lower layer in most biphasic systems.

Perfluorinated alkanes such as FC-72 have extremely low polarities, being even considerably less polar than the corresponding hydrocarbons. Trifluoromethylbenzene (α,α,α-trifluorotoluene, benzotrifluoride, BTF) has a polarity in the range of benzene or methylene chloride. Owing to the strong electron-withdrawing effect of the fluorinated substituents, the oxygen lone pairs of perfluorinated polyethers (PFPEs) are usually not available for bonding interactions and these solvents do not show coordinating properties like diethylether or THF. A number of perfluoroalkyl-substituted

Table 5.1. Selected Fluorous Compounds with Potential Use as Solvents in Organic Synthesis[a]

Compound	Trade Name/Acronym	bp (°C)	mp (°C)	*d* (g/mL)
Perfluoromethylcyclohexane	PFMC	76.1	−44.7	—
Perfluorohexane	FC-72	57.1	−87.1	—
Perfluorooctanes	FC-77	103–105	—	—
Perfluorobutyltetrahedrofuran	FC-75	102	—	1.78
C_6F_6	Perfluorobenzene	81.5	—	1.60
α, α, α-Trifluorotoluene	Benzotrifluoride (BTF)	102	−29	1.18
Perfluorodecalin		142	−10	1.95
1-Bromoperfluorooctane		140.5	—	1.89
Perfluoropolyether	PFPE (M_W ca. 410)	70	—	—
Perfluoropolyether	PFPE (M_W ca. 580)	110	—	—
Perfluorotributylamine	FC-43	178–180	−50	1.90

[a] Data taken from Betzemeier and Knochel (1999) and Barthel-Rosa and Gladysz (1999).

alcohols are available, which are, of course, highly polar and provide a protic fluorous environment. However, such liquids have found comparably less attention as solvents up to now, probably because their solvent behavior is largely dominated by the alcohol function, making the utilization of the unique properties of the fluorous part less obvious.

As outlined in detail in other chapters of this volume and elsewhere (Jessop and Leitner, 1999), the critical point of carbon dioxide is defined by its critical temperature $T_c = 31.0\,°C$ and critical pressure $P_c = 73.8$ bar. Strictly speaking, the supercritical state of CO_2 extends beyond these data, but below the very high pressures required for solidification. The bulk density of CO_2 at the critical point ($D_c = 0.466$ g mL^{-1}) is the mean value of the density of the liquid and gaseous phase just before entering the supercritical region. It must be noted, however, that the local density of a supercritical fluid (SCF) is subject to large fluctuations and may differ considerably from the bulk density, especially around solute molecules. Many attempts have been made to utilize "chemical probes" like radical reactions or Diels–Alder additions to investigate these so-called clustering or augmentation phenomena. For most practical applications, however, it seems more important that the bulk density of sc CO_2 can be varied continuously with relatively small variations in temperature and/or pressure, owing to the high compressibility of the SCF. For example, the density of sc CO_2 at 37 °C is only 0.33 g mL^{-1} at 80 bar, but it rises to 0.80 g mL^{-1} at 150 bar. These variations can have dramatic effects on the outcome of a chemical reaction and the density of sc CO_2 therefore provides an additional and unique reaction parameter. In favorable cases, transition-state theory can help to rationalize such phenomena.

The solubility of solid or liquid solutes also depends strongly on the bulk density of a supercritical medium, and appreciable solubilities are generally observed only at densities greater than D_c. The pressure required to achieve such densities increases rapidly with increasing temperature and this sets a practical limitation to the upper temperatures applicable in this medium. Typical laboratory equipment is generally rated for use below 500 bar, in most cases below 300 bar. Under these conditions, temperatures above 100 °C will not allow the medium to reach sufficient densities to conduct typical solution chemistry. The lower limit of the temperature range for the use of sc CO_2 is naturally set by the critical temperature T_c. However, many of the potential benefits associated with the use of CO_2 as a solvent are retained in the liquid state as well, and temperatures down to 0 °C or even −10 °C are certainly practical in these cases.

Compressed carbon dioxide is a very nonpolar reaction medium, albeit that small variations of the dielectric constant ε can be achieved with pressure and temperature. At temperatures close to T_c, ε increases slightly from 1.3 to just above 1.6 in the pressure range between P_c and $P = 300$ bar. The polarity can be influenced much more dramatically by even small amounts of cosolvents (modifiers or entrainers) like water, methanol, or THF. It should be noted that dissolved substrate(s) and product(s) can, of course, also influence the polarity of the medium considerably. In many cases, polarity changes resulting from the chemical transformation during the reaction will exceed those obtained by variation of external parameters like pressure or temperature. The possibility to generate aqueous emulsions or microemulsions with sc CO_2 opens an exciting new approach to highly polar media based on sc CO_2, which has been barely explored in synthetic chemistry up to now (Clarke et al., 1997; Jacobson et al., 1999a, 1999b; Johnston et al., 1996).

Toxicology, Environmental Impact, and Safety Aspects

Ecological and toxicological considerations are a major theme in the search for new reaction media. Fluorous solvents have sometimes been met with some skepticism in this respect (Cornils, 1997), mainly because fluor-containing molecules have a bad record for their ozone depletion potential. However, the detrimental impact of fluorochemicals on the world's ozone layer is largely connected with the carbon–chlorine bond in chlorinated fluorocarbons (CFCs) and fluorinated hydrocarbons are far less critical. In fact, a number of fluorohydrocarbons have received, or are being examined for, exemptions from VOC regulations (VOC = volatile organic compound = organic compounds that participate in atmospheric photoreactions). The persistence of fluorinated compounds in the environment varies widely, ranging from full atmospheric lifetimes in the order of months for BTF to environmental half-lives of several thousand years for perfluorinated alkanes (Ravishankara et al., 1993). In accord with their widespread applications in the food and health sectors, most perfluorinated solvents have very low or negligible acute and chronic toxicity. The use of perfluorinated liquids as an artificial blood substitute may serve as an illustrative example (Riess and Le Blanc, 1982).

Obviously, acute or chronic toxicity is not an issue with the use of CO_2. Similarly, the ecological impact of carbon dioxide is negligible when used as a solvent, and CO_2 is—by definition—not regulated as a VOC. The application of CO_2 as a solvent in chemical synthesis would not generate any additional carbon dioxide and therefore would not contribute to greenhouse gas emission. Even for large-scale applications, no toxic or environmental risk would arise in case of any accidental contamination of the immediate environment with the solvent CO_2 if sufficient ventilation is provided.

Perfluorinated solvents can be handled like any other conventional liquids, and in some cases additional safety benefits may arise from their non-existant or very low flammability. Similarly, the use of nonflammable CO_2 will result in largely reduced risks of ignition or explosion. These factors make both media particularly attractive for oxidation chemistry. The handling of compressed CO_2 requires suitable high-pressure equipment and—as with all compressed gases—appropriate safety precautions have to be taken in order to avoid any potential hazards arising from catastrophic failure of the equipment. Such measures are, however, readily met on a laboratory or industrial scale and standard equipment is commercially available from many vendors (Jessop and Leitner, 1999).

Solubility and Separation Schemes

The most striking feature of fluorous liquids is their restricted miscibility with a vast majority of organic compounds, allowing the design of highly efficient separation schemes in many cases. Obviously, the degree of miscibility or solubility will critically depend on the degree of fluorination. Solvents with very high fluor contents such as FC-72 or PFPEs can dissolve almost exclusively fluorinated materials, whereas partially fluorinated liquids like BTF may act as "hybrid solvents" for both organic and fluorinated reagents (Maul et al., 1999).

A strong temperature dependence is often observed for the miscibility of highly fluorinated liquids with organic substances, allowing one to switch from monophasic to biphasic conditions by simple heating or cooling, respectively. A typical separation scheme utilizing such a fluorous biphasic system (FBS) is outlined in Figure 5.1, exemplified for the recycling of a homogeneous catalyst (Horváth and Rabai, 1994a, 1994b). The catalyst is "fluorophilic"; that is, it contains a sufficient degree of fluorination to dissolve preferably in the fluorous solvent. The reaction can be carried out at a temperature where both liquid phases are miscible, but biphasic reactions are also possible. At the end of the reaction, the mixture is brought to temperatures at which the substrates and/or products are immiscible with

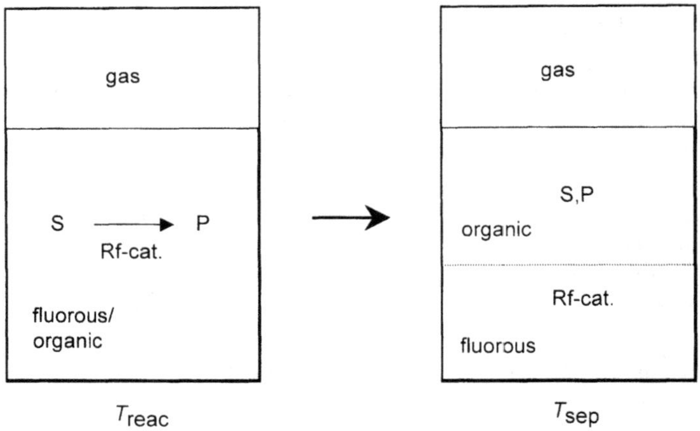

Figure 5.1. Separation scheme to recycle a homogeneous metal catalyst in a typical fluorous biphasic system (FBS).

the fluorous phase and the upper organic layer is then separated by simple decanting. An additional organic solvent may be used to facilitate this separation step. This principle is, of course, not restricted to catalysts; substrates, reagents, or products can be "tagged" with fluorophilic substituents and immobilized or separated in an analogous manner (Curran, 1998; Horváth, 1998; Studer et al., 1997).

The successful application of this separation strategy relies on the highly selective partitioning of the individual components between two liquid phases. In particular, the fluorinated components must have an almost exclusive preference for the fluorous phase to avoid any leaching and/or contamination of the product. Typically, the fluorophilicity is achieved by tagging the reagent or catalyst with a long perfluoroalkyl substituent or "ponytail" of general formula $(CH_2)_x(CF_2)_yF$ (Betzemeier and Knochel, 1999; Curran, 1998; Horváth, 1998). Partition coefficients between perfluoroalkanes and typical organic solvents have been collected for a number of organic and organometallic reagents containing these groups (Barthel-Rosa and Gladysz, 1999). It seems to emerge from these studies that at least three substituents with $y \geq 6$ are required to achieve partition coefficients of potential utility, and even higher degrees of fluorination are often necessary in practical systems.

Another elegant technique in fluorous synthesis is the use of reverse-phase fluorinated silica gel as a solid separation phase. Such fluorinated solids allow the separation of fluorous from nonfluorous materials by selective adsorption, and can even be used for chromatographic separations by degree of fluorination (Kainz et al., 1998). These techniques require a considerably lower content of fluorine for efficient separation and have been referred to as "light fluorous synthesis" (Curran and Luo, 1999).

The phase behavior of multicomponent mixtures with sc CO_2 can be highly complex and a detailed discussion is beyond the scope of this account. Exact numerical descriptions or simulations are often impossible for practical reaction systems containing substrate(s), reagent(s), and product(s) in varying amounts. The following generalizations are, however, possible. The solubility of organic compounds in sc CO_2 is largely determined by their polarity, their vapor pressure, and by the presence of specific functional groups. Most organic molecules of molecular weights below 1000 and with low to medium polarity are moderately soluble in sc CO_2, and their solubility increases greatly with increasing volatility of the compound. The presence of highly polar functional groups like carboxylic acids or their salts strongly reduces the solubility. Aromatic rings also suppress the solubility considerably. The tendency of a molecule to dissolve in sc CO_2 can be enhanced by orders of magnitude if "CO_2-philic" substituents are introduced (DeSimone et al., 1994; Harrison et al., 1994; Lin et al., 1993; Yazdi and Beckman, 1996). For example, it was shown that CO_2-insoluble catalysts containing aryl phosphine ligands become highly soluble in this medium upon attaching the perfluoroalkyl chains $(CH_2)_x(CF_2)_yF$ (Kainz et al., 1997). Hence, there is a direct overlap between the requirements for solubility in fluorous systems and in sc CO_2. All compounds that have a reasonable degree of solubility in fluorous solvents can be expected also to be soluble in sc CO_2, but not vice versa.

Owing to the wide utility of the perfluoroalkyl chains $(CH_2)_x(CF_2)_yF$, the shortcut Rf_yh_x has been introduced to symbolize these substituents in chemical formulae (Curran, 1996). For compounds with established acronyms, it has been proposed to use the prefix z-H^xF^y to designate a substituent of this type in z position relative to the main functional group (Kainz et al., 1997). Both types of nomenclatures will be used in this chapter if appropriate.

Figure 5.2 shows a possible reaction/separation sequence utilizing CO_2 as the solvent and mass separation agent. As in Figure 5.1, a homogeneously catalyzed process is used to illustrate the principle. After the reaction proceeded under homogeneous conditions, phase separation can be induced by variation of either concentration, temperature, and/or pressure. Subsequent supercritical fluid extraction (SFE) allows separation of the more "CO_2-philic" parts from the remaining material. The term CESS (catalysis and extraction using supercritical solvents) was introduced for processes of this type designed to recycle homogeneous catalysts (Kainz et al., 1999; Leitner, 1999), but application may again be more general. More sophisticated separation processes based on sc CO_2 could replace the SFE step (Jessop and Leitner, 1999), but practical utility has yet to be demonstrated. In addition, one can envisage the use of sc CO_2 in biphasic systems with an immiscible liquid phase. Water (Bhanage et al., 1999) or ionic liquids (Blanchard et al., 1999) have been suggested as the liquid phases for such systems and very promising initial results have been described.

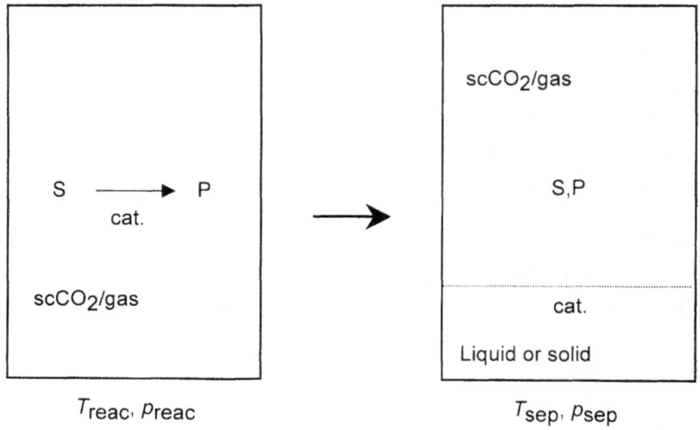

Figure 5.2. Separation scheme to recycle homogeneous catalysts using the CESS process.

A major difference between the sequences outlined in Figure 5.1 and Figure 5.2 is the number of phases during the individual steps, if gaseous reactants are involved. Typically, the gas phase is neglected in descriptions of the separation schemes based on two immiscible liquids. Thus, a hydrogenation or oxidation reaction in a fluorous "biphasic" system comprises actually three phases: the fluorous phase, the organic phase, and the gas phase. As described earlier, the liquid/liquid separation can be eliminated in thermoregulated mixtures, but the gas/liquid phase boundary will always remain and constitutes a potential origin for mass transport limitation. In contrast, a gas/liquid phase boundary is impossible in the supercritical medium and truly homogeneous single-phase conditions can be established. Obviously, such conditions provide an ideal environment for chemical transformations.

Chemical Behavior and Reactivity

In general, perfluorinated compounds can be considered as rather inert and thermally robust solvents. The oxidation stability of perfluoroalkanes and -ethers has stimulated considerable efforts to use them as solvents for reactions involving molecular oxygen (see section "Oxidation Using Molecular Oxygen"). The C–H bonds of partly fluorinated compounds can be quite acidic and such materials are not compatible with strongly basic reagents. The carbon–fluorine bond itself is among the strongest bonds in organic molecules. Some low-valent late transition-metal complexes have been designed to activate and cleave C–F bonds via oxidative addition (leading references: Burdenuic et al., 1997; van der Boom et al., 1999). Up to now, however, there have been no reports that such reactions might interfere with synthetic reactions when typical transition-metal catalysts are used in fluorous solvents.

Despite its reputation of being a "nonreactive" molecule, carbon dioxide undergoes a number of chemical interactions with various organic functionalities. Usually, the carbon center of CO_2 is the subject of nucleophilic attack in such reactions. The most well-known example is the reaction with H_2O to yield carbonic acid, but other O, C, or N nucleophiles will undergo similar reactions. The formation of carbamates from N–H groups and CO_2 has been exploited to utilize sc CO_2 simultaneously as solvent and protecting group in reactions involving catalysts that would otherwise be deactivated in the presence of primary or secondary amines (Fürstner et al., 1997).

The very few currently available mechanistic studies on metal-catalyzed reactions in sc CO_2 have shown that intermediates containing M–H and M–C units can be formed during hydroformylation (Rathke et al., 1991) and hydrogenation (Lange et al., 2000) in sc CO_2 without reacting with the solvent. Consequently, these reactions occur readily in CO_2 and may benefit from the unique properties of the SCF. Nevertheless, it is important to keep in mind that CO_2 itself can interact directly with many catalytically active transition-metal compounds (Leitner, 1996). Among the best-known processes are insertion reactions into metal-element bonds or deoxygenation of CO_2 to give CO. Such interactions were identified as major catalyst deactivation pathways in several cases (Kreher et al., 1998; Mason and Ibers, 1982; Six et al., 1999), but, of course, the interactions must not always be detrimental. Among the possible benefits associated with such processes is the highly attractive concept of using CO_2 as a solvent and a feedstock at the same time (Jessop et al., 1994; Reetz et al., 1993). The best-known example for this approach is the metal-catalyzed reaction of H_2 and CO_2 to yield formic acid and its derivatives, which involves the insertion of CO_2 into an M–H bond as one of the productive steps (Jessop et al., 1995b; Leitner, 1995) and was one of the first homogeneously catalyzed processes to be conducted with high efficiency in the supercritical phase (Jessop et al., 1994).

Selected Successful Applications of Fluorous Phases and Carbon Dioxide in Chemical Synthesis

Radical Reactions with Tin Hydride Reagents

Radical reactions based on tin hydrides have found a large number of important applications in synthetic organic chemistry, with $(n\text{-Bu})_3SnH$ being the most widely used example (Davies, 1997). The application of this reagent is, however, often accompanied by severe practical problems arising from difficulties separating toxic tin residues from the products and from the lack of methods to reuse the tin reagent. In an attempt to overcome these limitations, a series of perfluoroalkyl-substituted tin hydrides, **2**, were synthesized and tested as alternative reagents for synthesis in fluorous phases (Curran and Hadida, 1996; Curran et al., 1999) and in sc CO_2 (Curran and Hadida, 1997). Like many other syntheses of fluorinated reagents and catalysts, the preparations of **2** were synthesized from the readily available iodides **1** as starting materials (eq. 5.1).

$$\underset{\mathbf{1}}{F(F_2C)_x(H_2C)_yI} \xrightarrow[\text{3. } Br_2 \quad \text{4. } LiAlH_4]{\text{1. Mg} \quad \text{2. } PhSnCl_3} \underset{Rf_xh_y\text{-}\mathbf{2}}{[F(F_2C)_x(H_2C)_y]_3SnH} \qquad (5.1)$$

$$x = 4,6,8,10 \quad y = 2,3$$

Compounds **2** were found to partition preferably into the fluorous phase in mixtures of FC-72 and organic solvents like acetonitrile and benzene. As one would expect, the preference for the fluorous phase decreased when the ethylene spacer ($y = 2$) was replaced with a propylene spacer ($y = 3$) and it increased considerably with increasing values of x. The application of a reagent is, however, also related to its absolute solubility, and Rf_6h_2-**2** was found to provide the optimum solubility behavior. Rf_6h_2-**2** was also found to be much more soluble in sc CO_2 than $(n\text{-Bu})_3SnH$, but no systematic variation was performed in this case.

The perfluorinated tin hydride Rf_6h_2-**2** was used successfully in a large number of radical reductions of alkyl halides under either stoichiometric or catalytic conditions. The performance was very similar in either fluorous liquids or sc CO_2, as exemplified in eq. 5.2. Using basically the strategy shown in Figure 5.1, the products were isolated from the organic phases in high purity and all tin compounds were recovered in the fluorous phase. Most significantly, a three-phase (fluorous/organic/water) separation protocol could be established in the case of the catalytic procedure, and efficient recycling of Rf_6h_2-**2** was possible. The selective separation between the tin hydrides and the products might also be possible with sc CO_2, but the experimentally described separation with this medium was carried out using fluorous-phase extractions after venting the CO_2 from the reaction mixture.

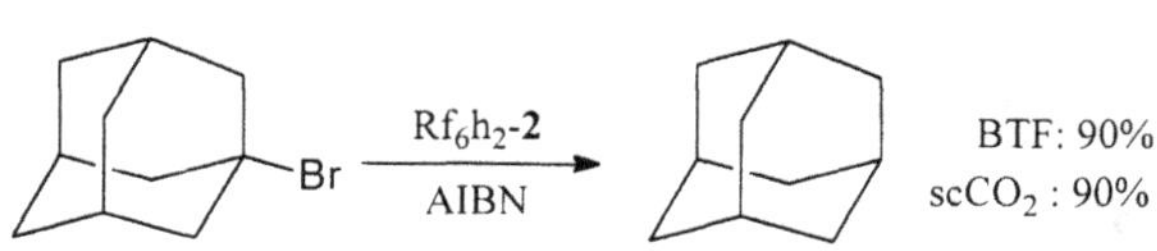

(5.2)

3 **4**

The combination of the catalytic method and the three-phase separation could be applied to the parallel rapid synthesis of a small library of nine compounds of type **7** using the Giese reaction (eq. 5.3). The attractive prospect from this simple experiment arises from the very efficient workup protocol, allowing facile and quantitative separation even if excess reagents are used. The potential for automation and the possible application of such strategies in combinatorial chemistry is immediately apparent (Curran, 1996; Studer et al., 1997). Although the reaction itself works also very well in sc CO_2 (eq. 5.3), workup procedures of similar efficiency are not obvious in this case.

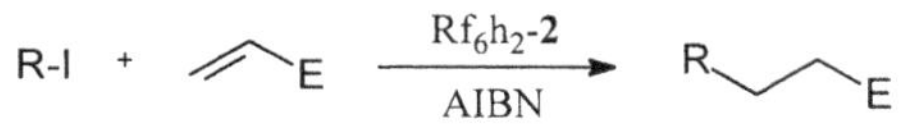

5 **6** **7**

BTF 67-94% $R = C_{15}H_{31}$, Cy, Ad; E = CN, CO_2Me, COMe

$scCO_2$ 81% R = Ad; E = CN

(5.3)

Cyclization reactions (eq. 5.4) could be realized in a similar manner in the fluorous phase and occurred with relative rates and selectivities similar to those using $(n\text{-Bu})_3SnH$ in typical organic liquids (Horner et al., 1997). In contrast, the hydrocarbon **10** was isolated as a by-product from the cyclization of **8** in sc CO_2 (eq. 5.5). This finding was explained with the higher diffusion rate in the SCF, which might facilitate the bimolecular reduction step relative to the unimolecular cyclization in these radical reactions. Another remarkable side reaction occurred when the iodo anthracene **11** was reduced with Rf_6h_2-**2** in sc CO_2 (eq. 5.5). The carboxylic acid **13** was isolated in 10% yield and it was speculated that the aryl radical intermediate was intercepted with CO_2 before reduction. Usually, dense-phase CO_2 is considered a very inert reaction medium for radical reactions (DeSimone et al., 1994) and further clarification is needed to elucidate the pathway leading to **13** and to devise eventual synthetic transformations based on this observation.

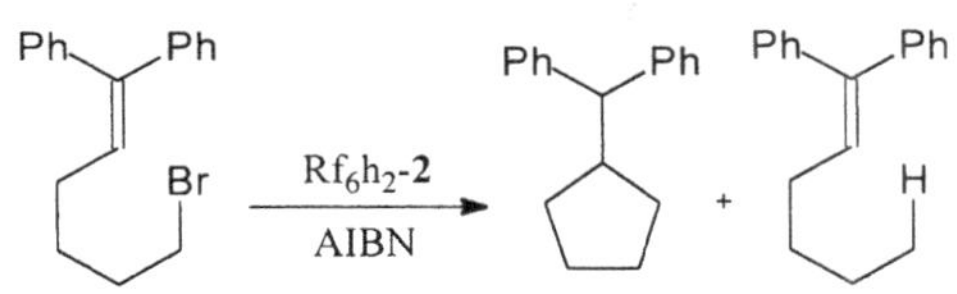

8 **9** **10**

	9	**10**
BTF:	72%	---
$scCO_2$:	87%	7%

(5.4)

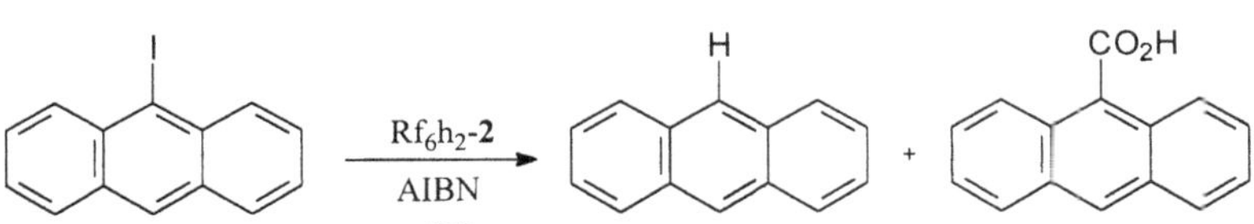

11 **12**, 71% **13**, 10%

(5.5)

Metal-Catalyzed Hydrogenation Reactions

The obvious advantage of a single solution phase with very high concentrations of H_2, gaslike diffusion, and low viscosity has stimulated a large number of studies devoted to the use of sc CO_2 as a medium for hydrogenation reactions. Heterogeneously catalyzed hydrogenations in SCFs are already finding serious industrial attention and various possible applications are considered (Härröd et al., 1997; Hitzler et al., 1998; Pickel and Steiner, 1994; Tacke et al., 1996). In homogeneous catalysis, examples range from hydrogenation of CO_2 itself to give formic acid and its derivatives (Jessop et al., 1994, 1996; Kröcher et al., 1998) to highly enantioselective hydrogenations of C=C double bonds (Burk et al., 1995; Lange et al., 2000; Xiao et al., 1996) and C=N double bonds (Kainz et al., 1999) with chiral metal complex catalysts. For example, the secondary amine **16** was obtained with an enantioselectivity of 81% enantiomeric excess (ee) for the (*R*)-isomer (eq. 5.6), which compares well to the selectivity achieved in methylene chloride under otherwise identical conditions. At the same time, the use of sc CO_2 led to an approximately 20 times higher catalytic efficiency and allowed for the separation and recycling of the iridium catalyst in active and selective form by means of the CESS process.

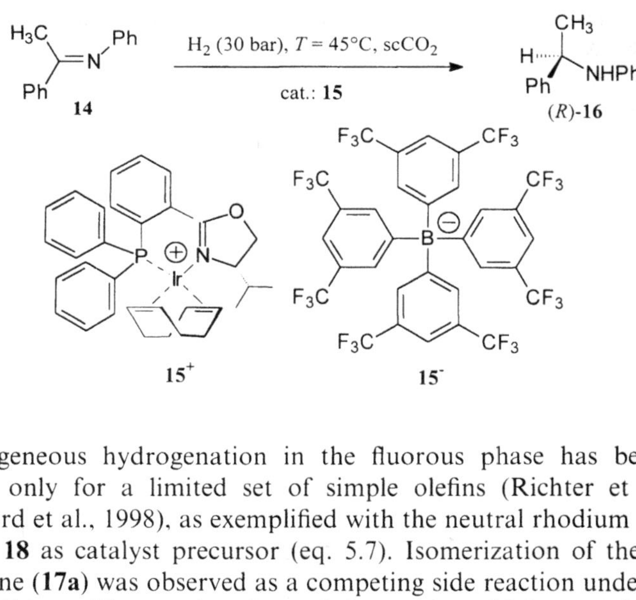

(5.6)

Homogeneous hydrogenation in the fluorous phase has been so far reported only for a limited set of simple olefins (Richter et al., 1999, Rutherford et al., 1998), as exemplified with the neutral rhodium phosphine complex **18** as catalyst precursor (eq. 5.7). Isomerization of the substrate 1-dodecene (**17a**) was observed as a competing side reaction under the reaction conditions. The catalyst formed from **18** could be recycled using a typical FBS protocol, but deactivation under formation of metal deposits limited the catalyst lifetime.

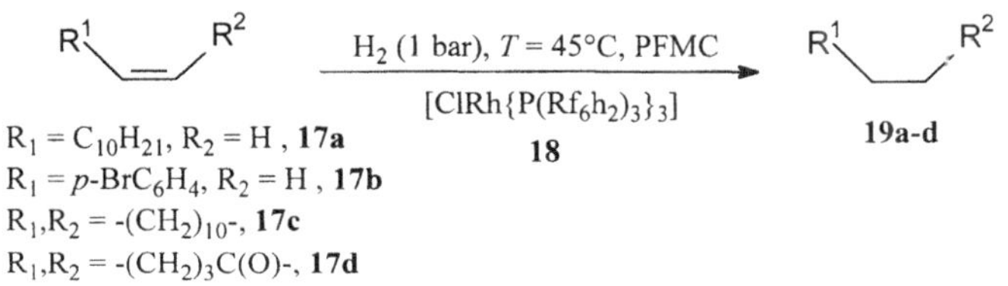

(5.7)

The discrepancy in the scope of hydrogenation in the two media can be attributed to the difficulty to adjust a sufficiently high degree of fluorination in chiral homogeneous hydrogenation catalysts to use them under FBS conditions, whereas the solubility issue is more readily addressed in the case of sc CO_2. For example, the modification of established chiral phosphorus ligands with up to four fluorinated ponytails Rf_6h_2 is now synthetically relatively straightforward and allows reasonable "CO_2-philicity" even for fairly complex ligand systems (Franciò and Leitner 1999; Hope et al., 1999; Kainz et al., 1997, 1998, 1999; Lange et al., 2000). A similar degree of fluorination will, however, often be insufficient for FBS systems owing to the large molecular weight and the polarity of the ligands and the metal complex catalysts (Klose and Gladysz, 1999; Richter et al., 1999). The cationic nature of many homogeneous hydrogenation catalysts (e.g., **15**$^+$) suggests that even more extensive fluorination will not always result in absolute solubilities that are high enough for practical application in the extremely nonpolar fluorous solvents. On the other hand, the application of tetrakis(3,5-trifluoromethylphenyl)borate (BARF, **15**$^-$) or related fluorinated anions is often sufficient to yield highly efficient and recyclable cationic catalysts for asymmetric catalysis in sc CO_2 (Burk et al., 1995; Kainz et al., 1999; Wegner and Leitner, 1999).

Hydroformylation of Long-Chain Olefins

The hydroformylation (oxo-reaction) is one of the most important applications of transition-metal catalysis in the synthesis of fine chemicals and commodities. It is therefore not surprising that some of the pioneering studies on the use of fluorous phases (Horváth and Rabai, 1994a, 1994b) and sc CO_2 (Kainz et al., 1997; Koch and Leitner, 1998; Rathke et al., 1991) in homogeneous catalysis have focused on this process. Later, highly enantioselective and regioselective hydroformylation of vinyl arenes was achieved using the chiral ligand 3-H^2F^6-BINPAHOS in sc CO_2 (Franciò and Leitner, 1999; Kainz and Leitner, 1998). Two detailed reports on the use of phosphine-modified rhodium catalysts for hydroformylation of long-chain olefins (Horváth et al., 1998; Koch and Leitner, 1998) allow an informative comparison of the two media (Figure 5.3).

The maximum turnover frequency (TOF_{max} = maximum rate given as mole product per mole Rh per hour) is an order of magnitude higher with the catalyst used in sc CO_2 than in the FBS system. These values

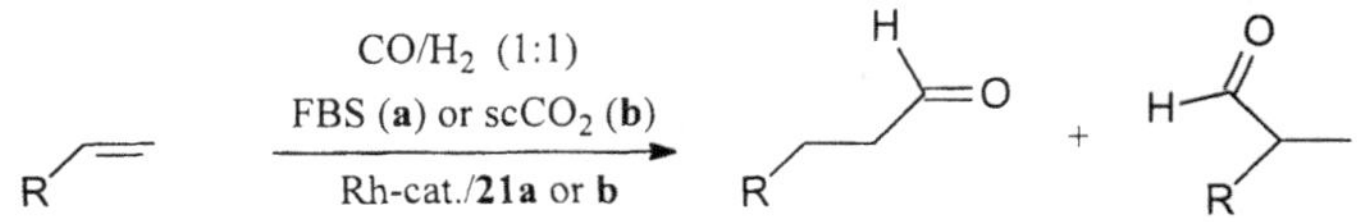

$R = C_8H_{17}$, **20a**　　*n*-**22a**　　*iso*-**22a**

$R = C_6H_{13}$, **20b**　　*n*-**22b**　　*iso*-**22b**

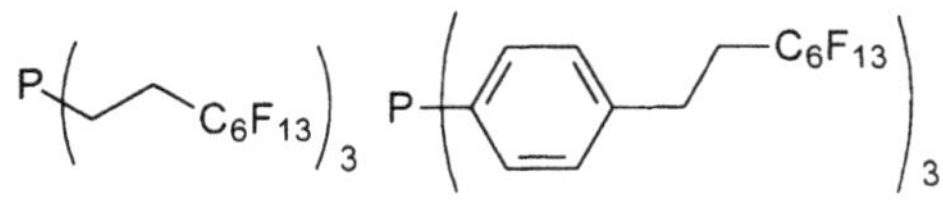

21a　　**21b**

conditions

	Ref.	Solvent	P-ligand	Rh-cat.	P:Rh	c(Rh) [mM]	C^0(**20**) [M]	T [°C]	$p(CO/H_2)$ [bar]
a	41	PFMC/ toluene	**21a**	[(H(CO)Rh(**21a**)$_3$]	27:1	0.81	1.0	100	10
b	42	Sc CO_2	**21b**	[(cod)Rh(hfacac)]	10:1	0.13	0.28	65	20

results

	conv. [%]	TON	TOF_{max} [h^{-1}]	*n*-**22**/ *iso*-**22**	Isomer. [%]	hydrog. [%]	Number of runs	Rh-loss [%]
a	80-95%	1100	36 (initially)	3.25	10%	< 1%	9	4.2
b	93-99%	2100	430 (cv = 25%)	4.5	< 1%	< 1%	5	2.4

Figure 5.3. Rhodium-catalyzed hydroformylation of long-chain olefins under FBS conditions (a) and in sc CO_2 (b).

correspond most likely to an even larger difference in catalytic activity considering the different reaction temperatures and substrate concentrations; the reaction is first-order in substrate in the FBS (Horváth et al., 1998), but was found to be half-order in substrate with a related catalyst in sc CO_2 (Palo and Erkey, 1999). Isomerization of **20a** to form internal olefins is observed as a significant side reaction in the FBS, but not in sc CO_2. The regioselectivity (*n*/*iso* ratio) is slightly higher with the catalytic system operating in sc CO_2 compared with that used in the fluorous phase.

The differences in rate and selectivity observed under comparable conditions can be assigned largely to the different ligand types used in the two studies (Moser et al., 1987). The ligand **21a** has the fluorinated substituent Rf_6h_2 attached directly to the phosphorus donor atom and therefore bears some resemblance in its steric and electronic properties to an alkyl phosphine. Ligand **21b**, on the other hand, is an Rf_5h_2-substituted analogue of triphenylphosphine (4-H^2F^6-tpp) and behaves more like a typical aryl phosphine. However, an influence of the reaction medium may well be operating at the same time, because a beneficial effect of sc CO_2 compared with classical organic solvents has been clearly established for unmodified rhodium (Koch and Leitner, 1998) and cobalt (Rathke et al., 1991) catalysts.

One of the most attractive features of both systems is the highly efficient isolation of the aldehydes **22** with simultaneous recycling of the catalyst, which is a major difficulty and cost factor in the current technology hydroformylation technology of long-chain olefins using conventional solvents (Frohning and Kohlpaintner, 1996). Catalyst recycling was achieved using either the FBS or CESS approach, whereby the overall rhodium loss was similar in both cases. A slight decrease in the *n*/*iso* ratio indicated some phosphine leaching in the FBS system, whereas the regioselectivity remained stable within experimental error in the CESS process. The principle technical feasibility of the fluorous-phase system was demonstrated for the hydroformylation of ethylene by continuous operation over 64 days. In this case, the separation was based on the high volatility of the product aldehyde rather than using a classical FBS procedure.

Pd-catalyzed C–C Coupling Reactions

A variety of Pd-catalyzed coupling reactions of aryl halides or triflates (**23**) with vinylic compounds (**24**) has been successfully applied to the generation of new carbon–carbon bonds in sc CO_2 as the reaction medium (eq. 5.8). The Heck (LG = H), Stille ($LG = SnR_3$), and Suzuki [$LG = B(OR)_2$] couplings were found to proceed smoothly under these conditions, whereby the design of soluble Pd-catalysts played again a crucial role for the stabilization of the active intermediates to control the activity and the selectivity of the reaction. The Stille coupling of PhI with vinyl(tributyl)tin gave almost quantitative yields of styrene using $[Pd_2(dba)_3]$ [dba = di(benzylidene)acetone] as the Pd(0) source in the presence of various fluorinated phosphine ligands, but initial rates were approximately two times lower than in toluene (Carrol and Holmes, 1998; Morita et al., 1998). It is noteworthy that even nonfluorinated tris(2-furyl)phosphine led to apparently soluble and active catalysts under similar conditions, especially when used together with only lightly fluorinated Pd-sources like palladium trifluoroacetate (Shezad et al., 1999). No reports are available so far on the possibility to separate the products of type **25** from the inorganic halides **26** or to recycle the catalysts in these reactions.

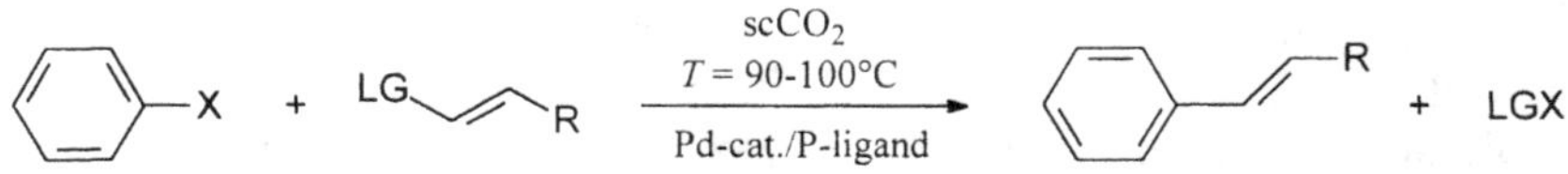

23 24 25 26

X = I, Br

LG = leaving group = H, Bu_3Sn, $(RO)_2B$

(5.8)

Replacement of the three *n*-Bu rests in the tributyl tin fragment, with the Rf_6h_2 substituent allowed to device a highly interesting enhanced protocol for the Stille coupling based on fluorous-phase separation (eq. 5.9). Aryl tin compounds of general formula **28** were coupled in high yields with various aryl or benzyl halides (**27**) using a nonfluorinated catalyst like **29** in dimethylformamide/tetrahydrofuran (DMF/THF) as a solvent either by thermal (Curran and Hoshino, 1996) or microwave (Olofsson et al., 1999) activation. After completion of the reaction, the solvent was evaporated and the nonvolatile reaction mixtures were partitioned between a three-phase system consisting of a fluorous phase (FC-72), an organic phase (CH_2Cl_2), and water. This resulted in a very efficient and yet simple separation of the coupling products (organic layer) from the salt additives (LiCl; aqueous layer) and especially from tin-containing by-products or excess tin reagent (fluorous phase). Isolation of the desired cross-coupling products **30** from the organic phase in pure form is straightforward after this initial separation, whereas traces of tin are very tedious to remove from Stille products obtained according to conventional procedures. Such traces of tin can be a severe problem for biologically active compounds and, consequently, the new fluorous protocol has already found interest in the pharmaceutical industry (Mulholland, 1998).

$$\underset{\mathbf{27}}{Ar^1Br} + \underset{\mathbf{28}}{Ar^2Sn(Rf_6h_2)_3} \xrightarrow[\text{[(Ph}_3\text{P)}_2\text{PdCl}_2\text{], } \mathbf{29}\text{, LiCl}]{\text{DMF/THF}} \underset{\mathbf{30}}{Ar^1\text{-}Ar^2} + \underset{\mathbf{31}}{BrSn(Rf_6h_2)_3} \quad (5.9)$$

Bisaryl coupling processes related to the Stille reaction have also been performed using a perfluorinated Pd-catalyst (Betzemeier and Knochel, 1997). Nucleophilic allylic substitutions are another class of important C–C bond-forming reactions using palladium catalysts. The use of $4\text{-}H^2F^6$-tpp, **21b** (Kling et al., 1998), or related highly fluorinated aryl phosphine ligands (Hope et al., 1999; Sinou et al., 1999), allow these reactions to be carried out in a typical FBS system consisting of THF and PFMC (perfluoromethylcyclohexane). Reaction rates were uniformly high and the products were collected in excellent yields by simple decantation of the organic phase. The fluorous phase could be reused effectively up to seven times, even under nonoptimized conditions.

Oxidation Using Molecular Oxygen

The high stability of fluorous phases and sc CO_2 toward oxidation has stimulated a number of activities to use these media as solvents for selective oxidation processes using various oxidating reagents (for FBS see Betzemeier et al., 1998; Pozzi et al., 1996, 1997a; van Vliet et al., 1999; for sc CO_2 see Leitner, 1999 and refs. therein). Most notably, however, the physicochemical properties of fluorinated liquids and sc CO_2 suggest a large potential of these media with molecular oxygen as the primary oxidant. The high solubility of gases in fluorinated phases has often been quoted as an additional potential advantage for such processes, although the net effect may not be so important if molar concentrations of O_2 are compared rather than mole fractions (Barthel-Rosa and Gladysz, 1999). In sc CO_2, the availability of O_2 is clearly at its optimum and the presence of CO_2 can add considerably to the safety of the process, as much higher O_2/substrate ratios become possible without reaching the explosion limits. The higher diffusion rates in the SCF can be beneficial for oxidations occurring via free-radical reaction pathways.

Porphyrins and other nitrogen ligands have been rendered "fluoro-philic" and were successfully used as ligands for cobalt (Horváth and Rabai, 1994b; Pozzi et al., 1997b) and manganese (Vincent et al., 1997) catalysts or as photosensitizers (DiMagno et al., 1996) for the aerobic oxidation of alkanes or alkenes. Ligand-free catalytic systems based on manganese were also employed in two-phase systems comprising fluorinated solvents (Ravikumar et al., 1998). High enantioselectivity was reported for the epoxidation of indene under FBS conditions when Mn^{2+} complexes of fluorinated chiral salen-type ligands were used (Pozzi et al., 1998). The ruthenium complex **33** and a related Ni^{2+} complex containing a highly fluorinated 1,3-diketonate ligand were developed as efficient and recyclable catalysts for the oxidation of aldehydes to carboxylic acids and of sulfides to sulfoxides or sulfones (Klement et al., 1997).

Complex **33** was found to also catalyze the epoxidation of various disubstituted olefins with O_2 as the primary oxidant and isobutyraldehyde as the sacrificial co-oxidant (Mukayama conditions; Klement et al., 1997). Yields of 70–85% were obtained after heating the olefins and **33** under oxygen atmosphere in a mixture of toluene and 1-bromoperfluorooctane to 50 °C for 5–12 h, as exemplified in eq. 5.10 for the synthesis of cycloocteneoxide **34**. The ruthenium was held back almost quantitatively in the fluorous phase and could be separated after cooling to 0 °C and reused several times.

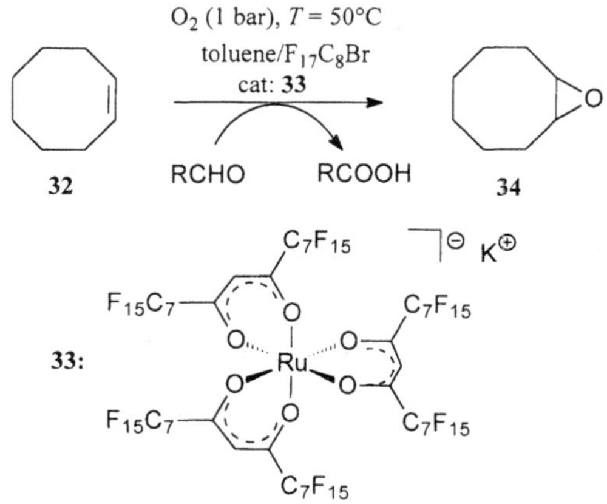

(5.10)

The cyclic substrate **32** and other disubstituted olefins such as **35a** were oxidized in sc CO_2 to give the corresponding epoxides with reasonable rates (>95% conversion in less than 18 h) and excellent selectivities (>98%) under otherwise similar reaction conditions (Loeker and Leitner, 2000). It is important to note, however, that no addition of a metal catalyst was required in the supercritical reaction medium. Detailed control experiments revealed that the stainless steel of the reactor walls served as efficient initiator for the epoxidation under these conditions. Terminal olefins **35b,c** were oxidized with somewhat reduced rates and either epoxidation or vinylic oxidation occurred as the major reaction pathway depending on the substrate (eq. 5.11). Apart from providing the first examples for efficient and highly selective oxidation with O_2 in sc CO_2 (earlier attempts: Birnbaum et al., 1999; Loeker et al., 1998; Wu et al., 1997), this study points to the possible importance of wall effects during catalytic reactions in this medium (see also Christian et al., 1999; Suppes et al., 1989).

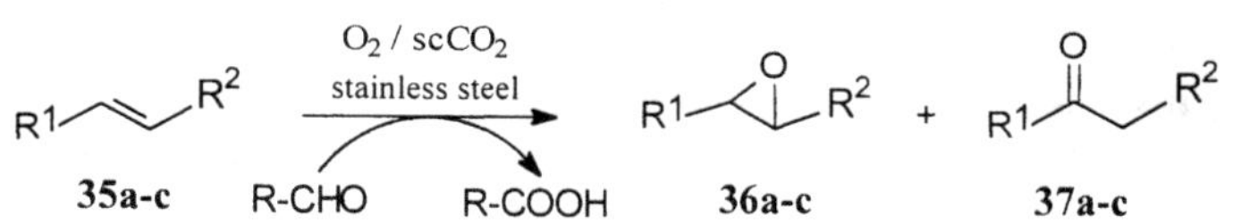

(5.11)

		36a-c	37a-c
a: $R^1 = R^2 = n\text{-}C_2H_5$	cv = 96%	>98%	---
b: $R^1 = n\text{-}C_6H_{13}$, $R^2 = H$	cv = 46%	94%	---
c: $R^1 = Me$, $R^2 = H$	cv = 73%	14%	67%

Conclusion and Outlook

The general discussion and the examples presented in this account demonstrate the remarkable potential of fluorous phases and sc CO_2 as alternative reaction media for modern chemical synthesis. These two solvent systems are in many ways complementary and can find broad application as toxicologically and environmentally benign replacements for conventional organic liquids in stoichiometric or catalytic conversions. At the same time, their unique properties can allow for additional benefits for chemical reactions, and we are just beginning to understand how to exploit these features efficiently.

The fluorous phases offer various possibilities for innovative separation schemes that are remarkably simple and effective. It is likely that we will see an increasing number of applications of this approach in combinatorial chemistry and high-throughput parallel synthesis. These separation processes are also attractive for industrial production, albeit the price of fluorous liquids and reagents seems to limit the technique to high-value products. However, it is important to note that comparisons between processes based on fluorous phases or organic liquids are not trivial and simply considering the price of the "solvents" may often not be sufficient. If designed properly, the fluorous-phase process will be operated with minimum loss of the fluorous components and considerable savings are possible in the downstream processing, including cuts on costs for wastewater treatment, waste disposal, or solvent regeneration.

Similarly, an economic evaluation of the use of sc CO_2 must take into account downstream processing as well. Although the high pressure required for the handling of sc CO_2 will be a very important factor during reactor and process design, other benefits like simple isolation of solvent-free products, catalyst recycling, higher space–time velocities, or increased selectivities may help to compensate for these additional investments. Increasing costs on CO_2 emissions would also influence the analysis considerably. Even now, the various technical applications of sc CO_2 in the food industry illustrate the feasibility of this technology, even for moderately costly products. The recent announcement of DuPont to build a pilot plant to evaluate the large-scale production of fluorinated polymers of the Teflon brand in sc CO_2 lends further support to this view (McCoy, 1999).

In addition to these questions concerning practical applications of fluorous phases and sc CO_2, there is also a great challenge and opportunity for basic science. Although recent research efforts have been fairly successful in transferring known chemistry into these media with various levels of added benefits, the potential for new reactions has barely been tapped up to now. Shifting equilibria by phase separation, mimicking dilution by controlled solubility, and using CO_2 as protecting group or building block are just some of the directions where novel chemistry may lay ahead. The rapid development that chemistry in "unconventional media" has seen since the mid-1990s forms a strong basis for the exploration of this prospering field.

References

Barthel-Rosa, L. P.; Gladysz, J. A. *Coord. Chem. Rev.* **1999**, *190–192*, 587.
Betzemeier, B.; Knochel, P. *Angew. Chem. Int. Ed. Engl.* **1997**, *36*, 2643.
Betzemeier, B.; Knochel, P. *Top. Curr. Chem.* **1999**, *206*, 61.
Betzemeier, B.; Lhermitte, F.; Knochel, P. *Tetrahedron Lett.* **1998**, *39*, 6667.
Bhanage, B. M.; Ikushima, B. M.; Shirai, M.; Arai, M. *Chem. Commun.* **1999**, 1277.
Birnbaum, E. R.; Le Lacheur, R. M.; Horton, A. C.; Tumas, W. *J. Mol. Catal. A: Chem.* **1999**, *139*, 11.
Blanchard, L. A.; Hancu, D.; Beckman, E. J.; Brenecke, J. F. *Nature* **1999**, *399*, 28.
Burdenuic, J.; Jedlicka, B.; Crabtree, R. H. *Chem. Ber./Recl.* **1997**, *130*, 145.
Burk, M. J.; Feng, S.; Gross, M. F.; Tumas, W. *J. Am. Chem. Soc.* **1995**, *117*, 8277.
Carrol, M. A.; Holmes, A. B. *Chem. Commun.* **1998**, 1395.
Christian, P.; Giles, M. R.; Howdle, S. M.; Major, R. C.; Hay, J. N.; Winder R. *Proceedings of the 6th Meeting on Supercritical Fluids*, Nottingham, U.K., 1999; p. 349.
Clarke, M. J.; Harrison, K. L.; Johnston, K. P.; Howdle, S. M. *J. Am. Chem. Soc.* **1997**, *119*, 6399.
Cornils, B. *Angew. Chem., Int. Ed. Engl.* **1997**, *36*, 2057.
Curran, D. P. *Chemtracts: Org. Chem.* **1996**, *9*, 75.
Curran, D. P. *Angew. Chem., Int. Ed. Engl.* **1998**, *37*, 1174.
Curran, D. P.; Hadida, S. *J. Am. Chem. Soc.* **1996**, *118*, 6480.
Curran, D. P.; Hadida, S. *J. Am. Chem. Soc.* **1997**, *119*, 7406.
Curran, D. P.; Hoshino, M. *J. Org. Chem.* **1996**, *61*, 6480.
Curran, D. P.; Luo, Z. *J. Am. Chem. Soc.* **1999**, *121*, 9069.
Curran, D. P.; Hadida, S.; Kim, S.-Y.; Luo, Z. *J. Am. Chem. Soc.* **1999**, *121*, 6607.
Davies, A. G. *Organotin Chemistry*; VCH: Weinheim, 1997.
DeSimone, J. M.; Maury, E. E.; Menceloglu, Y. Z.; McClain, J. B.; Romack, T. J.; Combs, J. R. *Science* **1994**, *265*, 356.
DiMagno, S. G.; Dussault, P. H.; Schultz, J. A. *J. Am. Chem. Soc.* **1996**, *118*, 5312.
Dinjus, E.; Fornika, R.; Scholz, M. *Chemistry Under Extreme or Non-Classical Conditions*; van Eldik, R., Hubbard, C. D., Eds.; Wiley: New York, 1996; pp. 219.
Franciò, G.; Leitner, W. *Chem. Commun.* **1999**, 1663.
Frohning, C. D.; Kohlpaintner C. W. *Applied Homogeneous Catalysis with Organometallic Compounds Vol. 1*; Cornils, B.; Herrmann, W. A., Eds.; VCH: Weinheim, 1996; pp. 47.
Fürstner, A.; Koch, D.; Langemann, K.; Leitner, W.; Six, C. *Angew. Chem., Int. Ed. Engl.* **1997**, *36*, 2466.
Harrison, K.; Goveas, J.; Johnston, K. P.; O'Rear III, E. A. *Langmuir* **1994**, *10*, 3536.
Härröd, M.; Macher, M.-B.; Högberg, J.; Møller, P. *Proceedings of the 4th Italian Conference on Supercritical Fluids and Their Applications*; Capri, Italy, 1997; pp. 319.
Hitzler, M. G.; Smail, F. R.; Ross, S. K.; Poliakoff, M. Org. Process Res. Dev. **1998**, *2*, 137.
Hope, E. G.; Kemmitt, R. D. W.; Paige, D. R.; Stuart, A. M.; Wood, D. R. W. *Polyhedron* **1999**, *18*, 2913.

Horner, J. H.; Martinez, F. N.; Newcomb, M.; Hadida, D.; Curran, P. *Tetrahedron Lett.* **1997**, *38*, 2783.

Horváth, I. T. *Acc. Chem. Res.* **1998**, *31*, 641.

Horváth, I. T.; Rábai, J. *Science* **1994a**, *266*, 72.

Horváth, I. T.; Rabai, J. EP 0 633 062, A1, **1994b**, 4 July 1994.

Horváth, I. T.; Kiss, G.; Cook, R. A.; Bond, J. E.; Stevens, P. A.; Rabai, J.; et al. *J. Am. Chem. Soc.* **1998**, *120*, 3133.

Jacobson, G. B.; Lee Jr., C. T.; Johnston, K. P. *J. Org. Chem.* **1999a**, *64*, 1201.

Jacobson, G. B.; Lee Jr., C. T.; daRocha, S. R. P.; Johnston, K. P. *J. Org. Chem.* **1999b**, *64*, 1207.

Jessop, P.G; Leitner, W., Eds. *Chemical Synthesis Using Supercritical Fluids*; Wiley-VCH: Weinheim, 1999.

Jessop, P. G.; Ikariya, T.; Noyori, R. *Nature* **1994**, *368*, 231.

Jessop, P. G.; Ikariya, T.; Noyori, R. *Science* **1995a**, *269*, 1065.

Jessop, P. G.; Ikariya, T.; Noyori, R. *Chem. Rev.* **1995b**, *95*, 259.

Jessop, P. G.; Hsiao, Y.; Ikariya, T.; Noyori, R. *J. Am. Chem. Soc.* **1996**, *118*, 344.

Johnston, K. P.; Harrison, K. L.; Clarke, M. J.; Howdle, S. M.; Heitz, M. P.; Bright, F. V., et al. *Science* **1996**, *271*, 624.

Kainz, S.; Leitner, W. *Catal. Lett.* **1998**, *55*, 223.

Kainz, S.; Koch, D.; Baumann, W.; Leitner, W. *Angew. Chem. Int. Ed. Engl.* **1997**, *36*, 1628.

Kainz, S.; Luo, Z.; Curran, D. P.; Leitner, W. *Synthesis* **1998**, 1425.

Kainz, S.; Brinkmann, A.; Leitner, W.; Pfaltz, A. *J. Am. Chem. Soc.* **1999**, *121*, 6421.

Klement, I.; Lütjens, H.; Knochel, P. *Angew. Chem., Int. Ed. Engl* **1997**, *36*, 1454.

Kling, R.; Sinou, D.; Pozzi, G.; Choplin, A.; Quignard, F.; Busch, S. et al. *Tetrahedron Lett.* **1998**, *39*, 9439.

Klose, A.; Gladysz, J. A. *Tetrahedron: Asymmetry* **1999**, *10*, 2665.

Koch, D.; Leitner, W. *J. Am. Chem. Soc.* **1998**, *120*, 13398.

Kreher, U.; Schebesta, S.; Walther, D. *Z. Anorg. Allg. Chem.* **1998**, *624*, 602.

Kröcher, O.; Köppel, R. A.; Fröba, M.; Baiker, A. *J. Catal.* **1998**, *178*, 284.

Lange, S.; Brinkmann, A.; Trautner, P.; Woelk, K.; Bargon, J.; Leitner, W. *Chirality* **2000**, *12*, 450.

Leitner, W. *Angew. Chem., Int. Ed. Engl.* **1995**, *34*, 2207.

Leitner, W. *Coord. Chem. Rev.* **1996**, *153*, 257.

Leitner, W. *Top. Curr. Chem.* **1999**, *206*, 107.

Lin, Y. H.; Brauer, R. D.; Laintz, K. E.; Wai, C. M. *Anal. Chem.* **1993**, *65*, 2549.

Loeker, F.; Leitner W. *Chem.—Eur. J.* **2000**, *6*, 2011.

Loeker, F.; Koch, D.; Leitner, W. *Selective Oxidations in Petrochemistry, DGMK-Tagungsbericht 9803*; Emig, G.; Kohlpaintner, C.; Lücke, B., Eds.; DGMK: Hamburg, 1998; p. 209.

Mason, M. G.; Ibers, J. A. *J. Am. Chem. Soc.* **1982**, *104*, 5153.

Maul, J. J.; Ostrowski, P. J.; Ublacker, G. A.; Linclau, B.; Curran, D. P. *Top. Curr. Chem.* **1999**, *206*, 61.

McCoy, M. *Chem. Eng. News* **1999**; June 14, p. 11.

Morgenstern, D. A.; LeLacheur, R. M.; Morita, D. K.; Borkowsky, S. L.; Feng, S.; Brown, G. H., et al. *Green Chemistry*; Anastas, P. T., Williamson, T. C., Eds.; ACS Symp. Series 626; American Chemical Society, Washington DC, 1996; p. 132.

Morita, D. K.; Pesiri, D. R.; David, S. A.; Glaze, W. H.; Tumas, W. *Chem. Commun.* **1998**, 1397.

Moser, W. R.; Papile, C. J.; Brannon, D. A.; Duwell, R. A.; Weininger, S. J. *J. Mol. Catal.* **1987**, *41*, 271.

Mulholland, K. Paper presented at the 16th Process Development Symposium, SCI, London, December 3, 1998.

Noyori, R.; Guest ed. *Chem. Rev.* **1999**, *99* (2).

Olofsson, K.; Kim, S.-Y.; Larhed, M.; Curran, D. P.; Hallberg, A. *J. Org. Chem.* **1999**, *64*, 4539.

Palo, D. R.; Erkey, C. *Ind. Eng. Chem. Res.* **1999**, *38*, 3786.

Pickel, K. H.; Steiner, K. *Proceedings of the 3rd International Symposium on Supercritical Fluids* Strasbourg, France, 1994, p. 25.

Pozzi, G.; Banfi, S.; Manfredi, A.; Montanari, F.; Quici, S. *Tetrahedron* **1996**, *52*, 11879.

Pozzi, G.; Cavazzini, M.; Quici, S.; Fontana, S. *Tetrahedron Lett.* **1997a**, *38*, 7605.

Pozzi, G.; Montanari, F.; Quici, S. *Chem. Commun.* **1997b**, *69*.

Pozzi, G.; Cinato, F.; Montanari, F.; Quici, S. *Chem. Commun.* **1998**, *877*.

Rathke, J. W.; Klingler, R. J.; Krause, T. R. *Organometallics* **1991**, *10*, 1350.

Ravikumar, K. S.; Barbier, F.; Bégué, J.-P.; Bonnet-Delpon, D. *Tetrahedron* **1998**, *54*, 7457.

Ravishankara, A. R.; Solomon, S.; Turnipseed, A. A.; Warren, R. F. *Science* **1993**, *259*, 194.

Reetz, M. T.; Könen, W.; Strack, T. *Chimia* **1993**, *47*, 493.

Richter, B.; Deelman, B.-J.; van Koten, G. *J. Mol. Catal. A: Chem.* **1999**, *145*, 317.

Riess, J. G.; Le Blanc, M. *Pure Appl. Chem.* **1982**, *54*, 2383.

Rutherford, D.; Juliette, J. J. J.; Rocaboy, C.; Horváth, I. T.; Gladysz, J. A. *Catal. Today* **1998**, *42*, 381.

Savage, P. E.; Gopalan, S.; Mizan, T. I.; Martino, C. J.; Brock, E. E. *AIChE J.* **1995**, *41,* 1723.

Shezad, N.; Oakes, R. S.; Clifford, A. A.; Rayner, C. M. *Tetrahedron Lett.* **1999**, *40*, 2221.

Sinou, D.; Pozzi, G.; Hope, E. G.; Stuart, A. M. *Tetrahedron Lett.* **1999**, *40*, 849.

Six, C.; Gabor, B.; Görls, H.; Mynott, R.; Philipps, P.; Leitner, W. *Organometallics* **1999**, *18*, 3316.

Studer, A.; Hadida, S.; Ferritto, R.; Kim, S.-Y.; Jeger, P.; Wipf, P., et al. *Science* **1997**, *275*, 823.

Suppes, G. J.; Occhiogrosso, R. N.; McHugh, M. A. *Ind. Eng. Chem. Res.* **1989**, *28*, 1152–56.

Tacke, T.; Wieland, S.; Panster, P. *Process Technol. Proc.* **1996**, *12*, 17.

van der Boom, M.; Ben-David, Y.; Milstein, D. *J. Am. Chem. Soc.* **1999**, *121*, 6652.

van Vliet, M. C. A.; Arends, I. W. C. E.; Sheldon, R. A. *Chem. Commun.* **1999**, *263*.

Vincent, J.-M.; Rabion, A.; Yachandra, V. K.; Fish, R. H. *Angew. Chem., Int. Ed. Eng.* **1997**, *36*, 2346.

Wegner, A.; Leitner, W. *Chem. Commun.* **1999**, 1583.

Wu, X.-W.; Oshima, Y.; Koda, S. *Chem. Lett.* **1997**, 1045.

Xiao, J.; Nefkens, S. C. A.; Jessop, P. G.; Ikariya, T.; Noyori, R. *Tetrahedron Lett.* **1996**, *37*, 2813.

Yazdi, A. V.; Beckman, E. J. *Ind. Eng. Chem. Res.* **1996**, *35*, 3644.

6

Enzyme Chemistry in Carbon Dioxide

REBECCA L. RODNEY

ALAN J. RUSSELL

Enzymes are biocatalysts constructed of a folded chain of amino acids. They may be used under mild conditions for specific and selective reactions. While many enzymes have been found to be catalytically active in both aqueous and organic solutions, it was not until quite recently that enzymes were used to catalyze reactions in carbon dioxide when Randolph et al. (1985) performed the enzyme-catalyzed hydrolysis of disodium *p*-nitrophenol using alkaline phosphatase and Hammond et al. (1985) used polyphenol oxidase to catalyze the oxidation of *p*-cresol and *p*-chlorophenol. Since that time, more than 80 papers have been published concerning reactions in this medium. Enzymes can be 10–15 times more active in carbon dioxide than in organic solvents (Mori and Okahata, 1998). Reactions include hydrolysis, esterification, transesterification, and oxidation. Reactor configurations for these reactions were batch, semibatch, and continuous.

There are many factors that influence the outcome of enzymatic reactions in carbon dioxide. These include enzyme activity, enzyme stability, temperature, pH, pressure, diffusional limitations of a two-phase heterogeneous mixture, solubility of enzyme and/or substrates, water content of the reaction system, and flow rate of carbon dioxide (continuous and semibatch reactions). It is important to understand the aspects that control and limit biocatalysis in carbon dioxide if one wants to improve upon the process. This chapter serves as a brief introduction to enzyme chemistry in carbon dioxide. The advantages and disadvantages of running reactions in this medium, as well as the factors that influence reactions, are all presented. Many of the reactions studied in this area are summarized in a manner that is easy to read and referenced in Table 6.1.

Table 6.1. Biocatalysis in Carbon Dioxide

Reference	Batch/Continuous	Study	Enzyme
Hammond et al. (1985)	Batch/semibatch	Oxidation of *p*-cresol and *p*-chlorophenol to corresponding *o*-benzoquinones	Polyphenol oxidase
Randolph et al. (1985)	Batch	Hydrolysis of disodium *p*-nitrophenyl phosphate to *p*-nitrophenol	Alkaline phosphatase
Nakamura et al. (1986)	Batch	Esterification of triolein using stearic acid	Lipase from *Rhizopus delemar*
Kamihira et al. (1987)	Batch	Synthesis of aspartame precursors through esterification	Thermolysin
Chi et al. (1988)	Batch	Esterification of triolein using stearic acid	Lipases from *Rizopus delemar*, *Rhizopus japonicus*, *Alcaligenes* sp., and *Mucor miehei*
Randolph et al. (1988)	Batch/semibatch	Oxidation of cholesterol	Cholesterol oxidases *Geioeocysticum chrysocreas* and *Streptomyces* sp.
van Eijs et al. (1988)	Continuous	Synthesis of isoamyl acetate and nonyl acetate by transesterification of ethyl acetate with the corresponding alcohols	Lipozyme^{TM} IM
Nakamura (1989)	Batch	Esterification of triolein using stearic acid	Lipase from *Rhizopus delemar*
Pasta et al. (1989)	Batch	Transesterification of *n*-acetyl-phenylalanine chloroethyl ester and ethanol	Subtilisin Carlsberg
Erickson et al. (1990)	Batch	Esterification of triolein with palmitic acid	Lipases from *Rhizopus arrhizius* and *Mucor miehei*
Marty et al. (1990)	Batch	Esterification of oleic acid by ethanol	Lipase from *Mucor miehei*

Nakamura et al. (1990)	Batch	Esterification of triolein with stearic acid	Lipase from *Rhizopus delemar*
Aaltonen and Rantakylä (1991)	Batch/continuous	Resolution of ibuprofen	Lipase from *Mucor miehei*
Dumont et al. (1991)	Batch	Esterification of myristic acid with ethanol	Lipase from *Mucor miehei*
Martins et al. (1991)	Batch	Esterification of glycidol with butyric acid	Porcine pancreatic lipase
Miller et al. (1991)	Batch	Inter-esterification of triglycerides	Lipase from *Rhizopus arrhizus*
Steytler et al. (1991)	Continuous	Transesterification of lauric acid with butanol	Lipase B from *Candida antarctica*
Bernard et al. (1992)	Batch	Esterification of myristic acid with ethanol	Lipozyme IM
Bornscheuer et al. (1992)	Batch	Esterification, hydrolysis, and transesterification of 3-hydroxyhexanoic acid methylester	Lipase from *Pseudomonas cepacia*
Capewell et al. (1992)	Batch	Hydrolysis, esterification, and transesterification of β-hydroxycarbonic esters and acids	Lipase PS
Dumont et al. (1992)	Batch	Esterification of myristic acid by ethanol	Lipase from *Mucor miehei*
Janssens et al. (1992)	Batch	Transesterification of ethyl acetate and nonyl alcohol	Lipozyme IM
Kamat et al. (1992)	Batch	Transesterification of methyl methacrylate by 2-ethylhexanol	Lipase from *Candida cylindracea*
Knez and Habulin (1992)	Batch	Esterification of oleic acid with oleyl alcohol	Lipozyme IM
Martins et al. (1992)	Batch	Esterification of glycidol using butyric acid	Porcine pancreatic lipase
Nakamura and Hoshino (1992)	Batch	Esterification of triolein with stearic acid	Lipase from *Rhizopus delemar*
Shen et al. (1992)	Batch	Esterification of butyric acid with glycidol	Porcine pancreatic lipase
Vermüe et al. (1992)	Batch	Transesterification of nonanol and ethyl acetate	Lipozyme IM

(continued)

Table 6.1 (continued)

Reference	Batch/Continuous	Study	Enzyme
Yu et al. (1992)	Continuous	Esterification of oleic acid using ethanol to make ethyl oleate	Lipase from *Candida cylindracea*
Chulalaksananukul et al. (1993)	Batch	Synthesis of geranyl acetate by transesterification of propyl acetate with geraniol	Lipozyme IM
Berg et al. (1993)	Semibatch	Synthesis of methyl and butyl esters of fatty acids through transesterification of triglycerides	Lipozyme IM
Dumont et al. (1993)	Continuous	Esterification of myristic acid with ethanol	Lipozyme IM
Ikushima et al. (1993)	Continuous	Esterification of oleic acid by ±-citronellol	Lipase from *Candida cylindracea*
Lee et al. (1993)	Batch	Hydrolysis of starch to glucose	α-Amylase and glucoamylase
Endo and Fujimoto (1994)	Batch/continuous	Esterification between tricaprylin and methyl oleate	Lipozyme IM
Lee et al. (1994)	Batch	Hydrolysis of starch	α-Amylase and glucoamylase
Martins et al. (1994)	Batch	Esterification of glycidol with butyric acid	Porcine pancreatic lipase and Lipozyme IM
Rantakylä and Aaltonen (1994)	Batch	Esterification of ibuprofen with *n*-propanol	Lipozyme IM
Bernard and Barth (1995)	Batch	Esterification of myristic acid with ethanol	Lipozyme IM
Chrisochoou et al. (1995)	Batch	Transesterification of ethylacetate and isoamyl alcohol	
Gunnlaugsdottir and Sivik (1995)	Continuous	Alcoholysis of cod liver oil with ethanol	Novozym™ IM

Kamat et al. (1995)	Batch	Transesterification of methyl methacrylate with 2-ethylhexanol	Lipase from *Candida cylindracea*
Bernard and Barth (1996)	Batch	Esterification of myristic acid with ethanol	Lipozyme IM
Combes (1996)	Continuous	Esterification of oleic acid with ethanol	Lipozyme IM
Glowacz et al. (1996)	Batch	Hydrolysis of triolein and its partial glycerides	Porcine pancreatic lipase
Ikushima et al. (1996)	Batch	Ester synthesis from acyl donors and terpene alcohols	Lipase from *Candida cylindracea*
Michor et al. (1996a)	Batch	Transesterification of menthol with isopropenyl acetate	Esterase from *Pseudomonas marginata*
Michor et al. (1996b)	Batch	Transesterification of ±menthol and ±citronellol	Lipases from *Candida rugosa, Pseudomonas* sp., *Candida antarctica*, Lipozyme IM, and esterase from *Pseudomonas marginata*
Rantakylä et al. (1996)	Batch	Hydrolysis of 3-(4-methoxy-phenyl glycidic) ester	Lipozyme IM
Yoon et al. (1996)	Batch	Transesterification of triolein and ethyl behenate	Lipozyme IM
Zheng and Tsao (1996)	Batch	Hydrolysis of cellulosic material to produce glucose	Enzyme from cellulase
Jackson et al. (1997)	Continuous	Randomization of fats and iols	Novozym-435
Seretti et al. (1997)	Batch	Esterification of hexanoic acid with hexanol	Cutinase
Almeida et al. (1998)	Batch	Transesterification of *n*-butyl acetate by 1-hexanol	Novozym-435

(*continued*)

Table 6.1 (continued)

Reference	Batch/Continuous	Study	Enzyme
Fontes et al. (1998a)	Batch	Transesterification of *N*-acetyl-L-phenylalanine ethyl ester with 1-propanol	Subtilisin Carlsberg
Fontes et al. (1998b)	Batch	Hydrolysis of *p*-nitrophenyl butyrate	Cutinase
Gunnlaugsdottir et al. (1998a, 1998b)	Batch/continuous	Alcoholysis of cod liver oil with ethanol and glycerolysis	Lipase from *Candida antarctica*
Knez et al. (1998)	Continuous	Esterification of oleic acid with oleyl alcohol	Lipzoyme IM
Liang et al. (1998)	Continuous	Inter-esterification of palm oil by stearic acid	Lipase from Mucor miehei
Mensah et al. (1998)	Batch	Esterification of proionic acid with isoamyl alcohol	Lipozyme IM
Mori and Okahata (1998)	Batch	Transgalactosylation from 1-*O*-*p*-nitrophenyl-β-D-galactopyranoside to 5-phenylpentan-1-ol	β-D-galactosidase
Nakaya et al. (1998)	Batch	Transesterifications by triolein using stearic acid	Lipozyme IM
Tsitsimpikou et al. (1998)	Batch	Acylation of glucose with lauric acid	Lipases from *Mucor miehei* and *Candida antarctica*
Yoon et al. (1998)	Batch	Inter-esterification between triolein and behenic acid and ethyl behenate	Lipozyme IM
Caussette et al. (1999)	Continuous	Lysozyme inactivation	Lysozyme
Overmeyer et al. (1999)	Batch	Esterification of 1-phenyl ethanol and ibuprofen	Novozym-435

Advantages and Disadvantages

Carbon dioxide is cited as a good choice of solvents for a number of reasons. Some of the advantages of running reactions in carbon dioxide instead of the more traditional organic solvents include the low viscosity of the solvent, the convenient recovery of the products and non-reacted components, abundant availability, low cost, no solvent contamination of products, full miscibility with other gases, non-existent toxicity, low surface tension, non-flammability, and recyclability. The low mass-transfer limitations are an advantage because of the large diffusivity of reactants. Because the density of the fluid is tunable using pressure, the dielectric constant, viscosity, and Hildebrand solubility parameters may also be tuned to the specific needs of every reaction system. There is one large hindrance to running reactions in carbon dioxide and that obstacle is the solubility of both substrates and products in the carbon dioxide. Enzymes are not soluble in carbon dioxide, and many substrates possess limited solubility. Only non-polar solids are soluble in carbon dioxide at high concentrations. The solubility of polar compounds may increase with the addition of cosolvents. There are several advantages to using enzymes in carbon dioxide. These include the recovery and re-use of insoluble enzymes, no microbial contamination (the release of proteases that inhibit enzymes can occur in aqueous solutions), reduction of major side reactions such as hydrolysis, and easier product recovery. Carbon dioxide also serves as a good heat sink for thermally unstable compounds.

Factors Influencing Reactions

Temperature

In general, one expects the reaction rates to increase as the temperature increases; but, with the addition of enzymes to the system, an additional complication is introduced. Enzyme activity is related to temperature; yet, as temperature increases beyond the optimum, activity decreases. Enzymes have been found to be stable and active in carbon dioxide at temperatures up to 70 °C (Tsitsimpikou et al., 1998) and, surprisingly, at 136 °C (Overmeyer et al., 1999).

There are three distinct findings for the effect of temperature on enzyme activity and stability in carbon dioxide. The first is that as temperature increases, activity of the enzyme decreases. Nakamura et al. (1986), Chi et al. (1988), and Lee et al. (1993) all observed slight decreases in activity as temperature was increased slightly. Chulalaksananukul et al. (1993) observed a more marked decrease in activity and attribute the loss of activity to thermal denaturation of the enzyme.

Other researchers have found that as temperature increases, the activity passes through an optimum (Miller et al., 1991; Rantakaylä and Aaltonen, 1994; Steytler et al., 1991; Yoon et al., 1996; Zheng and Tsao, 1996). It is believed that this bell-shaped behavior is the result of increasing activity as temperature increases to the optimum, followed by a decrease in activity above the optimum because of thermal denaturation. In one instance, another contributing factor to the lower activity below the optimum temperature was that the carbon dioxide was subcritical (Miller et al., 1991). Yoon et al. (1996) observed optimal temperatures for activity only at certain concentrations of products. No explanation was provided.

An increase in activity as temperature increases is the third and final finding for the effect of temperature. Kamat et al. (1995) and Knez and Habulin (1992) also observed increases in activity as temperature increases. Kamat et al. (1995) suggest that this behavior is the result of an unstable or reduced-activity enzyme complex formed at low temperatures, and that this complex is disassociated at the elevated temperatures.

Pressure

Pressure is a very important experimental factor to consider for any reaction that one wishes to carry out in carbon dioxide because changes in the outcome of the reaction are the result of not just the change in pressure, but also changes in the properties of the carbon dioxide. When the pressure changes, so too do solvent properties such as density, solubility, and dielectric constant. In addition, other factors such as the rate-determining step and activation volumes must be considered when interpreting the effect of pressure on a reaction (Kamat et al., 1995). Fontes et al. (1998a) found that increasing the pressure reduced the efficiency of the catalyst and resulted in positive activation volumes.

In several instances, as pressure increases, so too does conversion (Ikushima et al., 1993; Steytler et al., 1991), initial rates (Ikushima et al., 1996; Miller et al., 1991), hydrolysis (Lee et al., 1993), and productivity (Lee et al., 1994). As mentioned earlier, it is difficult to determine whether these apparent increases are because of the change in pressure or the accompanying change in solvent properties. Equations of state should be used to predict the conditions and monitor these extra variables.

In several other instances, a direct relationship between the reaction outcome and pressure was not observed. In two cases, an optimum was found for the reaction rate (Ikushima et al., 1996; Yoon et al., 1996). As was the case for the effects of temperature, while many of the products became more soluble as pressure increases, sometimes the enzyme activity and stability are adversely affected by the increased pressure. Examples of enzymatic stabilities at various reaction conditions may be seen in Table 6.2.

Nakaya et al. (1998) divided their findings based on pressure into three sections. At low pressures (below 5 MPa) the reaction was slow, and these

Table 6.2. Examples of Enzymatic Stability in Carbon Dioxide

Enzyme	Temperature (°C)	Pressure (MPa)	Time (h)	Effect	Reference
Alkaline phosphatase	35	10	24	Full activity retained	Randolph et al. (1985)
Lipase from *Rhizopus delemar*	35–50	13.7–30	24	Slight reduction; reduction increases as temperature increases	Nakamura et al. (1986)
Thermolysin	40	30	1	No loss	Kamihira et al. (1987)
Lipozyme IM	60	10–15	4	No loss	van Eijs et al. (1988)
Cholesterol oxidases	35	10	72	No loss	Randolph et al. (1988)
Lipase from *Rhizopus arrhizus*	35	9.6	80	No loss	Miller et al. (1991)
Lipase from *Candida cylindracea*	40	13.6	168	75% residual retained	Yu et al. (1992)
Lipozyme IM	40	13–18	384	90% residual retained	Marty et al. (1992a, 1992b, 1992c)
Lipase from *Mucor meihei*	50	15	120	96% residual retained	Dumont et al. (1992)
Lipozyme IM	40-80	13–18	24	50% residual retained	Chulalaksananukul et al. (1993)
Porcine pancreatic lipase	35	20	35	No loss	Martins et al. (1994)
Lipozyme IM	30–60	13	144	80% residual retained	Combes (1996)
Lipases from *Candida rugosa*, *Pseudomonas* spp., *Pseudomonas marginata*, and *Candida antarctica*	35–70	15	21	No loss	Michor et al. (1996b)
Cutinase	45	13	144	90% residual retained	Sereti et al. (1997)

authors claim that this is because it is the nonsolvent region. For pressures between 5 and 10 MPa, the pressure is near critical and the reaction rate was highest at 5.9 MPa. Nakaya et al. claim that this is because the enzyme–substrate complex was stabilized in this region. For pressures above 10 MPa, the reaction rate increased linearly with pressure because the substrates were more soluble.

Water Content

Enzymes need some water to remain catalytically active, but this amount of water is small. Randolph et al. (1988) investigated the use of cholesterol oxidase from *Cleocysticum chrysocreas* by placing the immobilized enzyme in a packed bed and exposing it to bone-dry carbon dioxide. While the enzyme quickly lost activity, it was rapidly restored with only 1% v/v water in the system. Water acts not only to help the enzyme retain its native conformation, but also it can act as a solubility modifier. Organic solvents and other nonaqueous environments can remove the enzyme's essential water (Zaks and Klibanov 1988).

It has been shown that, in supercritical carbon dioxide, increases in water concentration result in increases in enzyme activity. The amount of added water needed for this increase varies and can depend on many factors, such as reaction type, enzyme utilized, and initial water content of the system. This is true until an optimal level is reached. For hydrolysis reactions, activity will either continue to increase or maintain its value. For esterification or transesterification reactions, once the optimal level of hydration has been reached, additional water will promote only side reactions such as hydrolysis. Dumont et al. (1992) suggests that additional water beyond the optimal level needed for enzyme hydration may also act as a barrier between the enzyme and the reaction medium and thereby reduce enzyme activity. Mensah et al. (1998) also observed that water above a concentration of 0.5 mmol/g enzyme led to lower catalytic activity and that the correlation between water content of the enzyme and reaction rate was independent of the substrate concentrations.

Miller et al. (1991) demonstrated the effect of increased hydrolysis once an optimal amount of water has been added when they studied the inter-esterification of the triglycerides trilaurin and myristic acid. The observed rate of inter-esterification decreased to approximately one-third of its original rate when the water content was reduced from 1.75 (grams water per kilogram carbon dioxide) to zero. They also showed that the amount of hydrolysis increased about 10% as the water content increased. Knez et al. (1998) also found that the addition of small amounts of water increased the conversion rate when they studied the esterification of oleic acid with oleyl alcohol.

Nakamura et al. (1990) studied the esterification of triolein with stearic acid using four different lipases: *Mucor miehei*, *Rhizpous delmar*, *Alcaligenes* sp. and *Rhizopus japonicus*. They varied the water content of the system from 0.1% to 50% w/w carrier. For three of the enzymes, the addition of water increased the extent of inter-esterification to a maximum, and then above that point the extent decreased. These maximums were ~4% for *Alcaligenes* spp., ~13% for *Rhizopus delemar*, and ~40% for *Rhizopus japonicus*. In contrast to this behavior, lipase from *Mucor miehei* showed little change over this large range of water content; extent of inter-esterification for this enzyme is always higher than for any of the other three.

Mass Transfer

There are two types of mass transfer that can be considered for biocatalysis in carbon dioxide. The first is external mass transfer, which is the diffusion of the substrate(s) from the solvent to the surface of the enzyme. The second is internal mass transfer, which is the diffusion of the substrate within the enzyme particle to the active site of the enzyme. External mass transfer may generally be eliminated through agitation in batch systems, and it is controlled through flow rate in continuous systems. Internal mass transfer is dependent on the morphology of the enzyme preparation, but in carbon dioxide the enzyme powders can undergo morphological changes. Supercritical carbon dioxide provides intrinsically higher diffusivity of substrates than conventional solvents and thus is a very good solvent for enzymatic reactions that are diffusionally limited.

The role of external diffusion has been studied by several groups using continuous-flow systems. Marty et al. (1992a) investigated the synthesis of ethyloleate from oleic acid and ethanol and varied the flow rate to observe the effect of residence time on conversion. They found that increasing the residence time (a lower flow rate) increased conversion. They modeled their reaction system using a simple plug flow reactor equation and found that it gave effective predictions for the carbon dioxide reactor data. Jackson et al. (1997) found a similar result when they studied the degree of randomization and the rate of triglyceride throughput controlled by the flow rate of carbon dioxide. They used carbon dioxide at flow rates of 6 and 13 L/min and found that the lower flow rate increased the extent of reaction more than the higher flow rate. They attribute this to the increased contact time between the oil and the enzyme in the reactor at the lower flow rate

In contrast to the findings of both Jackson et al. and Marty et al. several other groups have found that increasing the flow rate of carbon dioxide actually increases the reaction rates. Dumont et al. (1993) discovered that the production rate of product increased with increasing fluid flow rates, while the flow of reactants, pressure, temperature, and water content remained constant. They also found that when they increased the flow of carbon dioxide while maintaining the concentration of substrates, the conversion rate decreased exponentially. The reaction that they studied was the esterification of myristic acid with ethanol and they suggest that this behavior is likely the effect of ethanol inhibiting the reaction. Gunnlaugsdottir et al. (1998a) also found that ethanol inhibited the reaction and changed the effect of altering the flow rate of carbon dioxide. Both Randolph et al. (1988) and van Eijs et al. (1988) observed linear rate increases with a linear increase in flow rate. Their conclusions are that the diffusion of substrate to the enzyme is not the rate-limiting step and that the transfer from the bulk to the enzyme is dependent not only on the flow rate of the carbon dioxide, but also on the viscosity and diffusivity of the fluid.

Internal diffusional limitations are possible any time that a porous immobilized enzymatic preparation is used. Bernard et al. (1992) studied internal diffusional limitations in the esterification of myristic acid with ethanol, catalyzed by immobilized lipase from *Mucor miehei* (Lipozyme). No internal mass diffusion would exist if there was no change in the initial velocity of the reaction while the enzyme particle size was changed. Bernard found this was not the case, however, and the initial velocity decreased with increasing particle size. This corresponds to an efficiency of reaction decrease from 0.6 to 0.36 for a particle size increase from ~180 μm to ~480 μm. Using the Thiele modulus, they also determined that for a reaction efficiency of 90% a particle size of 30 μm would be necessary. While Bernard et al. found that their system was limited by internal diffusion, Steytler et al. (1991) found that when they investigated the effect of different sizes of glass bead, 1 mm and 3 mm, no change in reaction rate was observed.

Stereoselectivity and Enantioselectivity

There are many pharmaceutical applications for the modification of one enantiomer over another, and to this end, many have studied these selective reactions in carbon dioxide. Glowacz et al. (1996) studied the enzymatic hydrolysis of triolein and its partial glycerides and found that stereoselectivity depends on reaction time and enzyme water content. They suggest that the water content varies the local environment of the enzyme in carbon dioxide and changes the local pH value. Rantakylä et al. (1996) also found that the hydrolysis of one stereoisomer over another was water-dependent. They studied the hydrolysis of 3-(4-methoxyphenyl)glycidic acid methylester and found that the 2*S*,3*R* enantiomer hydrolyzed more than fivefold faster than the 2*R*,3*S* form.

In a study of ester synthesis, Ikushima et al. (1996) found that, in a limited pressure range near the critical point, the enzyme undergoes conformational changes that allow it to catalyze stereoselective synthesis. Taniguchi et al. (1994) showed spectroscopically, using high-pressure IR, that the structure of the enzyme changes with pressure. In contrast to the findings of Ikushima et al., Rantakylä and Aaltonen (1994) found that enantioselective esterification was possible at different pressures. They examined the enantioselective esterification of racemic ibuprofen with *n*-propanol and found that the enantioselectivity was not affected by pressure changes, although the initial reaction rates increased with increasing pressure.

Fontes et al. (1998b) studied the enantioselectivity of cutinase and found that it was very selective toward one enantiomer with an enantiomeric excess of almost 100%. They found that the enantioselectivity was very sensitive to changes in water content. Bornscheuer et al. (1992) studied hydrolysis, esterification, and transesterification in carbon dioxide to try to find the best method for producing enantiomerically pure substances in carbon dioxide. They found that the thermodynamically favored hydrolysis led to higher enantiomeric excess with less enzyme in the shortest time. Michor et al. (1996b) also examined more than one system to determine a better route to product and found that while the transesterification of ±-menthol was fast and resulted in high enantiomeric excess, resolution of ±-citronellol was not feasible. The reaction rate for the reaction of ±-citronellol was 10–20 times of that of ±-menthol, but was not selective.

Enzymes in Carbon Dioxide—Non Reaction Applications

Thus far, the bulk of this chapter has focused on the biocatalytic applications of enzymes in carbon dioxide. There are other applications and needs for enzymes in carbon dioxide, such as for deactivation, for extraction, and for the production of protein particles for their controlled release as drugs. Supercritical carbon dioxide is used to deactivate pectinesterase in orange juice so that it can no longer de-esterify high methoxyl pectin and cause the orange juice to lose its desirable cloudiness (Balaban et al., 1991). Here, carbon dioxide replaces the alternative treatments previously used to maintain cloudiness that required higher energy input or the addition of chemicals. Each of these prior methods affected aroma and flavor. It is suggested that the carbon dioxide works by temporarily lowering pH and thus deactivating the enzyme.

Another application of enzymes in the food industry is the separation of peroxidase (POD EC 1.11.1.7) from wasabi horseradish (Taniguchi et al., 1988). Using carbon dioxide, the enzyme was successfully separated from the flavor components so that the enzyme could be used as a label enzyme in clinical diagnosis and microanalysis with immunoassays, and the flavor components could be added to spices and foods.

For some applications, it is important to use protein particles of a specific size. When drugs are released within the body there is an initial boost followed by an exponential decay in the concentration over time. Sometimes it is desirable or necessary to minimize the variation of concentration within patients and one of the ways to accomplish this is by using microparticles distributed within a polymer carrier. Tom et al. (1993) used carbon dioxide to first process proteins (catalase and insulin) into micron-sized particles and then they incorporated the proteins into polymer-drug microspheres.

Conclusions and Suggestions for Future Research

There are many factors that control and limit biocatalysis in carbon dioxide, including, but not limited to, pressure, water content, temperature, and mass transfer, and their influence on each enzyme's activity and stability. It is also important to understand whether one is operating in a one-phase or two-phase reaction medium. To this end, solubility studies should be carried out regularly to determine if the reaction is one or two phases. Only if the reaction is one-phase can one determine the kinetic and thermodynamic parameters with certainty.

Carbon dioxide is an appealing solvent in which to conduct biocatalysis, mostly because of the desirable physical properties of the solvent. Two of the most influential solvent properties are its high diffusivity and low viscosity. Solvent properties and product solubility are easily varied, especially near the critical temperature and pressure.

Many enzymes are both active and stable in carbon dioxide and have been used to conduct a number of reactions. Several different types of reactions have been examined, including hydrolysis (Lee et al., 1993; Randolph et al., 1985; Zheng and Tsao, 1996), oxidation (Hammond et al., 1985; Randolph et al., 1988), and esterification/transesterification (Kamihira et al., 1987; Nakamura et al., 1986; Rantakylä and Aaltonen, 1994), but there are other types of reactions that would make worthwhile investigations in carbon dioxide. These include preparation of amides, reduction of ketones, preparation of cyanohydrins from aldehydes, aldol reactions, hydroxylation reactions, and Baeyer–Villiger oxidation.

References

Aaltonen, O.; Rantakylä, M. Lipase Catalyzed Reactions of Chiral Compounds in Supercritical Carbon Dioxide; In *Proceedings of the 2nd International Symposium on Supercritical Fluids*; Boston, MA, 1991; pp. 146–149.

Almeida, M. C.; Ruivo, R.; Maia, C.; Freire, L.; de Sampaio, T. C.; Barreiros, S. Novozym 435 Activity in Compressed Gases. Water Activity and Temperature Effects. *Enzyme Microb. Technol.* **1998**, *22*, 494–499.

Balaban, M. O.; Arreola, A. G.; Marshall, M.; Peplow, A.; Wei, C. I.; Cornell, J. Inactivation of Pectinesterase in Orange Juice by Supercritical Carbon Dioxide. *J. Food Sci.* **1991**, *56* (3), 743–747.

Berg, B. E.; Hansen, E. M.; Gjørven, S.; Greibrokk, T. On-Line Enzymatic Reaction, Extraction, and Chromatography of Fatty Acids and Triglycerides with Supercritical Carbon Dioxide. *J. High Resolut. Chromatogr.* **1993**, *16*, 358–363.

Bernard, P.; Barth, D. Internal Mass Transfer Limitation During Enzymatic Esterification in Supercritical Carbon Dioxide and Hexane. *Biocatal. Biotransform.* **1995**, *12*, 299–308.

Bernard, P.; Barth, D. Enzymatic Reaction in Supercritical Carbon Dioxide Internal Mass Transfer Limitation. *High Pressure Chem. Eng.* **1996**, *12*, 103–108.

Bernard, P.; Barth, D.; Perrut, M. Internal Mass Transfer Limitation on Enzymatic Esterification in Supercritical Carbon Dioxide. In *High Pressure and Biotechnology*; Balny, C., Hayashi, R., Heremans, K., Masson, P., Eds.; John Libbey; U.K., 1992; pp. 451–455.

Bornscheuer, U.; Capewell, A.; Scheper, T.; Meyer, H. H.; Kolisis, F. A Comparison of Enzymatic Reactions in Aqueous, Organic and Supercritical Phases. *Ann. NY. Acad. Sci.* **1992**, *672*, 336–342.

Capewell, A.; Bornscheuer, U.; Herar, A.; Scheper, T.; Meyer, H. H.; Kolisis, F. A Comparison of Enzymatic Reactions in Aqueous, Organic and Supercritical Phases. *Dechema Biotechnol. Conf.* **1992**, *5*, 57–60.

Caussette, M.; Gaunnand, A.; Planche, H.; Colombie, S.; Monsan, P.; Lindet, B. Lysozyme Inactivation by Inert Gas Bubbling: Kinetics in a Bubble Column Reactor. *Enzyme Microb. Technol.* **1999**, *24*, 412–418.

Chi, Y. M.; Nakamura, K.; Yano, T. Enzymatic Interestification in Supercritical Carbon Dioxide. *Agric. Biol. Chem.* **1988**, *52*, 1541–1550.

Chrisochoou, A.; Schaber, K.; Bolz, U. Phase Equilibria for Enzyme-Catalyzed Reactions in Supercritical Carbon Dioxide. *Fluid Phase Equilib.* **1995**, *108*, 1–14.

Chulalaksananukul, W.; Condoret, J. S.; Combes, D. Geranyl Acetate Synthesis by Lipase-Catalyzed Transesterification in Supercritical Carbon Dioxide. *Enzyme Microb. Technol.* **1993**, *15*, 691–698.

Combes, D. Reaction/Separation Process in Supercritical CO_3 Using Lipases. In *NATO ASI Series. Series E: Applied Science*, Vol. 317; Malcata, F. X., Ed.; 1996, pp. 613–618.

Dumont, T.; Barth, D.; Perrut, M. In *Proceedings of the 2nd International Symposium on Supercritical Fluids*; Boston, MA, 1991; pp. 150–153.

Dumont, T.; Barth, D.; Corbier, C.; Branlant, G.; Perrut, M. Enzymatic Reaction Kinetic: Comparison in an Organic Solvent and in Supercritical Carbon Dioxide. *Biotechnol. Bioeng.* **1992**, *39*, 329–333.

Dumont, T.; Barth, D.; Perrut, M. Continuous Synthesis of Ethyl Myristate by Enzymatic Reaction in Supercritical Carbon Dioxide. *J. Supercrit. Fluids* **1993**, *6*, 85–89.

Endo, Y.; Fujimoto, K. Enzyme Reaction in Supercritcal Fluid. In *Proceedings of the International Congress on Food Engineering, Part 2*; Yano, T., Matsuno, R., Nakamura, K, Eds.; Blackie: Glasgow, U.K., 1994; pp. 849–851.

Erickson, J. C.; Schyns, P.; Cooner, C. L. Effect of Pressure on an Enzymatic Reaction in a Supercritical fluid. *AIChE J.* **1990**, *36*, 299–301.

Fontes, N.; Nogueiro, E.; Elvas, A. M.; de Sampaio, T. C.; Barreiros, S. Effect of Pressure on the Catalytic Activity of Subtilisin Carlsberg Suspended in Compressed Gases. *Biochim. Biophys. Acta*, **1998a**, *1383*, 165–174.

Fontes, N.; Almeida, M. C.; Peres, C.; Garcia, S.; Grave, J.; Aires-Barros, M. R.; Soares, C. M.; Cabral, J. M. S.; Maycock, C. D.; Barreiros, S Cutinase Activity and Enantioselectivity in Supercritical Fluids. *Ind. Eng. Chem Res.* **1998b**, *37*, 3189–3194.

Glowacz, G.; Bariszlovich, M.; Linke, M.; Richter, P.; Fuchs, C.; Mörsel, J. T. Stereoselectivity of Lipases in Supercritical Carbon Dioxide. I. Dependence of the Regio- and Enantioselectivity of Porcine Pancrease Lipase on the Water Content During the Hydrolysis of Triolein and its Partial Glycerides. *Chem. Phys. Lipids* **1996**, *79*, 101–106.

Gunnlaugsdottir, H.; Sivik, B. Lipase-Catalyzed Alcoholysis of Cod Liver Oil in Supercritical Carbon Dioxide. *J. Am. Oil Chem. Soc.* **1995**, *77*, 399–405.

Gunnlaugsdottir, H.; Jaremo, M.; Sivik, B. Process Parameters Influencing Ethanolysis of Cod Liver Oil in Supercritical Carbon Dioxide. *J. Supercrit. Fluids* **1998a**, *12*, 85–93.

Gunnlaugsdottir, H.; Wannerberger, K.; Sivik, B. Alcoholysis and Glyceride Synthesis with Immobilized Lipase on Controlled-Pore Glass of Varying Hydrophobicity in Supercritical Carbon Dioxide. *Enzyme Microb. Technol.* **1998b**, *22*, 360–67.

Hammond, D. A.; Karel, M.; Klibanov, A. M. Enzymatic Reactions in Supercritical Gases. *Appl. Biochem. Biotechnol.* **1985**, *11*, 393–400.

Ikushima, Y.; Saito, N.; Yokoyama, T. Solvent Effects on an Enzymatic Ester Synthesis. *Chem. Lett.* **1993**, 109–112.

Ikushima, Y.; Saito, N.; Hatakeda, K.; Sato, O. Promotion of a Lipase-Catalyzed Esterification in Supercritical Carbon Dioxide in the Near-Critical Region. *Chem. Eng. Sci.* **1996**, *51*, 2817–2822.

Jackson, M. A.; King, J. W.; List, G. R.; Neff, W. E. Lipase-Catalyzed Randomization of Fats and Iols in Flowing Supercritical Carbon Dioxide. *J. Am. Oil Chem. Soc.* **1997**, *74*, 635–639.

Janssens, R. J. J.; van der Lugt, J. P.; Oostrom, W. H. M. The Integration of Biocatalysis and Downstream Processing in Supercritical Carbon Dioxide. In *High Pressure and Biotechnology*; Balny, C., Hayashi, R., Heremans, K., Masson, P., Eds. John Libbey: U.K., 1992; Vol. 224, pp. 447–449.

Kamat, S.; Barrera, J.; Beckman, E. J.; Russell, A. J. Biocatalytic Synthesis of Acrylates in Organic Solvents and Supercritical Fluids: I. Optimization of Enzyme Environment. *Biotechnol. Bioeng.* **1992**, *40*, 158–166.

Kamat, S.; Critchley, G.; Beckman, E. J.; Russell, A. J. Biocatalytic Synthesis of Acrylates in Organic Solvents and Supercritical Fluids: III. Does Carbon Dioxide Covalently Modify Enzymes? *Biotechnol. Bioeng.* **1995**, *46*, 610–620.

Kamihira, M.; Taniguchi, M.; Kobayashi, T. Synthesis of Aspartame Precursors by Enzymatic Reaction in Supercritical Carbon Dioxide. *Agric. Biol. Chem.* **1987**, *51*, 3427–3428.

Knez, Ž.; Habulin, M. Lipase Catalyzed Esterification in Supercritical Carbon Dioxide. In *Biocatalysis in Non-Conventional Media*; Tramper, J., Vermüe, M. H., Beeftink, H. H., Eds.; Elsevier Science: Amsterdam, 1992, pp. 401–406.

Knez, Ž.; Habulin, M.; Krmelj, V. Enzyme Catalyzed Reactions in Dense Gases. *J. Supercrit. Fluids* **1998**, *14*, 17–29.

Lee, H. S.; Lee, W. G.; Park, S. W.; Lee, H.; Chang, H. N. Starch Hydrolysis Using Enzyme in Supercritical Dioxide. *Biotechnol. Tech.* **1993**, *7*, 267–270.

Lee, H. S.; Yeon, W. R.; Kim, C. Hydrolysis of Starch by α-Amylase and Glucoamylase in Supercritical Carbon Dioxide. *J. Microbiol. Biotechnol.* **1994**, *4*, 230–232.

Liang, M. T.; Chen, C. H.; Liang, R. C. The Interesterification of Edible Palm Oil by Stearic acid in Supercritical Carbon Dioxide. *J. Supercrit. Fluids*. **1998**, *13*, 211–216.

Martins, J. F.; Barreiros, S. F.; Azevedo, E. G.; daPonte, N. In *Proceedings of the 2nd International Symposium on Supercritical Fluids*; Boston, MA, 1991; pp. 406–407.

Martins, J. F.; Sampaio, T. C.; Carvalho, I. B.; da Ponte, M. N.; Barreiros, S. Lipase Catalyzed Esterification of Flycidol in Chloroform and in Supercritical Carbon Dioxide. In *High Pressure and Biotechnology*; Balny, C., Hayashi, R., Heremans, K., Masson, P., Eds.; John Libbey; U.K., 1992; Vol. 224, pp. 411–415.

Martins, J. F.; de Carvalho, I.B.; de Sampaio, T. C.; Barreiros, S. Lipase-Catalyzed Enantioselective Esterification of Glycidol in Supercritical Carbon Dioxide. *Enzyme Microb. Technol.* **1994**, *16*, 785–790.

Marty, A.; Chulalaksananukul, Condoret, J. S.; Willemot, R. M.; Durand, G. Comparison of Lipase-Catalysed Esterification in Supercritical Carbon Dioxide and in *n*-Hexane. *Biotechnol. Lett.* **1990**, *12*, 11–16.

Marty, A.; Combes, D.; Condoret, J. S. Fatty Acid Esterification in Supercritical Carbon Dioxide. In *Biocatalysis in Non-Conventional Media*; Tramper, J.; Vermüe, M. H.; Beeftink, H. H., Eds.; Elsevier Science: Amsterdam, 1992a; pp. 425–432.

Marty, A.; Chulalaksananukul, W.; Willemot, R. M.; Condoret, J. S. Kinetics of Lipase-Catalyzed Esterification in Supercritical CO_2. *Biotechnol. Bioeng.* **1992b**, *39*, 273–280.

Marty, A.; Chulalaksananukul, W.; Condoret, J. S.; Combes, D. Transesterification and Esterification in Supercritical Carbon Dioxide. In *High Pressure and Biotechnology*; Balny, C., Hayashi, R., Heremans, K., Masson, P., Eds.; John Libbey: U.K., 1992c; Vol. 224, pp. 461–463.

Mensah, P.; Gainer, J. L.; Carta, G. Adsorptive Control of Water in Esterification with Immobilized Enzymes: I. Batch Reactor Behavior. *Biotechnol. Bioeng.* **1998**, *60* (4), 434–444.

Michor, H.; Marr, R.; Gamse, T. Enzymatic Catalysis in Supercritical Carbon Dioxide: Effect of Water Activity. *High Pressure Chem. Eng.* **1996a**, *12*, 115–120.

Michor, H.; Marr, R.; Gamse, T.; Schilling, T.; Klingsbichel, E.; Schwab, H. Enzymatic Catalysis in Supercritical Carbon Dioxide: Comparison of Different Lipases and a Novel Esterase. *Biotechnol. Lett.* **1996b**, *18*, 79–84.

Miller, D. A.; Blanch, H. W.; Prausnitz, J. M. Enzyme-Catalyzed Interesterification of Triglycerides in Supercritical Carbon Dioxide. *Ind. Eng. Chem. Res.* **1991**, *30*, 939–946.

Mori, T.; Okahata, Y. Effective Biocatalytic Transgalactosylation in a Supercritical Fluid Using a Lipid-Coated Enzyme. *Chem. Commun.* **1998**, *20*, 2215–2216.

Nakamura, K. Supercritical Fluid Bioreactor. In *Bioproducts and Bioprocesses*; Flechter, A., Okada, H., Tanner, Eds.; Springer-Verlag: Berlin, 1989; pp. 257–265.

Nakamura, K.; Hoshino, T. Novel Utilization of Supercritical Carbon Dioxide for Enzymatic Reaction in Food Processing. In *Advances in Food Engineering*; Singh, R. P., Wirakartakusumah, M. A., Eds.; CRC Press; Boca Raton, FL, 1992; pp. 257–262.

Nakamura, K.; Chi, Y. M.; Yamada, Y.; Yano, T. Lipase Activity and Stability in Supercritical Carbon Dioxide. *Chem. Eng. Commun.* **1986**, *45*, 207–212.

Nakamura, K.; Fujii, H.; Chi, Y. M.; Yano, T. A Novel Nonaqueous Medium to Integrate Enzymatic Reaction and Separation. *Ann. N.Y. Acad. Sci.* **1990**, *613*, 319–332.

Nakaya, H.; Miyawaki, O.; Nakamura, K. Transesterification Between Triolein and Stearic Acid Catalyzed by Lipase in CO_2 at Various Pressures. *Biotechnol. Tech.* **1998**, *12* (12), 881–884.

Overmeyer, A.; Schrader-Lippelt, S.; Kasche, V.; Brunner, G. Lipase-Catalysed Kinetic Resolution of Racemates at Temperatures from 40 °C to 160 °C in Supercritical CO_2. *Biotechnol. Lett.* **1999**, *21*, 65–69.

Pasta, P.; Mazzola, G.; Carrea, G.; Riva, S. Subtilisn-Catalyzed Transesterification in Supercritical Carbon Dioxide. *Biotechnol. Lett.* **1989**, *2*, 643–648.

Randolph, T. W.; Blanch, H. W.; Prausnitz, J. M.; Wilke, C. R. Enzymatic Catalysis in a Supercritical Fluid. *Biotechnol. Lett.* **1985**, *7*, 325–328.

Randolph, T. W.; Blanch, H. W.; Prasnitz, J. M. Enzyme-Catalyzed Oxidation of Cholesterol in Supercritical Carbon Dioxide. *AIChE J.* **1988**, *34*, 1354–1360.

Rantakylä, M.; Aaltonen, O. Enantioselective Esterification of Ibuprofen in Supercritical Carbon Dioxide by Immobilized Lipase. *Biotechnol. Lett.* **1994**, *16*, 825–830.

Rantakylä, M.; Alkio, M.; Aaltonen, O. Stereospecific Hydrolysis of 3-(4-Methoxyphenyl)glycidic Ester in Supercritical Carbon Dioxide by Immobilized Lipase. *Biotechnol. Lett.* **1996**, *18*, 1089–1094.

Seretti, V.; Stamatis, H.; Kolisis, F. N. Improved Stability and Reactivity of *Fusarius solani* Cutanase in Supercritical CO_2. *Biotechnol. Tech.* **1997**, *11*, 661–665.

Shen, X. M.; de Loos, T. W.; de Swaan Arons, J. Enzymatic Reaction in Organic Solvents and Supercritical Gases. In *Biocatalysis in Non-Conventional Media*; Tramper, J., M. H. Vermüe, M. H., Beeftink, H. H., Eds.; Elsevier Science: Amsterdam, 1992; pp. 417–423.

Steytler, D. C.; Moulson, P. S.; Reynolds, J. Biotransformations in Near-Critical Carbon Dioxide. *Enzyme Microb. Technol.* **1991**, *13*, 221–226.

Taniguchi, M.; Nomura, R.; Kamihiri, M.; Kijima, I.; Kobayashi, T. Effective Utilization of Horseradish and Wasabi by Treatment with Supercritical Carbon Dioxide. *J. Ferment. Technol.* **1988**, *66* (3), 347–353.

Taniguchi, Y.; Takeda, N.; Kato, M. *High Pressure Liquids and Solutions*; Elsevier Science and the Society of Materials Science, Amsterdam: 1994.

Tom, J. W.; Lim, G. B.; Debenedetti, P. G.; Prud'homme, R. K. Applications of Supercritical Fluids in the Controlled Release of Drugs. *Supercrit. Fluid Eng. Sci.* **1993**, *514*, 238–257.

Tsitsimpikou, C.; Stamatis, H.; Sereti, V.; Daflos, H.; Kolisis, F. N. Acylation of Glucose Catalyzed by Lipases in Supercritical Carbon Dioxide. *J. Chem. Technol. Biotechnol.* **1998**, *71*, 309–314.

Van Eijs, A. M. M.; de Jong, J. P. L.; Doddema, H. J.; Lindeboom, D. R. Enzymatic Transesterification in Supercritical Carbon Dioxide. In *Proceedings of the International Symposium on Supercritical Fluids*; Nice, France, 1988; pp. 933–942.

Vermüe, M. H.; Tramper, J.; de Jong, J. P. J.; Oostrom, W. H. M. Enzymic Transesterification in Near-Critical Carbon Dioxide: Effect of Pressure,

Hildebrand Solubility Parameter and Water Content. *Enzyme Microb. Technol.* **1992**, *14*, 649–654.

Yoon, S. H.; Miyawaki, Osata, Park, K. H.; Nakamura, K. Transesterification Between Triolein and Ethylbehenate by Immobilized Lipase in Supercritical Carbon Dioxide. *J. Ferment. Bioeng.* **1996**, *82*, 334–340.

Yoon, S. H.; Nakaya, H.; Ito, O.; Miyawaki, O.; Park, K. H.; Nakamura, K. Effects of Substrate Solubility in Interesterification with Riolein by Immobilized Lipase in Supercritical Carbon Dioxide. *Biosci. Biotechnol. Biochem.* **1998**, *62*, 170–172.

Yu, Z. R.; Rizvi, S. S. H.; Zollweg, J. A. Enzymatic Esterification of Fatty Acid Mixtures from Milk Fat and Anhydrous Milk Fat with Canola Oil in Supercritical Carbon Dioxide. *Biotechnol. Prog.* **1992**, *8*, 508–513.

Zaks, A.; Klibanov, A. M. The Effect of Water on Enzyme Action in Organic Media. *J. Biol. Chem.* **1988**, *263*, 3194.

Zheng, Y.; Tsao, G. T. Avicel Hydrolysis by Cellulase Enzyme in Supercritical CO_2. *Biotechnol. Lett.* **1996**, *18*, 451–454.

PART II

POLYMERS IN CARBON DIOXIDE

7

Solubility of Polymers in Supercritical Carbon Dioxide

MARK A. McHUGH

A great deal of information is known about the solvent character of CO_2 with a wide range of polymers and copolymers based on well-characterized and systematic solubility studies that are available in the literature (Kirby and McHugh, 1999). Nevertheless, the prediction of polymer solubility in CO_2, or any solvent for that matter, presents a formidable challenge since contemporary equations of state are still not facile enough to describe the unique characteristics of a long-chain polymer in solution. The difficulty resides in accounting for the intra- and intersegmental interactions of the many segments of the polymer connected to a single backbone relative to the small number of segments in a solvent molecule. An additional challenge exists to describe the density dependence of the intermolecular potential functions used in the calculations since SCF–polymer solutions (SCF, supercritical fluid) can be highly compressible mixtures. In this brief review, the solvent character of CO_2 is described using the principles of molecular thermodynamics and also using a select number of phase behavior studies to reveal the impact of polymer architecture on solubility.

Molecular Thermodynamics of CO_2—Polymer Mixtures

To form a stable polymer–SCF solvent solution at a given temperature and pressure, the Gibbs energy, shown in eq. 7.1, must be negative and at a minimum.

$$\Delta G_{\text{mix}} = \Delta H_{\text{mix}} - T\Delta S_{\text{mix}} \tag{7.1}$$

where ΔH_{mix} and ΔS_{mix} are the change of enthalpy and entropy, respectively, on mixing (Prausnitz et al., 1986). Enthalpic interactions depend predominantly on solution density and on polymer segment–segment, solvent–solvent, and polymer segment–solvent interaction energies. The value of ΔS_{mix} depends on both the combinatorial entropy of mixing and the noncombinatorial contribution associated with the volume change on mixing, a so-called equation-of-state effect (Patterson, 1982). The combinatorial entropy always promotes the mixing of a polymer with a solvent. However, the noncombinatorial contribution can have a negative impact on mixing as a result of monomer–monomer interactions that arise due to the connectivity of the segments in the backbone of the polymer chain.

For a dense SCF solution, ΔH_{mix} is expected to be approximately equal to the change in internal energy on mixing, ΔU_{mix}, which is shown in eq. 7.2

$$\Delta U_{\text{mix}} \approx \frac{2\pi \rho(P, T)}{kT} \sum_{i,j} x_j x_j \int \Gamma_{ij}(r, T) g_{ij}(r, \rho, T) r^2 dr \tag{7.2}$$

where x_i and x_j are mole fractions of components i and j, respectively, Γ_{ij} (r, T) is the intermolecular pair-potential energy of the solvent and the polymer segments, $g(r, \rho, T)$ is the radial distribution function, r is the distance between molecules, $\rho(P, T)$ is the solution density, and k is the Boltzmann constant (Lee, 1988). The radial distribution function describes the spatial positioning of molecules or segments of molecules with respect to one another, which has embedded in it information on the positioning of the segments of the polymer chain. Since the repeat units of a given chain are connected to one another, the polymer–SCF solution cannot be considered a random mixture of repeat units and SCF molecules. Nevertheless, important generalities can still be gleaned from an interpretation of eq. 7.2. For example, given that the internal energy of the mixture is roughly proportional to density, the solubility of a polymer is expected to improve by increasing the system pressure, by using a denser SCF solvent, or by adding a dense liquid cosolvent to an SCF solvent. However, the polymer dissolves only if the energetics of segment–solvent interactions outweigh segment–segment and solvent–solvent interactions. For certain polymer–SCF combinations, hydrostatic pressure alone will not overcome a mismatch in energetics in solution as described by the interchange energy, ω:

$$\omega = z\left\{\Gamma_{ij}(r, T) - \frac{1}{2}\left[\Gamma_{ii}(r, T) + \Gamma_{jj}(r, T)\right]\right\} \tag{7.3}$$

where z is the coordination number, or number of dissimilar pairs in solution (Prausnitz et al., 1986). Equation 7.4 presents an approximate form of the attractive part of the intermolecular potential energy, $\Gamma_{ij}(r, T)$, for *small, freely tumbling* molecules:

$$\Gamma_{ij}(r, T) \approx -\left[C_1 \frac{\alpha_i \alpha_j}{r^6} + C_2 \frac{\mu_i^2 \mu_j^2}{r^6 kT} + C_3 \frac{\mu_i^2 Q_j^2}{r^8 kT} + C_4 \frac{\mu_j^2 Q_i^2}{r^8 kT} + C_5 \frac{Q_i^2 Q_j^2}{r^{10} kT} + \text{Complex formation}\right] \tag{7.4}$$

where α is the polarizability, μ is the dipole moment, Q is the quadrupole moment, and C_{1-5} are constants (Prausnitz et al., 1986). Induction interactions are not shown in eq. 7.4 since their contribution tends to be much smaller than dispersion and polar interactions. Equation 7.4 does not describe rigorously the interaction of a polymer segment with another segment or with the solvent since segmental motion is constrained by chain connectivity and this architectural feature is not taken into account. Nonpolar dispersion interactions, the first term in eq. 7.4, depend only on the polarizability of the components in solution and not on temperature. Unfortunately, the polarizability of CO_2 is similar to that of methane, which is not a good solvent unless the system pressure is exceptionally high, or stated differently, unless the density of methane is increased considerably. By analogy, CO_2 is not expected to be a good solvent at high operating temperatures where dispersion interactions are dominant. The tenets of chemistry teach that CO_2 must have a polar moment because oxygen and carbon have different electron affinities. Due to structural symmetry, CO_2 does not have a dipole moment, but it does have a substantial quadrupole moment that operates over a much shorter distance than dipolar interactions. Carbon dioxide is a dense solvent at modest temperatures and pressures, which magnifies quadrupolar interactions that scale with molar density to the 5/6 power as indicated in eq. 7.4 if the distance between the center of mass of two CO_2 molecules is twice the radius of an effective spherical volume occupied by CO_2. The dipolar and quadrupolar interaction terms in eq. 7.4 are inversely proportional to temperature, which means that at elevated temperatures polar molecules behave as if they were nonpolar. Hence, it may be possible to dissolve a nonpolar polymer in CO_2 if the temperature is high enough to diminish CO_2–CO_2 quadrupolar interactions relative to CO_2–polymer segment nonpolar dispersion interactions. As will be shown, CO_2 is a weak solvent for polar polymers since dipole interactions outweigh quadrupole interactions, especially at low temperatures where polar interactions are magnified. The challenge that still remains today is to predict the level of polarity needed in the polymer to make it soluble in CO_2 at modest pressures and temperatures.

Temperature-sensitive specific interactions, such as complex or hydrogen-bond formation, also contribute to the attractive pair potential energy. Certain polymers that possess electron-donating groups, such as carbonyls, exhibit specific interactions with CO_2 where the carbon atom of CO_2 acts as an electron acceptor and the carbonyl oxygen in the polymer acts as an electron donor (Kazarian et al., 1996). The strength of the CO_2–segment complex is generally less than 1 kcal/mol, which makes it only slightly stronger than dispersion interactions; however, in a dense CO_2–polymer solution, this increase in interaction strength can be significant. It will be shown that replacing hydrogen with fluorine increases polymer solubility in CO_2. The basic carbon atom of CO_2 can interact with the fluorine in a C–F bond since fluorine has a higher electron affinity than does hydrogen. High-pressure NMR has been used to elucidate specific CO_2–fluorocarbon interactions in low-molecular-weight fluorocarbon solvents (Dardin et al.,

1998) and fluorocarbon repeat units in different fluorinated polymers and copolymers (Dardin et al., 1997).

Polymer solubility in a given solvent also depends on the free volume, or thermal expansivity, difference between the solvent and the polymer (Patterson, 1982). To dissolve a polymer, it is necessary for the solvent molecules to solvate the repeat units tethered to a given chain, which reduces the number of conformations available to the pure solvent, resulting in a negative entropy of mixing. If the free volume of a polymer is increased, it becomes easier to dissolve it in a given solvent. Polymer free volume increases as chain branches are added to the backbone or as the attractive potential between repeat segments is reduced. Hence, it is difficult to uncouple energetic and entropic contributions to polymer solubility, although several examples of this are offered in the following section that reveals the impact of polymer architecture on solubility. A more complete compilation of CO_2–polymer studies can be found elsewhere (Kirby and McHugh, 1999).

Phase Behavior Studies with CO_2

Carbon dioxide at or near room temperature readily dissolves many poly(dimethyl) and poly(phenylmethyl) silicones, perfluoroalkylpolyethers, and chloro- and bromotrifluoroethylene polymers (Hoefling et al., 1991; Krukonis, 1985; McHugh and Krukonis, 1984; Xiang and Kiran, 1995; Yilgor et al., 1984). In general, the polymers that dissolve in CO_2 possess some degree of polarity due to oxygen or other electronegative groups such as chlorine or bromine incorporated into the polymer backbone or on a side chain. The high solubility of the poly(dimethyl) and poly(phenylmethyl) silicones in CO_2 is also likely due to the very flexible character of these polymers that endows them with much larger free volumes as compared with other polymers.

Carbon dioxide does not dissolve polyolefins to any great extent unless the molecular weight is very low. For example, Gregg and co-workers show that a three-arm-star polyisobutylene with a molecular weight (M_w) of 4000 does not dissolve in CO_2 to temperatures of 200 °C and 2000 bar (Gregg et al., 1994); Rindfleisch and co-workers have shown that low-density polyethylene with an M_w of ~100,000 does not dissolve in CO_2 even to temperatures of 270 °C and 2,750 bar (Rindfleisch et al., 1996). Carbon dioxide can dissolve octane, hexadecane, and squalane, but these nonpolar solutes have molecular weights in the range 100–500 (Liphard and Schneider, 1975). Rindfleisch et al. (1996) argue that the interchange energy, given in eq. 7.3, is dominated by CO_2–CO_2 quadrupolar self-interactions rather than CO_2–polymer dispersion or induction cross interactions for mixtures of CO_2 with alkanes or polyolefins.

Very polar polymers or polymers that are water-soluble do not dissolve in CO_2 even at high temperatures. For example, poly(acrylic acid), $M_w = 450{,}000$, does not dissolve in CO_2 to temperatures of 272 °C and pressures of 2220 bar (Rindfleisch et al., 1996). Carbon dioxide can dissolve very-low-molecular-weight, slightly polar polymers, but solubility levels quickly drop as the molecular weight increases above 10,000 (Gregg et al., 1994; Hoefling et al., 1991; Krukonis, 1985; Liphard and Schneider, 1975; McHugh and Krukonis, 1994). Even polystyrene (PS) with a molecular weight slightly higher than 100,000 does not dissolve in CO_2, even though the aromatic ring in PS has a quadrupole moment that interacts favorably with the quadrupole of CO_2 (Rindfleisch et al., 1996). However, PS has a high glass transition temperature, ~ 103 °C, which suggests that there is a significant energy penalty for chain-segment rotations resulting in enhanced segment–segment interactions. Once again, the perplexing question arises: "How much polarity is needed to make a polymer soluble in CO_2?" Consider the behavior of a family of poly(acrylates) in CO_2 shown in Figure 7.1, where each line is a cloud-point curve that represents the boundary between a single fluid phase (F) and two liquid phases consisting of a polymer-rich and a CO_2-rich phase (L + L) (Rindfleisch et al., 1996). As the acrylate alkyl tail increases, the effective polarity decreases since dipole interactions scale inversely with the square root of the molar volume. Likewise, the sharp upturn in the cloud-point curve shifts to higher temperatures with increasing alkyl tail length. It is surprising that the poly(octadecyl acrylate) (PODA)–

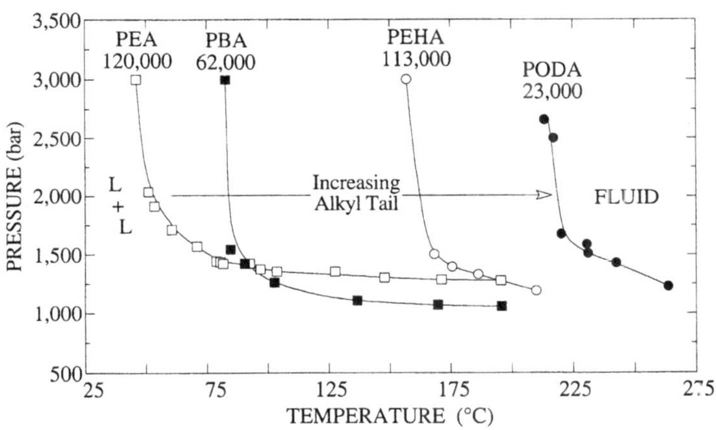

Figure 7.1. Comparison of the cloud-point curves of poly(ethyl acrylate) (PEA), poly(butyl acrylate) (PBA), poly(ethylhexyl acrylate) (PEHA), and poly(octadecyl acrylate) (PODA) in CO_2. The overall polymer concentration is ~ 5 wt % for each curve and the M_w of the polymer is given on each curve (Kirby and McHugh, 1999). The demarcations L + L and FLUID denote a two-phase and a one-phase region, respectively.

CO_2 curve turns up sharply in pressure at 215 °C since it suggests that, even at this very high temperature, quadrupolar CO_2–CO_2 interactions dominate. Although the free-volume difference between the poly(acrylates) and CO_2 decreases with increasing alkyl tail length, the gain in the conformational entropy of mixing does not outweigh the concomitant shift in the interchange energy toward CO_2–CO_2 and polymer–polymer self-interactions as opposed to CO_2–polymer cross interactions. Very high pressures are needed to dissolve the poly(acrylates), even at high temperatures where polar interactions are diminished, since CO_2 is a weak solvent with a very low polarizability.

Interpreting the phase behavior of polymers in CO_2 or any supercritical fluid solvent is a challenge since the connectivity of the monomer units and the conformation of the polymer in solution affects both the energy and entropy of mixing. For example, poly(methyl acrylate) (PMA) dissolves in CO_2 at pressures from 1800 bar at 200 °C to 2300 bar at 40 °C, although poly(methyl methacrylate) (PMMA) remains insoluble even though the strength of PMA–CO_2 intermolecular interactions is expected to be similar to PMMA–CO_2 interactions. However, the glass transition temperature of PMMA is ~ 100 °C higher than that of PMA, suggesting that the rotation of a methacrylate chain segment is more hindered than the rotation of an acrylate chain segment, which likely leads to enhanced PMMA segment–segment interactions. Carbon dioxide does dissolve poly(butyl methacrylate) (PBMA) at temperatures above 120 °C and pressures slightly higher than those needed to dissolved poly(butyl acrylate) (PBA), although in this case the T_g for PBMA is only 70 °C higher than that of PBA.

As further evidence of the challenge inherent in predicting polymer solubility in CO_2, consider a comparison of the PMA and poly(vinyl acetate) (PVAc) cloud-point curves (Rindfleisch et al., 1996). At 30 °C the PMA cloud-point curve is more than 1500 bar higher than the PVAc curve, even though the molecular weight of PVAc is four times greater than that of PMA. Both polymers are polar, but the T_g for PVAc is approximately 21 °C higher than the T_g of PMA, which is likely due to stronger polar interactions between vinyl acetate groups compared with methyl acrylate groups. Carbon dioxide can more easily access the carbonyl group in PVAc than in PMA, which facilitates the formation of a weak CO_2–acetate complex, especially at moderate temperatures.

A large body of work has been developed by DeSimone and co-workers on the solubility of fluorinated polymers, especially poly(1,1-dihydroperfluorooctylacrylate) (PFOA), in CO_2 (Hsiao et al., 1995; Luna-Barcenas et al., 1998). An excellent example of utilizing creative chemistry to design a CO_2–soluble polymer, PFOA is one of the very few fluoropolymers that dissolves in CO_2 at modest temperatures and pressures less than 300 bar. The characteristics needed to make a fluoropolymer soluble in CO_2 can be ascertained from Figure 7.2, which shows the difference in cloud-point curves for poly(vinylidene fluoride) (PDVF), a statistically random copoly-

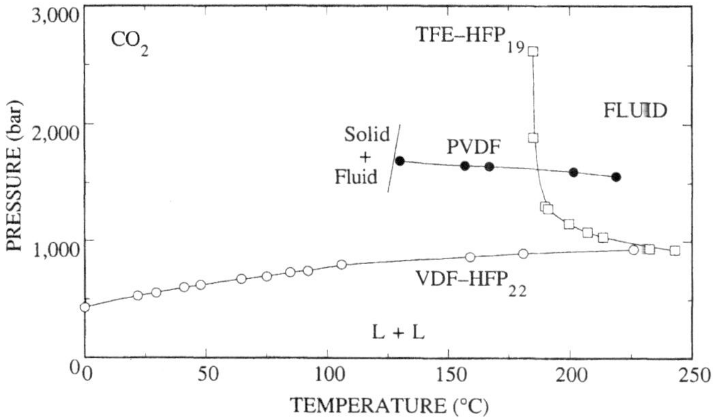

Figure 7.2. Comparison of the cloud-point behavior of poly(vinylidene fluoride) (PVDF), 78 mol % vinylidene fluoride and 22 mol % hexafluoropropylene (VDF–HFP_{22}), and 81 mol % tetrafluoroethylene and 19 mol % hexafluoropropylene (TFE–HFP_{19}) in CO_2. The polymer and copolymer concentrations are ~ 5 wt % in each case. The demarcations L + L and FLUID denote a two-phase and a one-phase region, respectively.

mer of 78 mol % vinylidene fluoride and 22 mol % hexafluoropropylene (VDF–HFP_{22}), and a nonpolar copolymer with 81 mol % tetrafluoroethylene and 19 mol % hexafluoropropylene (TFE–HFP_{19}). (DiNoia et al., 2000). The PVDF–CO_2 curve is at 1500 to 1700 bar and it terminates at a crystallization boundary at 130 °C. In contrast, the VDF–HFP_{22}–CO_2 curve is at pressures less than 1000 bar and it extends to 0 °C and 400 bar. The VDF–HFP_{22} polymer is less polar than PVDF and the copolymer has a higher free volume. Notice that the TFE–HFP_{19} curve exhibits a very steep increase in pressure at ~ 185 °C, which shows that nonpolar TFE–HFP_{19} does not dissolve in CO_2 until very high temperatures are obtained to reduce CO_2–CO_2 interactions. It should also be noted, however, that TFE–HFP_{19} does dissolve in CO_2 whereas polyolefins do not dissolve in CO_2, which lends credence to the observation that fluorinating a polyolefin makes it soluble in CO_2. However, it is obvious by comparison of the VDF–HFP_{22} and TFE–HFP_{19} curves that just fluorinating a polyolefin does not guarantee that it will dissolve at modest pressures and temperatures.

As previously noted, polymer free volume, or conversely thermal expansivity, is an important characteristic that affects solubility. Consider how the phase behavior of TFE–HFP changes as more HFP groups are added to the backbone to increase the free volume and decrease crystallinity (McHugh et al., 1998) (see Figure 7.3). Both cloud-point curves exhibit the same characteristic shapes even though TFE–HFP_{19} has a peak melting point at 147 °C and TFE–HFP_{48} is amorphous. At high temperatures, where quadrupolar CO_2–CO_2 interactions are diminished, it takes approximately 1000

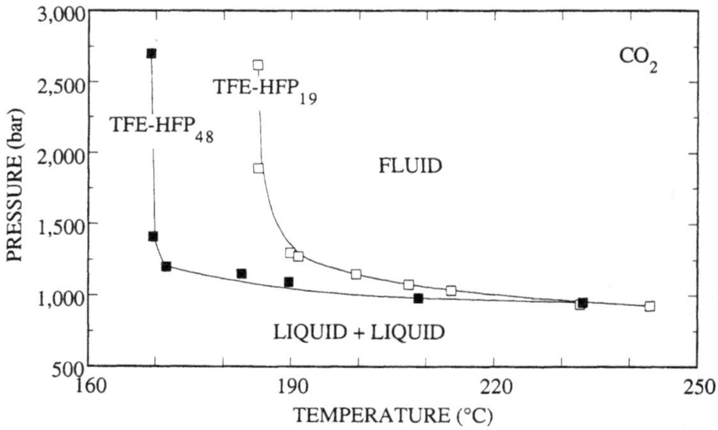

Figure 7.3. Impact of copolymer free volume on the solubility of 81 mol % tetrafluoroethylene and 19 mol % hexafluoropropylene (TFE–HFP_{19}) and (TFE–HFP_{48}) in CO_2. The copolymer concentration is ~ 5 wt % in each case (McHugh et al., 1998).

bar to dissolve either copolymer in CO_2, which is a consequence of the small polarizability of CO_2. Increasing the HFP content increases copolymer free volume and it causes a shift in the low-temperature portion of the cloud-point curve from 190 °C to 170 °C. The sharp upturn in cloud-point pressure in both cases is a consequence of quadrupolar interactions between CO_2 molecules that makes the interchange energy unfavorable for dissolving nonpolar TFE–HFP_x.

Conclusions

Carbon dioxide exhibits both polar and nonpolar character, although the nonpolar character of CO_2 is very modest and is similar to that of methane and perfluoromethane, which are weak SCF solvents. Carbon dioxide does not dissolve polyolefins as long as the molecular weight is greater than a few thousand. The larger the difference in polarity between CO_2 and the polymer, the higher the temperature needed to obtain a single phase. Complex formation between CO_2 and the polymer dramatically improves solvent quality that, in turn, lowers the pressures and temperatures needed to obtain a single phase. However, if the polymer self-associates, it is virtually impossible to dissolve it in CO_2. Certainly, fluorinating a polymer or a copolymer makes it CO_2-soluble, but the pressures needed to solubilize fluorinated copolymers can be very high and they can be extremely temperature-sensitive, depending on the polarity of the polymer. Relatively mild pressures and temperatures are needed to dissolve the fluorinated acrylates since they are partially fluorinated, they have bulky alkyl tails, and they have carboxyl groups that give them some polarity. All of these features enhance solubility in CO_2.

References

Dardin, A.; Cain, J. B.; DeSimone, J. M.; Johnson, C. S., Jr.; Samulski, E. T. *Macromolecules* **1997**, *30*, 3593.

Dardin, A.; DeSimone, J. M; Samulski, E. T. *J. Phys. Chem. B* **1998**, *102*, 1775.

DiNoia, T. P.; Conway, S. E.; Lim, S.-J.; McHugh, M. A. *J. Polym. Sci., Polym. Phys. Ed.* **2000**, *38*, 2832–2840.

Gregg, C. J.; Stein, F. P.; Radosz, M. *Macromolecules* **1994**, *27*, 4981–4985.

Hoefling, T. A.; Enick, R. M.; Beckman, E. J. *J. Phys. Chem.* **1991**, *95*, 7127.

Hoefling, T. A.; Stofesky, D.; Reid M.; Beckman, E. J.; Enick, R. M. *J. Supercrit. Fluids* **1992**, *5*, 237–241.

Hoefling, T. A.; Newman, D. A.; Enick, R. M.; Beckman, E. J. *J. Supercrit. Fluids* **1993**, *6*, 165.

Hsiao, Y. L.; Maury, E. E.; DeSimone, J. M.; Mawson, S.; Johnston, K. P. *Macromolecules* **1995**, *28*, 8159.

Kazarian, S. G.; Vincent, M. F.; Bright, F. V.; Liotta, C. L.; Eckert, C. A. *J. Am. Chem. Soc.* **1996**, *118*, 1729.

Kirby, C. F.; McHugh, M. A. *Chem. Rev.* **1999**. *99*, 565–602.

Krukonis, V. J. *Polym. News* **1985**, *11*, 7.

Lee, L. L. *Molecular Thermodynamics of Nonideal Fluids*; Butterworth: Stoneham, MA, 1988.

Liphard, K. G.; Schneider, G. M. *J. Chem. Thermodyn.* **1975**, *7*, 805.

Luna-Barcenas, G.; Mawson, S.; Takishima, S.; DeSimone, J. M.; Sanchez, I. C.; Johnston, K. P. *Fluid Phase Equilib.* **1998**, *146*, 325.

McHugh, M. A.; Krukonis, V. J. *Supercritical Fluid Extraction: Principles and Practice*; Butterworth: Stoneham, MA, 1994.

McHugh, M. A.; Mertdogan, C. A.; DiNoia, T. P.; Lee, S. H.; Anolick, C.; Tuminello, W. H.; Wheland, R. *Macromolecules* **1998**, *31*, 2252.

Patterson, D. *Polym. Eng. Sci.* **1982**, *22*, 64.

Prausnitz, J. M.; Lichtenthaler, R. N.; de Azevedo, E. G. *Molecular Thermodynamics of Fluid-Phase Equilibria*; Prentice-Hall: NJ, 1986.

Rindfleisch, F.; DiNoia, T. P.; McHugh, M. A. *J. Phys. Chem.* **1996**, *100*, 15581.

Xiang, Y.; Kiran, E. *Polymer* **1995**, *36*, 4817.

Yilgor, I.; McGrath, J. E.; Krukonis, V. J. *Polym. Bull.* **1984**, *12*, 499.

8

Interfacial Phenomena with Carbon Dioxide Soluble Surfactants

KEITH P. JOHNSTON
S. R. P. DA ROCHA
J. D. HOLMES
GUNILLA B. JACOBSON
C. T. LEE
M. Z. YATES

A fundamental understanding of colloid and interface science for surfactant design in CO_2-based systems is emerging on the basis of studies of interfacial tension and surfactant adsorption (da Rocha et al., 1999) along with complementary studies of colloid structure (Chillura-Martino et al., 1996; Meredith and Johnston, 1999; Wignall, 1999) and stability (Meredith and Johnston, 1999; O'Neill, 1997; Yates et al., 1997). The interfacial tension, γ, between a supercritical fluid (SCF) phase and a hydrophilic or lipophilic liquid or solid, along with surfactant adsorption, play a key role in a variety of processes including nucleation, coalescense and growth of dispersed phases, formation of microemulsions and emulsions (Johnston et al., 1999), particle and fiber formation, atomization, foaming (Goel and Beckman, 1995), wetting, adhesion, lubrication, and the morphology of blends and composites (Watkins et al., 1999).

The first generation of research involving surfactants in SCFs addressed water/oil (w/o) microemulsions (Fulton and Smith, 1988; Johnston et al., 1989) and polymer latexes (Everett and Stageman, 1978) in ethane and propane (Bartscherer et al., 1995; Fulton, 1999; McFann and Johnston, 1999). This work provided a foundation for studies in CO_2, which has modestly weaker van der Waals forces (polarizability per volume) than ethane. Consequently, polymers with low cohesive energy densities and thus low surface tensions are the most soluble in CO_2: for example, fluoroacrylates (DeSimone et al., 1992), fluorocarbons, fluoroethers (Singley et al., 1997), siloxanes, and to a lesser extent propylene oxide. Since CO_2 is nonpolar (unlike water) and has weak van der Waals forces (unlike lipophilic phases), it may be considered to be a third type of condensed phase. Surfactants with the above types of "CO_2-philic" segments and a "CO_2-phobic" segment have been used to form microemulsions (Harrison et al., 1994; Johnston et al., 1996), emulsions (da Rocha et al., 1999; Jacobson et al., 1999a; Lee et al., 1999b), and organic polymer latexes (DeSimone et al., 1994) in CO_2. Microemulsion droplets are typically 2–10 nm in diameter, making them optically transparent and thermodynamically stable, whereas kinetically stable emulsion droplets and latexes in the range of 200 nm to 10 μm are opaque and thermodynamically unstable.

The environmentally benign, nontoxic, and nonflammable fluids water and carbon dioxide (CO_2) are the two most abundant and inexpensive solvents on Earth. Water-in-CO_2 (w/c) or CO_2-in-water (c/w) dispersions in the form of microemulsions and emulsions offer new possibilities in waste minimization for the replacement of organic solvents in separations, reactions, and materials formation processes. Whereas the solvent strength of CO_2 is limited, these dispersions have the ability to function as a "universal" solvent medium by solubilizing high concentrations of polar, ionic, and nonpolar molecules within their dispersed and continuous phases. These emulsions may be phase-separated easily for product recovery (unlike the case for conventional emulsions) simply by depressurization.

The following discussion begins with studies of surfactant adsorption and its effect on interfacial tension for various CO_2-based interfaces, particularly water–CO_2 interfaces. These properties are utilized to understand how to design surfactants to form w/c microemulsions and emulsions with the desired curvature and stability. A variety of practical applications of these emulsions are examined, including pH buffers, organic reactions with and without catalysts, enzymatic catalysis, and synthesis of metal and metal oxide nanoparticles. The section on polymer colloids presents theoretical and light-scattering studies of the mechanism of steric stabilization. The last sections discuss in situ studies of the mechanism of steric stabilization in dispersion polymerization and novel ambidextrous surfactants for stabilizing both organic–CO_2 and organic–water interfaces.

Interfacial Tension and Surfactant Adsorption

Recently, surfactant adsorption and γ have been measured at CO_2–water and CO_2–organic interfaces with a tandem variable-volume tensiometer (Harrison, 1996). A pendant drop of an aqueous or organic phase, saturated with CO_2, may be suspended in CO_2 or a CO_2-surfactant mixture and equilibrated. From the digitized droplet shape and density difference between the phases, γ may be calculated from the Laplace equation. In Figure 8.1, γ of the binary CO_2–water (da Rocha et al., 1999), –polyethylene

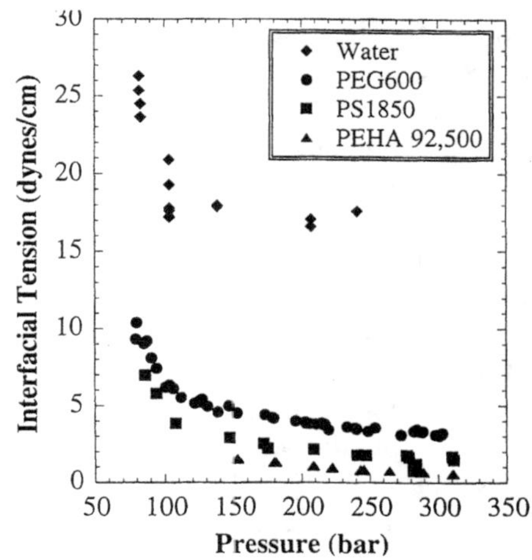

Figure 8.1. Liquid–CO_2 interfacial tension at 45 °C determined by pendant drop tensiometry (Harrison, 1996).

glycol (PEG, 0.6 kg/mol) (Harrison et al., 1996), –polystyrene (PS, 1.85 kg/mol) (Harrison et al., 1998), and –poly-2-ethylhexylacrylate (PEHA, 92.5 kg/mol) systems are plotted as a function of pressure. For each system, γ decreases with pressure (neglecting the critical region) as the intermolecular forces of CO_2 become closer to that of the other component, as has been predicted theoretically with a gradient model along with lattice-fluid theory (Harrison, 1996). As expected, changes are larger at low pressures where the CO_2 density variation is more pronounced. At a given pressure, γ decreases for the above series from CO_2–water to CO_2–PEHA in the same order as the surface tension of the pure condensed phase, as expected.

In Figure 8.2, γ is plotted versus surfactant concentration at a constant CO_2 density of 0.842 g/mL for a perfluoropolyether ammonium carboxylate (PFPE–$COO^-NH_4^+$) surfactant (2500 g/mol) at the CO_2–water interface (da Rocha and Johnston, 1999). A shallow break in the curve is observed at the critical microemulsion concentration. The surfactant adsorption, Γ may be determined by the slope according to the Gibbs adsorption equation, $d\gamma = -RT\Gamma d\ln c$. The area per surfactant molecule, given by $1/\Gamma$ at 25 °C, 45 °C, and 65 °C is 76, 96, and 107 A^2, respectively. This surfactant occupies a somewhat greater area than the smaller 695 g/mol PFPE–$COO^-NH_4^+$ at 35 °C, which occupies 120 A^2 according to small-angle neutron scattering (SANS) measurements of the droplet radius (Zielinsky et al., 1997). Because of the high adsorption of these surfactants, resulting in low interfacial tensions, they stabilize w/c microemulsions. It is interesting that 710 g/mol PFPE–$COO^-NH_4^+$ occupies only 50 A^2/molecule at the water–PFPE oil (900 g/mol) interface (Baglioni et al., 1996). The fact that surfactant areas for w/c microemulsions appear to be larger than for w/o microemulsions is an intriguing and presently unexplained phenomena. The same approach has been used to determine the area for block copolymer

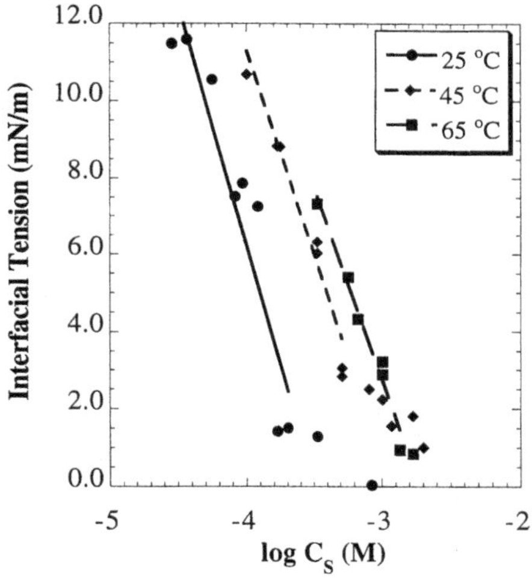

Figure 8.2. Surfactant adsorption at water–CO_2 interface from slope of γ versus concentration for 2500 g/mol PFPE $COO^-NH_4^+$.

surfactants adsorbed at CO_2–organic oligomer interfaces to understand the stability of organic emulsions and latexes (Harrison et al., 1996, 1998).

The interfacial tension is a key property for describing the formation of emulsions and microemulsions (Aveyard et al., 1990), including those in supercritical fluids (da Rocha et al., 1999), as shown in Figure 8.3, where the x-axis represents a variety of formulation variables. A minimum in γ is observed at the phase inversion point where the system is balanced with respect to the partitioning of the surfactant between the phases. Here, a middle-phase emulsion is present in equilibrium with excess CO_2-rich (top) and aqueous-rich (bottom) phases. Upon changing any of the formulation variables away from this point—for example, the hydrophilic/CO_2-philic balance (HCB) in the surfactant structure—the surfactant will migrate toward one of the phases. This phase usually becomes the external phase, according to the Bancroft rule. For example, a surfactant with a low HCB, such as PFPE $COO^-NH_4^+$ (2500 g/mol), favors the upper CO_2 phase and forms w/c microemulsions with an excess water phase. Likewise, a shift in formulation variable to the left would drive the surfactant toward water to form a c/w emulsion. Studies of γ versus HCB for block copolymers of propylene oxide, and ethylene oxide, and polydimethylsiloxane (PDMS) and ethylene oxide, have been used to understand microemulsion and emulsion formation, curvature, and stability (da Rocha et al., 1999).

Water-in-CO_2 Microemulsions

In order to form a w/c microemulsion, a relatively high surfactant concentration is required to cover the extremely large interfacial area resulting

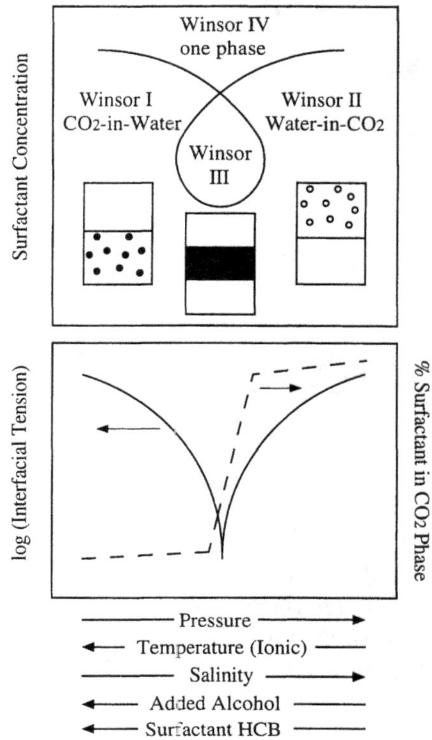

Figure 8.3. Schematic representation of the relationship between the phase behavior and interfacial tension for water/CO_2/ionic surfactant mixtures as a function of formulation variables.

from the extremely small droplets (< 10 nm). For PFPE $COO^-NH_4^+$w/c microemulsions, Fourier transform infrared (FTIR), UV-vis, fluorescence, and electron paramagnetic resonance (EPR) experiments have demonstrated the existence of an aqueous microdomain in CO_2 with a polarity approaching that of bulk water (Johnston et al., 1996). From SANS, the water droplet radius is 3.5 nm for a molar water-to-surfactant ratio of 30 (Zielinsky et al., 1997). Surfactants that have been reported to stabilize w/c microemulsions include PFPE $COO^-NH_4^+$ (Johnston et al., 1996; Niemeyer and Bright, 1998; Zielinsky et al., 1997), the hybrid–hydrocarbon-fluorocarbon surfactant $C_7F_{15}CH(OSO_3^-Na^+)C_7H_{15}$ (Harrison et al., 1994), and the surfactant di(1*H*, 1*H*, 5*H*-octafluoro-*n*-pentyl) sodium sulfosuccinate (di-HCF_4) (Holmes et al., 1998). Organic-in-CO_2 microemulsions have also been formed for 600 g/mol poly(ethylene glycol) (PEG600) and for polystyrene oligomers (Harrison et al., 1996; McClain et al., 1996).

For microemulsions with low water volume fractions (typically $\phi < 0.02$), the absolute amount of dissolved hydrophiles can be somewhat small, and this can limit phase-transfer reactions between CO_2 and aqueous phases (Jacobson et al., 1999b). Recently, highly concentrated w/c microemulsions ($\phi \sim 0.5$) have been formed (Lee et al., 1999a), providing for much greater

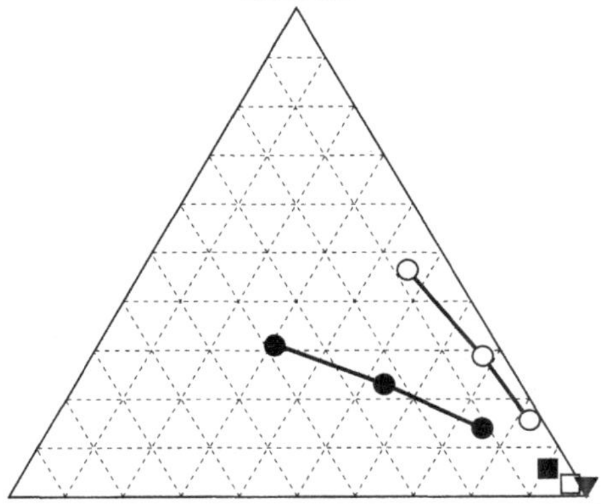

Figure 8.4. Phase behavior of water/CO_2/surfactant systems studied to date. ● $T = 35\,°C$, $P = 414$ bar and ○ $T = 35\,°C$, $P = 138$ bar. ▼ PFPE $COO^-NH_4^+$ (Johnston et al., 1996); □ PFPE $COO^-NH_4^+$ (Zielinsky et al., 1997); ■ di-HCF_4 (Holmes et al., 1998). The one-phase microemulsion region is to the right of each curve.

solubilization. As seen in Figure 8.4, the one-phase microemulsion region (the region to the right of each curve) increases in size as the pressure of the continuous CO_2-phase increases. The increase in solvation of the surfactant tails reduces attractive interdroplet interactions between surfactant tails (Zielinsky et al., 1997). The electrical conductivity of these microemulsions increases nearly three orders of magnitude due to changes in the interdroplet interactions with droplet concentration or temperature. Scaling analysis of the conductivity data supports a dynamic percolation model (Grest et al., 1986) for this system, whereby the attractive interdroplet interactions result in the formation of clusters of discrete droplets and not bicontinuous structures.

The pH inside microemulsion droplets is typically pH 3, as determined with fluorescence (Niemeyer and Bright, 1998) and absorbance (Holmes et al., 1998, 1999b) probes. Inorganic and organic bases and buffers, such as NaOH, can be used to control the aqueous pH in PFPE $COO^-NH_4^+$-stabilized microemulsions from pH 3 to values from pH 5 to 7 as shown in Figure 8.5. The titration of carbonic acid with 0.3 M alkaline buffers raises the pH to 5.5. Large amounts of base (e.g., ~ 1.5 mol kg^{-1} NaOH) are required to further increase pH to 7 due to the buffer capacity of the carbonic acid–bicarbonate equilibrium. The pH may be predicted from the relevant equilibrium constants and initial concentrations, along with small corrections due to the waters of hydration for the surfactant head group. The pH results are similar for the microemulsions and biphasic water–CO_2 systems without surfactant.

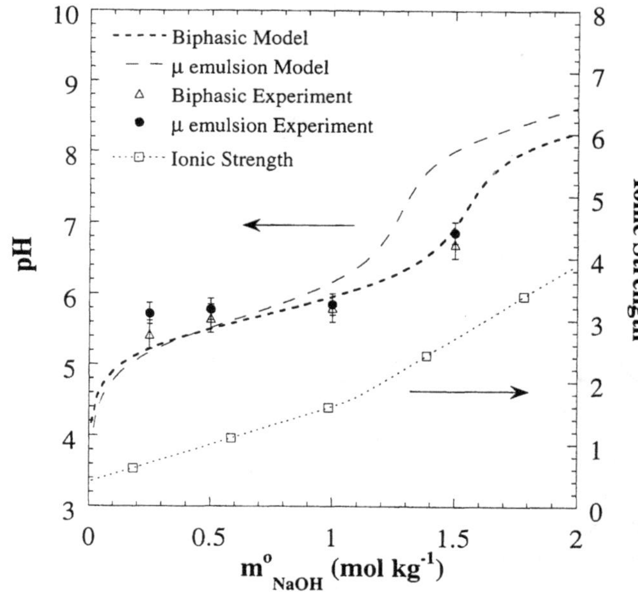

Figure 8.5. Experimentally and theoretically predicted aqueous pH results for w/c PFPE $COO^-NH_4^+$ microemulsions and biphasic systems (no surfactant) buffered with sodium hydroxide (Holmes et al., 1999b) ($T = 35\,^{\circ}C$, $P = 345$ bar).

Water-in-CO_2 Emulsions

The formation of emulsion droplets requires lower surfactant concentrations than in the case of microemulsions because of the much larger droplets. Stable w/c emulsions, for either liquid or supercritical CO_2, have been formed with the surfactants (PFPE $COO^-)_2$ Mn^{+2} (Johnston et al., 1999), PFPE $COO^-NH_4^+$ (672–7500 g/mol) (Lee et al., 1999b), and block copolymer surfactants composed of PDMS and poly(acrylic acid) or poly(methacrylic acid). The emulsions were formed by shear through a 100-μm capillary or a small orifice. The ratio of water to CO_2 has been varied from 9:1 to 1:9 with less than 1 wt % surfactant. In contrast, w/c microemulsions contain less than 5 wt % water for this level of surfactant because of the much higher interfacial area. A stability plot is shown for the surfactant PFPE $COO^-NH_4^+$ (2500 g/mol) in Figure 8.6. Similar behavior was seen for the other types of surfactants. Above the inversion pressure, the specific conductivity was less than 0.1 μS/cm, indicating water droplets in a CO_2-continuous phase. Water droplet sizes ranging from 3 to 10 μm were determined by optical microscopy. Emulsion stability is a maximum of 1 day, at low temperature and high pressure, and decreases with either an increase in temperature or a decrease in the pressure. Eventually, an inver-

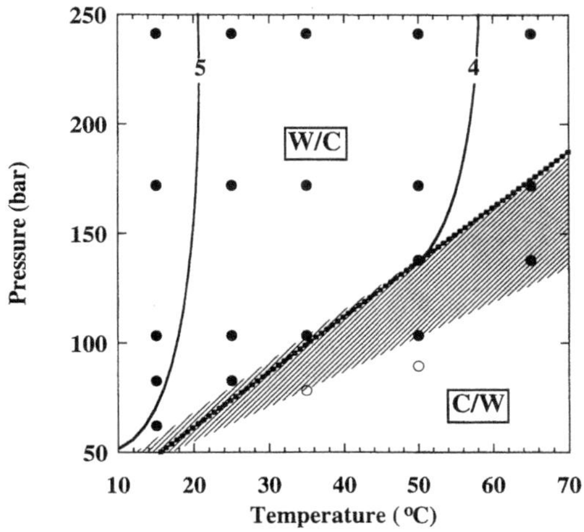

Figure 8.6. Emulsion stability contours (log of emulsion stability in seconds) for 50/50 (mass) CO_2–water (0.01 M NaCl) systems with 10.7 mM PFPE $COO^-NH_4^+$ (g/mol = 2500) surfactant. ● w/c emulsions; ○ (c/w) emulsions. Dotted line indicates the phase boundary of the surfactant in CO_2; cross-hatched region indicates highly flocculated emulsions (Lee et al., 1999b).

sion to CO_2-in-water (c/w) emulsions occurs. This inversion also takes place with a decrease in the PFPE molecular weight (increase in HCB). Near the inversion point, as defined in Figure 8.3, w/c emulsion stability decreases rapidly. For example, in the cross-hatched region the emulsion is highly flocculated. Emulsions are destabilized by flocculation due to attraction between droplets and by coalescence of the flocs due to thinning of the continuous phase between the droplets. Emulsions become more stable with a change in any of the formulation variables away from the inversion point (see Figure 8.3). The resulting increase in interfacial tensions, and thus interfacial tension gradients, resist film drainage (Marangoni–Gibbs stabilization) (Lee et al., 1999b). For PFPE-based surfactants, stability is enhanced with an increase in the molecular weight of the surfactant tails that increases steric stabilization and the thickness of the stabilizing films between droplets.

Reactions in Microemulsions and Emulsions

Reactions of CO_2-soluble hydrophobic organic substrates across an interface with hydrophilic nucleophiles have been performed in both w/c or c/w microemulsions (Jacobson et al., 1999b) and emulsions (Jacobson et al., 1999a). For the reaction of benzyl chloride (CO_2-soluble) with KBr, the yield of benzyl bromide and the rate constant were found to be an order of magnitude higher in a w/c microemulsion as compared with w/o microemulsions. This enhancement is related to a difference in interfacial structure resulting from the lower interfacial tension between water and CO_2. Even higher yields were obtained in w/c emulsions since the much larger amount of water in the emulsions (50 wt % versus 1 wt %) allowed for a greater excess of KBr. The reaction rates for the hydrolysis of water-insoluble benzoyl chloride and *p*-nitrophenyl chloroformate were also shown to be an order of magnitude higher in w/c microemulsions as compared with w/o microemulsions. It was found that changing the droplet size of the micelle, thereby changing the polarity and nucleophilicity of the water droplet, could affect the reaction rates due to the different transition states of these substrates.

Holmes et al. (1998) performed two enzymatic reactions, the lipase-catalyzed hydrolysis of *p*-nitrophenol butyrate and lipoxygenase-catalyzed peroxidation of linoleic acid, in w/c microemulsions stabilized by a fluorinated two-chained sulfosuccinate surfactant (di-HCF_4). The activity of both enzymes in the w/c microemulsion environment was found to be essentially equivalent to that in a water/heptane microemulsion stabilized by Aerosol OT, a surfactant with the same headgroup as di-HCF_4. The buffer 2-(*N*-morpholino)ethanesulfonic acid (MES) was used to fix the pH in the range 5–6.

Synthesis of Metal and Metal Oxide Nanoparticles

A relatively recent development is the exploitation of w/c microemulsions for the synthesis of metallic and semiconductor nanoparticles. By reducing silver nitrate, Ji et al. (1999) were able to harvest silver nanoparticles from a w/c microemulsion. Analysis of the plasmon resonance peak at 400 nm indicated that samples collected at intervals of 20 and 10 min were ~ 4 nm in diameter. A subsequent decrease in the intensity of the plasmon band, over a period of 1 h, was attributed to the slow flocculation of nanoparticles.

Holmes et al. (1999a) synthesized semiconductor nanoparticles of cadmium sulfide, in PFPE $COO^-NH_4^+$-stabilized w/c microemulsions. The exciton peaks are shown in Figure 8.7. At $W_0 = 5$ and 10, the exciton energies are 3.86 and 3.09 eV, corresponding to mean particle radii of approximately 0.9 and 1.8 nm, respectively. These radii are consistent with measurements by transmission electron microscopy (TEM), as well as the microemulsion droplet diameters from SANS (Zielinsky et al., 1997). Despite the low viscosity of CO_2, resulting in higher droplet collision frequencies, agglomeration is minimal and the particles remain restricted by the water core radius (r_w) after a reaction period of 10 min. This templating effect possibly arises from the difficulty in containing two particles in one micelle at a particle diameter approaching r_w. Also, particles of an equivalent size to r_w presumably become surrounded by surfactant molecules, which act as a protective agent. Therefore, w/c microemulsions serve as an effective medium for compartmentalized growth of nanoparticles with controllable size and relatively narrow size distributions.

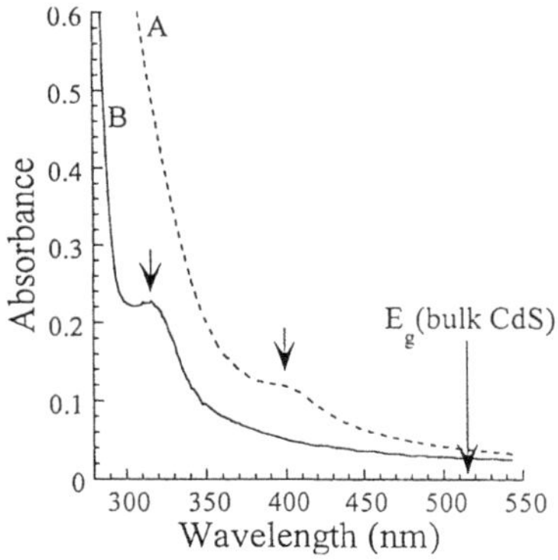

Figure 8.7. UV-visible absorbance spectra of CdS particles prepared in PFPE COO^- NH_4^+ (3 wt %) stabilized w/c microemulsions: (A) $W_0 = 10$ and (B) $W_0 = 5$ (Holmes et al., 1999a)

Polymer Colloids

Organic substances including polymers may also be emulsified in CO_2. Copolymers containing CO_2-philic segments such as poly(1,1-dihydroperfluorooctyl acrylate) (PFOA) (DeSimone et al., 1994), poly(perfluoropropylene oxide) (Lepilleur and Beckman, 1997), and PDMS have been used to sterically stabilize latexes that form during dispersion polymerization. Steric stabilization may prevent flocculation if (1) the adsorbed or grafted surfactant provides high surface coverage, (2) the length of the solvated CO_2-philic segment is sufficient to overcome the van der Waals attraction between the particle cores, and (3) the solvation of the CO_2-philic segment is sufficient to prevent interparticle attractive interactions between these segments. With theory (Meredith and Johnston, 1998, 1999) and computer simulation (Meredith et al., 1998), an analogy has been observed between polymer–supercritical fluid phase separation in bulk and the flocculation of surfaces with grafted polymer (see Figure 8.8). As the density is lowered, the polymer and the supercritical fluid phase separate at the upper critical solution density (UCSD). At this same density, the force between surfaces with grafted chains becomes attractive, indicating flocculation. This point is called the critical flocculation density (CFD). In each case, the solvent expands away from the polymer chains, resulting in either precipitation of a polymer-rich phase (phase behavior) or flocculation between surfaces. This expansion is entropically driven. To date, the simulations have been performed only for symmetric systems where the segment–segment interactions on the polymer match the segment–solvent interactions.

Steric stabilization of poly(2-ethyl hexyl acrylate) (PEHA) emulsions in CO_2 was studied with static and dynamic light scattering (DLS) (O'Neill et al., 1997; Yates et al., 1997). These emulsions were stabilized with PFOA-based surfactants. Figure 8.9 shows the average droplet size (hydrodynamic radius) measured by DLS for a liquid PEHA emulsion in CO_2 stabilized

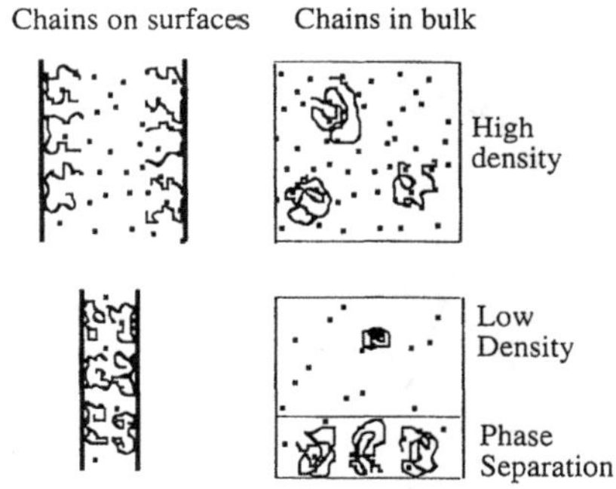

Figure 8.8. Schematic illustrating the analogy between colloid flocculation behavior and phase behavior of the stabilizer in bulk solution. As density is lowered, separation of solvent from chains in bulk solution resembles separation of solvent from chains on surfaces, which produces flocculation.

with poly(styrene)-*b*-PFOA at 25 °C. The emulsion was initially formed in CO_2 with a density of 0.896 g/mL. The droplet size changed very little as the density was lowered from 0.896 to 0.838 g/mL. However, a very large increase in the droplet size is observed as the density is reduced from 0.838 to 0.824 g/mL due to flocculation at the CFD. The CFD agreed very closely with the UCSD of PFOA in CO_2 at the same temperature, in agreement with theory and simulation.

For sub-micron silica particles with grafted PDMS (up to 22 K), a different result was obtained (Yates and Johnston, 1999). The particles were unstable and flocculated well above the UCSD of the PDMS–CO_2 binary system. These results may suggest that it is necessary to raise the density to the UCSD for PDMS at infinite molecular weight (theta density). Another possibility is that the parameters used in the theory and simulation are not applicable to PDMS, since the polymer–polymer interactions are far stronger than the polymer–CO_2 interaction, unlike the case for PFOA.

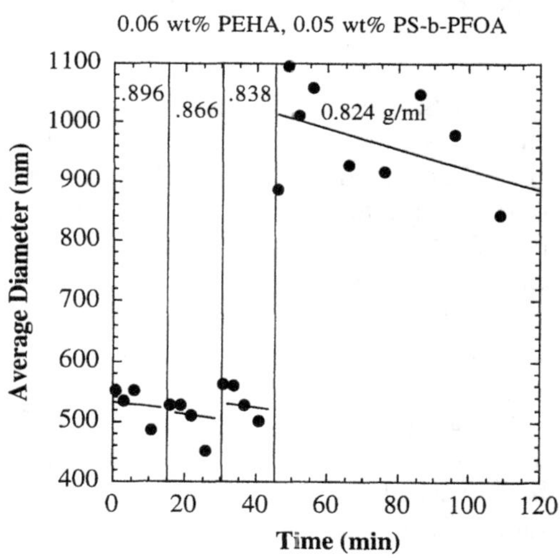

Figure 8.9. Critical flocculation density of a 0.04 wt % PEHA emulsion stabilized with 0.03 wt % PS-*b*-PFOA as determined by DLS from the average hydrodynamic radius (Yates et al., 1997).

In Situ Studies of the Mechanism of Dispersion Polymerization

Turbidimetry has been used to study dispersion polymerization of PMMA (O'Neill et al., 1998) and poly(vinyl acetate) in CO_2 (Canelas et al., 1998). The average particle size, particle number density, and overall surface area were measured versus time during particle formation. Coagulative nucleation and controlled coagulation regions were governed by the amount of stabilizer available relative to the total surface area of the dispersion. At the end of the controlled coagulation region, which can last tens of minutes, the particle number density approaches the final value. For the dispersion polymerization of vinyl acetate, the particle number density was obtained for the stabilizers: PDMS homopolymer, PDMS-*b*-PVAc, PFOA homopolymer, and PFOA-*b*-PVAc. The measured particle number densities after 40 min agreed with those for diluted samples at the completion of the reaction, again suggesting that the number density is fixed early in the reaction. The PDMS-based surfactants produce larger particles than the PFOA-based surfactants, suggesting that they provide less resistance to flocculation and coagulation, as was discussed earlier.

Ambidextrous Surfactants

Recently, the concept of an "ambidextrous" surfactant, which is active at an organic–CO_2 and organic–water interface, was demonstrated as shown in Figure 8.10. The surfactants included a PDMS block, a poly(methacrylic acid) (PMA) or poly(acrylic acid) block, and, in some cases, a third PMMA

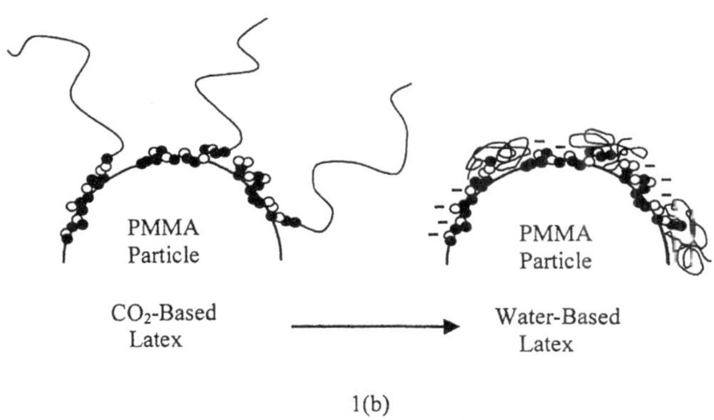

Figure 8.10. Trifunctional "ambidextrous" surfactant for steric stabilization of organic–CO_2 latexes and electrostatic stabilization of organic–water latexes (Yates et al., 1999).

block (Yates et al., 1999). In CO_2, PMMA particles have been produced with these surfactants by dispersion polymerization. These particles were sterically stabilized by the PDMS segments. Upon transferring the surfactant-coated particles to water (or buffered water) at up to 40% by weight, the PDMS block collapsed onto the latex surface, and some of the PMA groups are ionized, resulting in electrostatic stabilization. The surfactant is ambidextrous in that it stabilizes two different types of interfaces, each by a different stabilization mechanism. With this concept, it is possible to produce latexes in CO_2 with minimal waste, which can be vented, shipped as dry powder, and resuspended in water to form an aqueous latex for coating applications.

In conclusion, a fundamental understanding of colloid and interface science in CO_2-based systems is emerging from studies of interfacial tension, surfactant adsorption, and colloid interactions. This knowledge is being used to design surfactants for the stabilization of microemulsions, emulsions, latexes, and suspensions of inorganic and organic substances in CO_2. A variety of cleaning, separation, reaction, and materials formation processes are being developed on the basis of surfactants and novel colloids in CO_2.

References

Aveyard, R.; Binks, B. P.; Clark, S.; Fletcher, P. D. I. *J. Chem. Tech. Biotechnol.* **1990**, *48*, 161–171.

Baglioni, P.; Gambi, C. M. C.; Giordano, R.; Senatra, D. *J. Mol. Struct.* **1996**, *383*, 165–169.

Bartscherer, K. A.; Minier, M.; Renon, H. *Fluid Phase Equilib.* **1995**, *107*, 93–150.

Canelas, D. A.; Betts, D. E.; DeSimone, J. M.; Yates, M. Z.; Johnston, K. P. *Macromolecules* **1998**, *31*, 6794–6805.

Chillura-Martino, D.; Triolo, R.; McClain, J. B.; Combes, J. R.; Betts, D. E.; Canelas, D. A.; DeSimone, J. M.; Samulski, E. T.; Cochran, H. D.; Londono, J. D.; Wignall, G. D. *J. Mol. Struct.* **1996**, *383*, 3–10.

da Rocha, S. R. P.; Johnston, K. P. *Langmuir* **2000**, *16*(8), 3690–3695.

da Rocha, S. R. P.; Harrison, K. L.; Johnston, K. P. *Langmuir* **1999**, *15*, 419–428.

DeSimone, J. M.; Guan, Z.; Elsbernd, C. S.; *Science* **1992**, *257*, 945–947.

DeSimone, J. M.; Maury, E. E.; Menceloglu, Y. Z.; McClain, J. B.; Romack, T. J.; Combes, J. R. *Science* **1994**, *265*, 356.

Everett, D. H.; Stageman, J. F. *Faraday Discuss. Chem. Soc.* **1978**, *65*, 230–241.

Fulton, J. L. In *Microemulsions: Fundamentals and Applied Aspects*; Kumar, P., Ed.; Marcel Dekker: New York, 1999.

Fulton, J. L.; Smith, R. D.; *J. Phys. Chem.* **1988**, *92*, 2903–2907.

Goel, S. K.; Beckman, J. *AIChE J.* **1995**, *41*, 357–367.

Grest, G. S.; Webman, I.; Safran, S. A.; Bug, A. L. R. *Phys. Rev. A* **1986**, *33*, 2842–2845.

Harrison, K. Chemical Engineering Thesis; University of Texas at Austin, Austin, 1996.

Harrison, K. L.; Goveas, J.; Johnston, K. P.; O'Rear III, E. A. *Langmuir* **1994**, *10*, 3536–3541.

Harrison, K. L.; Johnston, K. P.; Sanchez, I. C. *Langmuir* **1996**, *12*, 2637–2644.

Harrison, K. L.; da Rocha, S. R. P.; Yates, M. Z.; Johnston, K. P. *Langmuir* **1998**, *14*, 6855–6863.

Holmes, J. D.; Steytler, D. C.; Rees, G. D.; Robinson, B. H. *Langmuir* **1998**, *14*, 6371–6376.

Holmes, J. D.; Bhargava, P. A.; Korgel, B. A.; Johnston, K. P. *Langmuir* **1999a**, *15*, 6613–6615.

Holmes, J. D.; Ziegler, K. J.; Audriani, M.; Lee, C. T.; Bhargava, P. A.; Steytler, D. C; Johnson, K. P. *J. Phys. Chem. B.* **1999b**, *103*, 5703–11.

Jacobson, G. B.; Lee, C. T.; da Rocha, S. R. P.; Johnston, K. P. *J. Org. Chem.* **1999a**, *64*, 1207–1210.

Jacobson, G. B.; Lee, C. T.; Johnston, K. P. *J. Org. Chem.* **1999b**, *64*, 1201–1206.

Ji, M.; Chen, X.; Wai, C. M.; Fulton, J. L. *J. Am. Chem. Soc.* **1999**, *121*, 2631–2632.

Johnston, K. P.; McFann, G.; Lemert, R. M. *Am. Chem. Soc. Symp. Ser.* **1989**, *406*, 140–164.

Johnston, K. P.; Harrison, K. L.; Clarke, M. J.; Howdle, S. M.; Heitz, M. P.; Bright, F. V.; Carlier, C.; Randolph, T. W. *Science* **1996**, *271*, 624–626.

Johnston, K. P.; Jacobson, G. B.; Lee, C. T.; Meredith, C.; da Rocha, S. R. P.; Yates, M. Z.; DeGrazia, J.; Randolph, T. W. In *Chemical Synthesis in Supercritical Fluids*; Jessop, P., Leitner, W., Eds.; Wiley-VCH: Weinheim, 1999.

Lee, C. T.; Bhargava, P.; Johnston, K. P. *J. Phys. Chem.* **2000**, *104* (18), 4448–4456.

Lee, T. C.; Psathas, P. A.; Johnston, K. P.; deGrazia, J.; Randolph, T. W. *Langmuir* **1999b**, *15*, 6781–91.

Lepilleur, C.; Beckman, E. J. *Macromolecules* **1997**, *30*, 745–756.

McClain, J. B.; Betts, D. E.; Canelas, D. A.; Samulski, E. T.; DeSimone, J. M.; Londono, J. D.; Cochran, H. D.; Wignall, G. D.; Chillura-Martino, D.; Triolo, R. *Science* **1996**, *274*, 2049.

McFann, G. J.; Johnston, K. P. In *Microemulsions: Fundamental and Applied Aspects*; Kumar, P., Ed.; Dekker: New York, 1999; pp. 281–307.

Meredith, J. C.; Johnston, K. P. *Macromolecules* **1998**, *31*, 5518–28.

Meredith, J. C.; Johnston, K. P. In *Supercritical Fluids: Fundamentals and Applications*; Kiran, E., Peters, P. D. C., Eds.; Kluwer Academic Publishers: Dordrecht, 2000, pp. 211–228.

Meredith, J. C.; Sanchez, I. C.; Johnston, K. P.; Pablo, J. J. D. *J. Chem. Phys.* **1998**, *109*, 6424–34.

Niemeyer, E. D.; Bright, F. V. *J. Phys. Chem.* **1998**, *102*, 1474.

O'Neill, M.; Yates, M. Z.; Harrison, K. L.; Johnston, P. K.; Canelas, D. A.; Betts, D. E.; DeSimone, J. M.; Wilkinson, S. P. *Macromolecules* **1997**, *30*, 5050–5059.

O'Neill, M. L.; Yates, M. Z.; Johnston, K. P.; Smith, C. D.; Wilkinson, S. P. *Macromolecules* **1998**, *31*, 2848–2856.

Singley, E. J.; Liu, W.; Beckman, E. J. *Fluid Phase Equilib.* **1997**, *128*, 199–219.

Watkins, J. J.; D. B. G.; Rao, V. S.; Pollard, M. A.; Russell, T. P. In *NATO Advanced Study Institute Series*; Kiran, E., Debenedetti, P. G., Peters, C. J., Eds.; Kluwer Academic Publishers, Dordrecht, 1999.

Wignall, G. D. *J. Phys.: Condens. Mat.* **1999**, *11* (15), R157–R177.
Yates, M. Z.; Shah, P. S.; Johnston, K. P.; Lim, K. T.; Webber, S. E. *J. Colloid Interface Sci.* **2000**, *227*, 176–184.
Yates, M. Z.; O'Neill, M. L.; Johnston, K. P.; Webber, S.; Canelas, D. A.; Betts, D. E.; DeSimone, J. M. *Macromolecules* **1997**, *30*, 5060–5067.
Yates, M. Z.; Li, G.; Shim, J. J.; Maniar, S.; Johnston, K. P. *Macromolecules* **1999**, *32*, 1018–1026.
Zielinsky, R. G.; Kline, S. R.; Kaler, E. W.; Rosov, N. *Langmuir* **1997**, *13*, 3934–3937.

9

Synthesis and Characterization of Polymers: From Polymeric Micelles to Step-Growth Polymerizations

JENNIFER L. YOUNG

JOSEPH M. DeSIMONE

The benefits of using CO_2 in polymer synthesis are numerous, ranging from environmental responsibility to improved materials properties. Carbon dioxide is an inert, nontoxic, nonflammable, and inexpensive reaction and processing medium that is an environmentally benign alternative to the organic solvents or water typically used today. Although the often toxic, carcinogenic, and environmentally hazardous organic solvents are recycled, some release to the environment is inevitable. Replacement of organic solvents with water still requires the costly purification of the wastewater prior to disposal and/or an energy-intensive drying process to remove the water. On the other hand, CO_2 can be easily separated from other chemical components and recycled through depressurization and recompression. Although CO_2 is a greenhouse gas, the CO_2 used as a solvent does not contribute to the greenhouse gases since it is acquired from natural reservoirs or recovered as a by-product from other industrial chemical processes. The more specific environmental benefits of using liquid or supercritical carbon dioxide as a solvent vary depending on the polymerization being considered.

The synthesis of fluoropolymers in CO_2 is of particular interest since these polymers have historically been prepared in chlorofluorocarbons (CFCs) and other fluorinated solvents, as well as in water. Due to the association of CFCs with ozone-layer depletion, these solvents have been banned and replacement solvents must be found. Alternative fluorinated solvents are expensive and also have environmental concerns.

In heterogeneous polymerizations, many polymer latexes produced by emulsion or dispersion polymerization in water or organic solvents can be produced in CO_2. To eliminate volatile organic compound (VOC) emissions, more polymer latexes are being synthesized in water. However, for dry polymer applications, the latexes must undergo energy-intensive drying by vacuum or heat to remove the water. For polymer latexes produced in CO_2, there are no VOC emissions and the energy-intensive drying step can be significantly reduced since the CO_2 has a much lower heat of vaporization in the liquid state and, in fact, has a zero heat of vaporization in the supercritical state. Additionally, the polymer can be shipped dry at 100% solids, thus saving energy and money in shipping the heavy water latex. With carefully designed surfactants, the polymer particles can be redispersed in water once the destination is reached.

In step-growth polymerizations conducted by melt-phase or solid-state polymerization, the glass transition temperature (T_g) depression of the polymer by CO_2 can reduce polymer viscosity, and thus can allow for use of lower reaction temperatures and can facilitate the removal of condenstate (which is essential for attaining high-molecular-weight polymer). Such effects of CO_2 on the step-growth polymerization can result in energy savings in the polymer production process and can lead to better materials with higher color quality because of the lower temperatures associated with CO_2-mediated processes.

Most types of polymerizations can be performed in liquid and supercritical CO_2. The two major types of polymerizations, chain-growth and step-growth, have been demonstrated in CO_2. Reviews in the literature (Canelas and DeSimone, 1997b; Kendall et al., 1999) have described numerous polymerizations in CO_2, many of which will not be discussed in this chapter. Since only amorphous or low-melting fluoropolymers and silicones show appreciable solubility at relatively mild temperatures and pressures ($T < 100\,°C$, $P < 400$ bar), only these two classes of polymers can be synthesized by a homogeneous polymerization in CO_2. All other types of polymers, including semicrystalline fluoropolymers and lipophilic or hydrophilic polymers, must be made by heterogeneous methods, such as precipitation, dispersion, emulsion, and suspension, since the polymers are insoluble in CO_2 (when $T < 100\,°C$ and $P < 400$ bar). Some semicrystalline fluoropolymers and hydrocarbon polymers can be dissolved at more extreme temperatures and pressures and are discussed in Chapter 7 of this book.

This chapter is divided into three main sections: fluoropolymer synthesis, nonfluorinated polymer synthesis, and polymer characterization in CO_2. In the fluoropolymer synthesis section, solution, precipitation, biphasic, and continuous polymerizations will be described. Several examples of nonfluorinated heterogeneous chain-growth polymer syntheses will be followed by step-growth polymerizations. A brief summary of polymer characterization methods in CO_2 will conclude the chapter.

Fluoropolymer Synthesis

The topic of fluoropolymers synthesized in CO_2 has been reviewed by DeYoung et al., (1999). While amorphous, low-melting fluoropolymers, such as highly fluorinated (meth)acrylates, are soluble in CO_2 and can be polymerized by homogeneous solution polymerizations, the semicrystalline, high-melting fluoropolymers, such as polytetrafluoroethylene (PTFE) and poly(vinylidene fluoride) (PVDF), are insoluble in CO_2 and must be polymerized by heterogeneous processes. Carbon dioxide is a suitable solvent for such polymerizations because there is no chain transfer to CO_2 since there are no hydrogens to abstract. A further advantage to conducting tetrafluoroethylene (TFE) polymerizations in CO_2 is the safety associated with this monomer in CO_2. However, TFE is a dangerous monomer in that it forms explosive mixtures with air, can violently disproportionate in the absence of air, and may form shock-sensitive polymeric peroxides in the presence of oxygen (Gangal, 1988). Fortunately, TFE forms a pseudo-azeotrope with CO_2 and can be safely stored and handled in this manner (Van Bramer et al., 1994).

Highly fluorinated acrylate polymers can be easily synthesized in supercritical CO_2 (DeSimone et al., 1992). High-molecular-weight poly(1,1-dihydroperfluorooctyl acrylate) (PFOA) was prepared at 59.4 °C and 207 bar initiated by azobis(isobutyronitile) (AIBN) (see Scheme 9.1). The polymer was identical to that polymerized in CFCs as confirmed by Fourier transform infrared (FTIR) and NMR analysis. Even high-molecular-weight polymer is soluble in CO_2. It was also shown that statistical copolymers that contain a fluoroacrylate monomer and either methyl methacrylate, styrene, butyl acrylate, or ethylene could be polymerized under homogeneous solution polymerization conditions. The homopolymers of the nonfluorinated monomers are insoluble in CO_2; however, the copolymerizations remained homogeneous in CO_2 even with comonomer incorporations of 50%.

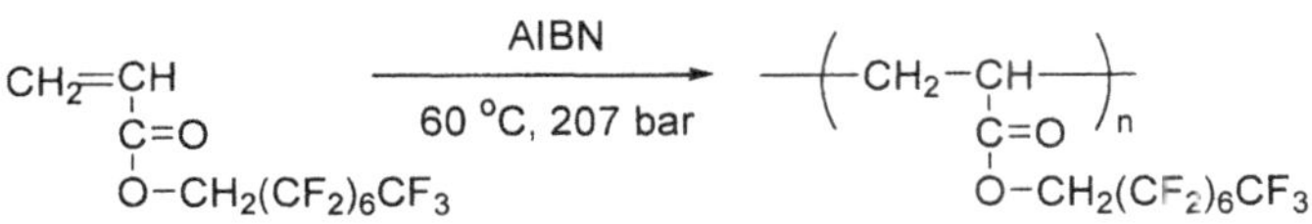

Scheme 9.1. Synthesis of PFOA.

While high-molecular-weight PTFE and PVDF are not soluble in CO_2 because of their crystallinity, low-molecular-weight oligomers are soluble up to some limited molecular weight and can be synthesized by homogeneous telomerization reactions in CO_2 (see Scheme 9.2) (Combes et al., 1994; Romack et al., 1995a). As commercially important intermediates, perfluoroalkyl iodides are used for the synthesis of fluorinated olefins, alcohols, and fluorinated acrylate and methacrylate monomers. The telomerization of TFE can form these important intermediates. Using perfluorobutyl iodide

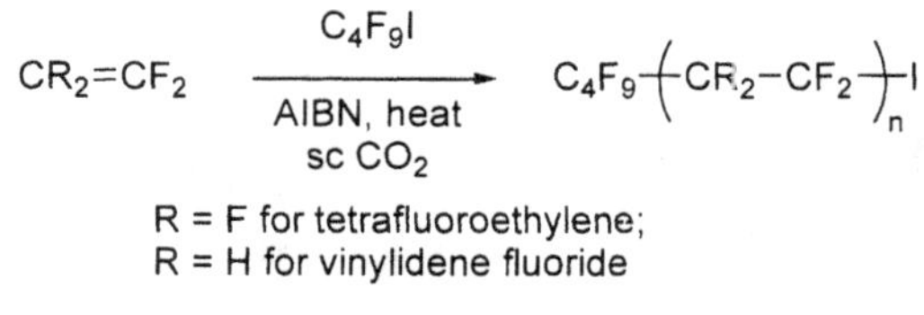

Scheme 9.2. Synthesis of PTFE and PVDF.

as the telogen, TFE was telomerized by AIBN initiation at 68 °C, 345 bar and by thermal initiation at 180 °C, 165–220 bar (Romack et al., 1995a). The number-averaged molecular weights were in the range 620–960 g/mol for AIBN initiation and 570–650 g/mol for thermal initiation. The average chain-transfer constants calculated for both initiation conditions were comparable to previously reported values. The use of CO_2 as a solvent prevents the unwanted chain-transfer reactions that would occur in hydrocarbon solvents. In a related study, telomerizations of vinylidene fluoride by perfluorobutyl iodide initiated by AIBN at 60 °C remained homogeneous throughout the reaction, producing weight-averaged molecular weights of 588–608 g/mol (Combes et al., 1994). A small number of head-to-head addition defects were detected, but analytical techniques confirmed that no CO_2 was incorporated into the product.

The studies of TFE polymerizations in CO_2 were extended to the synthesis of commercially important TFE-based melt-processable copolymers. Such TFE-based polymers synthesized in aqueous dispersions or emulsions have drawbacks as the result of using ionic initiators that produce carboxylic acid end groups, which can lead to polymer decomposition and discoloration at high temperatures. Thus, postpolymerization, high-temperature fluorination is often required to remove these unwanted functionalities. The initiator contribution to the acid end-group problem has been overcome by synthesizing "nonaqueous-grade" TFE-based copolymers in CFC solvents such as 1,1,2-trichlorotrifluoroethane. The recent ban on the use of CFCs has required the development of new solvent systems. Since fluoroolefin-derived radicals readily chain-transfer to hydrogen-containing solvents, only solvents such as perfluorocarbons and some hydrofluorocarbons are suitable alternatives. However, CO_2 has been shown to be an excellent solvent alternative for the synthesis of nonaqueous-grade fluoropolymers. Romack and DeSimone (1995) reported the precipitation polymerization synthesis of TFE copolymers of perfluoro(propyl vinyl ether) (PPVE) and hexafluoropropylene (HFP) (see Scheme 9.3) with compositions and molecular weights consistent with existing commercial materials. One interesting result from these studies was the unexpectedly low levels of acid end groups found in the TFE–PPVE copolymers synthesized in CO_2, 0–3 end groups per 10^6 carbon atoms. These numbers are comparable to copolymers synthesized in CFCs only after high-temperature, postpolymerization fluorination.

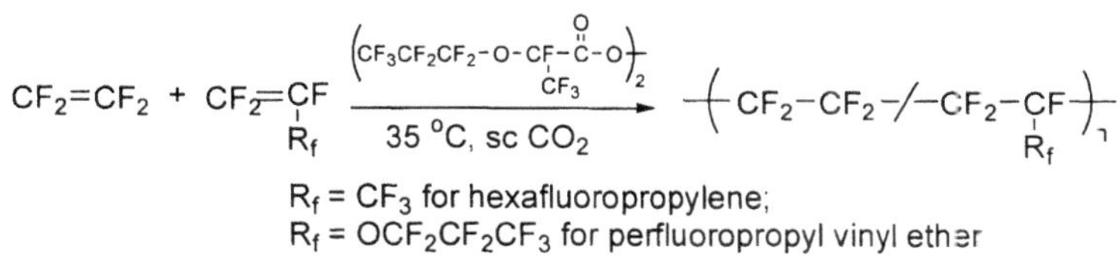

Scheme 9.3. Synthesis of PTFE copolymers.

Typically, PTFE is manufactured by heterogeneous suspension or dispersion polymerization methods. To mimic the commercial processes, the polymerization of TFE in the biphasic mixture of CO_2 and water was studied with and without surfactant using a persulfate initiator (Romack et al., 1995b). These reaction conditions take advantage of the safety issues associated with handling TFE in CO_2, the heat dissipation by the water during the highly exothermic polymerization, and the compartmentalization of the reagents between the water and CO_2 phases. The reaction temperatures ranged from 80 °C to 100 °C and the pressures ranged from 50 bar to 120 bar (50–65 bar for large scale). High-molecular-weight polymer was achieved (> 200 kg/mol). Scanning electron microscopy images showed that the morphology of particles was similar to commercial polymers. The polymerization without surfactant produced aggregates of 0.5–3 μm. This morphology is comparable to that of commercial granular PTFE formed by a suspension polymerization in water. The surfactant-containing polymerizations in the hybrid system produced 100–200 nm particles with intermittent fibrils. This is very similar to the morphology of a commercial sample made by an aqueous dispersion polymerization.

Perfluoropolyethers can also be synthesized in CO_2 (Bunyard et al., 1999). Such polymers have utility as high-performance lubricants and heat-transfer fluids. This class of fluoropolymer is soluble in CO_2. In this study, the perfluoropolyethers were prepared by the photooxidation of HFP in liquid CO_2 at −40 °C (see Scheme 9.4). The composition and molecular weights were comparable to those of perfluoropolyethers synthesized in perfluoro-

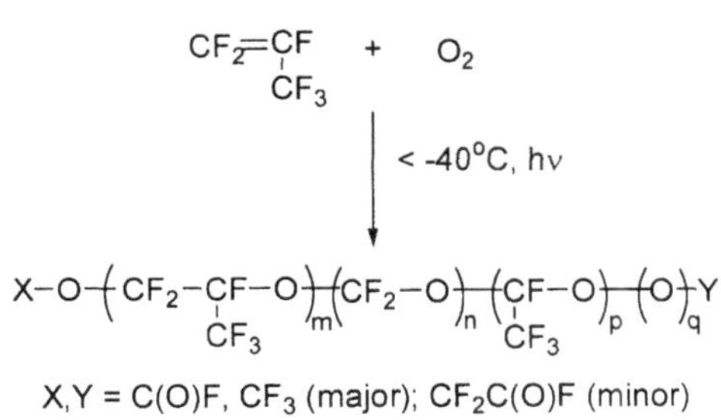

Scheme 9.4. Synthesis of perfluoropolyethers.

cyclobutane. This study showed that CO_2 is a viable solvent replacement for the dichlorodifluoromethane currently used in some commercial fluoroolefin photooxidations.

As in any type of polymerization, a batch reaction is not as commercially attractive as a continuous polymerization process that can produce larger quantities of polymer in the same amount of time. The first continuous polymerizations in CO_2 were reported (Charpentier et al., 1999) with the monomers acrylic acid and vinylidene fluoride. The vinylidene fluoride polymerization was extensively studied at 75 °C, 275 bar. The polymerizations were run with residence times that varied between 15 and 40 min in a continuous-stirred-tank reactor before collection in a filter. The maximum rate of polymerization was determined to be 19×10^{-5} mol $L^{-1}s^{-1}$. Future research will move toward continuous removal of polymer, recycling of unreacted monomer and CO_2, and expansion to other monomers.

Nonfluorinated Polymer Synthesis

Most polymers are essentially insoluble in CO_2 at relatively mild conditions at $T < 100$ °C and $P < 400$ bar—low-melting or amorphous fluoropolymers and silicones are the exceptions. As a result, most nonfluorinated polymers are synthesized in CO_2 by heterogeneous chain-growth or step-growth processes. Since most vinyl monomers and free-radical initiators are soluble in CO_2, precipitation and dispersion polymerizations are the most appropriate heterogeneous chain-growth techniques in CO_2. In a step-growth polymerization, condensate removal is essential for attaining high-molecular-weight polymer. This is achieved either by vacuum or by passing an inert carrier gas through the reaction to remove the condensate. Since CO_2 dissolves some small-molecule condensates (such as phenol) and plasticizes and swells many polymers, this benign solvent may facilitate step-growth polymerizations.

Most dispersion polymerizations in CO_2, including the monomers methyl methacrylate, styrene, and vinyl acetate, have been summarized elsewhere (Canelas and DeSimone, 1997b; Kendall et al., 1999) and will not be covered in this chapter. In a dispersion polymerization, the insoluble polymer is sterically stabilized as colloidal polymer particles by the surfactant that is adsorbed or chemically grafted to the particles. Effective surfactants in the dispersion polymerizations include CO_2-soluble homopolymers, block and random copolymers, and reactive macromonomers. Polymeric surfactants for CO_2 have been designed by combining CO_2-soluble (CO_2-philic) polymers, such as polydimethylsiloxane (PDMS) or PFOA, with CO_2-insoluble (CO_2-phobic) polymers, such as hydrophilic or lipophilic polymers (Betts et al., 1996, 1998; Guan and DeSimone, 1994). Several advances in CO_2-based dispersion polymerizations will be reviewed in the following section.

It was previously reported that the homopolymer surfactant PFOA successfully stabilized poly(methyl methacrylate) (PMMA) dispersion polymerizations (DeSimone et al., 1994; Hsiao et al., 1995), but was not successful for styrene dispersion polymerizations (Canelas et al., 1996). In these styrene polymerizations, the CO_2 pressure used was 204 bar. However, later studies showed that both PFOA and poly(1,1-dihydroperfluorooctyl methacrylate) (PFOMA) could stabilize polystyrene (PS) particles (Shiho and DeSimone, 1999) when a higher pressure was used. These polymerizations were conducted at 370 bar, 65 °C, and the particle size could be varied from 3 to 10 μm by varying the concentration of stabilizer. These homopolymer surfactants are less expensive and easier to synthesize than block copolymer surfactants and provide access to a large range of particle sizes.

A new concept for water-dispersible polymer particles has been developed by Johnston (Yates et al., 1999), in which water-insoluble polymer particles of PMMA are synthesized in CO_2 and then redispersed in water. An "ambidextrous" surfactant was used in which the hydrophilic block acts as the anchor and the PDMS acts as the steric stabilizer in CO_2, while in water the hydrophilic block becomes the electrostatic stabilizer and the PDMS collapses onto the particles. Two block copolymers were investigated for the dispersion polymerization of MMA. The PDMS-*b*-poly(methacrylic acid) surfactant (PDMS = polydimethylsiloxane) (Scheme 9.5A) was less effective during the CO_2 dispersion polymerization, but proved to be the best for redispersing the PMMA particles in water. Up to 10 wt % PMMA was redispersed in water at pH 8 or pH 11 with this surfactant containing approximately nine ionizable groups per surfactant chain. On the other hand, the PDMS-*g*-poly(pyrrolidonecarboxylic acid) (Scheme 9.5B) was more effective in the CO_2 dispersion polymerizations but did not redisperse the PMMA particles in water, containing only approximately two ionizable groups per surfactant chain.

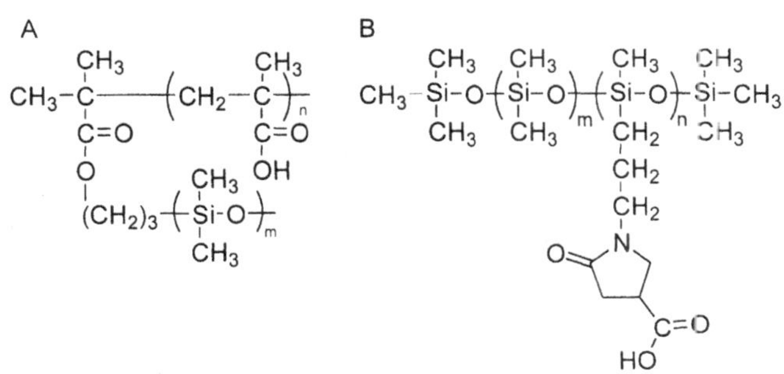

Scheme 9.5. Examples of ambidextrous surfactants.

A report by Cooper et al. (1999a) illustrates the ability to prepare highly cross-linked polymer particles in supercritical CO_2. Cross-linked polymers, due to their insolubility in solvents and water, have a variety of applications ranging from solid-phase synthesis and packing for chromatography to polymer-supported reagents and molecular imprinting. Cooper reported that, in the absence of stabilizer, particles containing a mixture of divinylbenzene (Scheme 9.6A) and ethylvinylbenzene (Scheme 9.6B) (56/43% or 78/20%) precipitated in the form of partially aggregated microspheres in the size range 1–5 μm, possibly due to the rigidity of the cross-linked particles. Other cross-linking monomers, ethylene glycol dimethacrylate (EGDM) (Scheme 9.6C) and trimethylolpropane trimethacrylate (TRM) (Scheme 9.6D), formed random agglomerates of smaller, irregular primary particles. Addition of 0.25 or 1% w/w (based on monomer) of a block copolymer consisting of PMMA and a fluorinated methacrylate block resulted in smaller (0.4–2.5 μm), less aggregated particles, and addition of 3% w/w stabilizer (based on monomer) produced nonaggregated, <0.5 μm sized monodisperse particles. Surface area and porosimetry measurements indicated that the particles were nonporous. In similar studies with the cross-linking monomers TRM and EGDM at high monomer concentrations (40–60%), continuous macroporous polymer monoliths, which conform to the shape of the reaction vessel, were produced in supercritical CO_2 (Cooper and Holmes, 1999). This technique could be particularly useful for the formation of macroporous polymers molded within narrow-bore chromatography capillaries where solvent removal would be difficult for solvents other than CO_2. The pore diameter could be varied from 7880 to 20 nm by varying the monomer concentration and the CO_2 pressure. The copolymerization of methacrylic acid and TRM shows the potential for molecular imprinting monoliths.

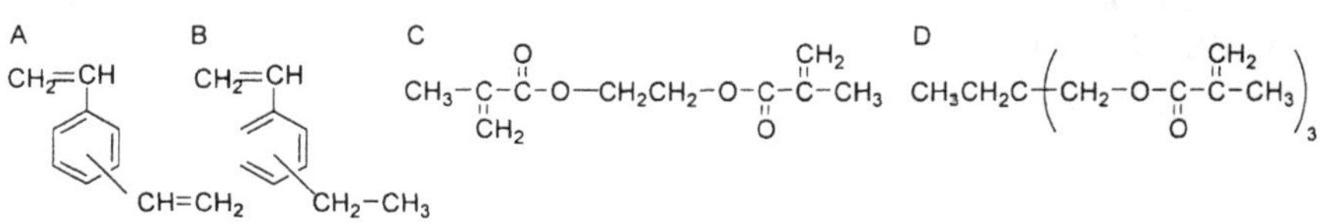

Scheme 9.6. Examples of cross-linking monomers.

"Controlled" free-radical polymerization methods, like atom-transfer radical polymerization (ATRP), can yield polymer chains that have a very narrow molecular-weight distribution and allow the synthesis of block copolymers. In a collaboration between Matyjaszewski and DeSimone (Xia et al., 1999), ATRP was performed in CO_2 for the first time. PFOMA-*b*-PMMA, PFOMA-*b*-PDMAEMA [DMAEMA = 2-(dimethylamino)ethyl methacrylate], and PMMA-*b*-PFOA-*b*-PMMA copolymers were synthesized in CO_2 using Cu(0), CuCl, a functionalized bipyridine ligand, and an alkyl halide initiator. The ATRP method was also conducted as a dispersion polymerization of MMA in CO_2 with PFOA as the stabilizer, generating a kinetically stable colloidal dispersion of high-molecular-weight PMMA with a relatively narrow molecular-weight distribution ($M_w/M_n = 1.41$). Similar dispersion polymerizations in CO_2 of MMA initiated by AIBN result in a broader molecular-weight distribution (polydispersity index, PDI = 2–3) (Hsiao et al., 1995). Future studies will seek to optimize the particle-forming dispersion polymerizations by ATRP in CO_2.

Step-growth polymerizations have been conducted in CO_2 to make polycarbonate (Gross et al., 1998, 1999, 2000). Polycarbonate is a tough, optically clear plastic that has many practical uses. In the synthesis of polycarbonates, one of the two primary methods is an interfacial polymerization that involves the environmentally unfriendly methylene chloride and the highly toxic phosgene. The other method is a melt polymerization, which involves no solvent but has limitations due to the high viscosities and high temperatures of polymerization and processing. Using CO_2 in a melt polymerization would plasticize the polymer and reduce the viscosity, allowing for easy removal of the condensate by increasing the diffusivity of the condensate and allowing for higher molecular weights to be achieved. Furthermore, phenol (the by-product of the melt-phase process) is soluble in sc CO_2. Therefore, CO_2 can be used to more efficiently extract the condensate from the polymer melt by supercritical fluid extraction. In work by Gross et al. (1998), bisphenol A and diphenyl carbonate were polymerized with tetraphenylphosphonium tetraphenylborate catalysis and the condensate was removed by supercritical CO_2 or argon (Scheme 9.7). The polymerizations yielded molecular weights up to 13 kg/mol. Under each reaction condition, higher molecular weight was achieved by supercritical CO_2 extraction as opposed to condensate removal by argon. Swelling experiments confirmed that polycarbonate is highly plasticized by sc CO_2 and showed swelling up to 55% at 235 °C and 350 bar.

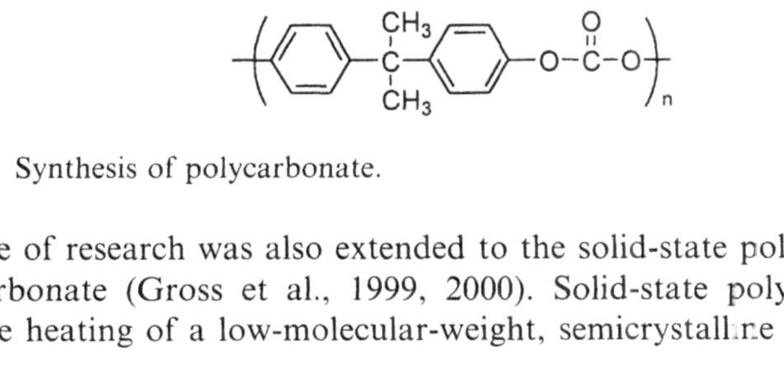

Scheme 9.7. Synthesis of polycarbonate.

This line of research was also extended to the solid-state polymerization of polycarbonate (Gross et al., 1999, 2000). Solid-state polymerizations involve the heating of a low-molecular-weight, semicrystalline step-growth polymer above its T_g but below its T_m to allow for condensate and chain mobility for further chain-extension reactions without particles coalescence. Polycarbonate, unlike poly(ethylene terephthalate), does not undergo thermal crystallization on a reasonable time-scale, so solvent-induced crystallization is necessary for solid-state polymerization to become a reality for polycarbonate. Fortunately, it was shown that thin films of polycarbonate can be crystallized in the presence of supercritical CO_2 (Beckman and Porter, 1987). This was extended to include the crystallization of granules and beads of polycarbonate suitable for solid-state polymerization. Following the synthesis of low-molecular-weight prepolymer polycarbonate beads, CO_2 was used to induce crystallinity of the polymer at elevated temperatures. The crystallized prepolymer beads underwent a solid-state polymerization with condensate removal facilitated by either nitrogen or carbon dioxide. It was found that the polymerization rates in the solid state were always greater in the presence of sc CO_2, as opposed to N_2 under comparable conditions. In the presence of sc CO_2, the T_g of polycarbonate is decreased by over 70 °C. Thus, the enhanced reaction rates could be due to enhanced condensate solubility and diffusivity, along with increased end-group mobility in the amorphous phases of the beads. Therefore, the use of CO_2 as a sweep fluid offers the advantage of allowing the polymerization to be run at much lower temperatures, suppressing the side reactions that lead to discoloration.

Characterization in CO_2

Many analytical techniques have been modified to accommodate high pressure, such as IR, UV-vis, and NMR spectroscopies and scattering techniques. Such high-pressure analyses can be used to probe polymer interactions in CO_2 and polymerization mechanisms. This chapter will focus on scattering techniques, including high-pressure small-angle neutron scattering (SANS), small-angle X-ray scattering (SAXS), and light scattering.

Both SANS and SAXS are powerful tools for examining micelle formation of block copolymers. Both of these scattering techniques can be used to determine the radius of gyration (R_g) and the second virial coefficient (A_2) of both homopolymers and copolymers, as well as the aggregation number, or the number of polymer chains forming a micelle. For example, using SAXS measurements, the aggregation behavior of three different amphiphiles in CO_2 was determined (Fulton et al., 1995). The polymer with a PFOA backbone and poly(ethylene oxide) (PEO) grafts formed micelles that stabilized water, the PTFE-*b*-PEO formed reverse micelles, and the diblock structure $F(CF_2)_{10}(CH_2)_{10}H$ formed small aggregates of about four chains per aggregate. In high-pressure SANS studies of homopolymers (Chillura-Martino et al., 1996; McClain et al., 1996a), PFOA ($14 < 10^3$ $M_w < 1000$) was determined to have a positive A_2 value, which increases as molecular weight decreases, indicating that CO_2 is a good solvent for this polymer. It was also shown that at 65 °C and 340 bar, CO_2 is a theta solvent for the CO_2-philic polymer poly(hexafluoropropylene oxide) ($M_w = 1.3 \times 10^4$ g/mol), by the A_2 value of zero, while CO_2 appears to be a poor solvent for polydimethylsiloxane (PDMS; $M_w = 1.3 \times 10^4$ g/mol), causing aggregation of chains (Chillura-Martino et al., 1996). Later SANS experiments show that PDMS has a temperature- and pressure-induced theta point (Melnichenko et al., 1999). The theta temperature at a constant CO_2 density of 0.95 g/cm^3 is 65 ± 5 °C and the theta pressure at 50 °C is 520 ± 40 bar. At conditions above the theta point, CO_2 becomes a good solvent for PDMS.

It was shown by McClain et al. that a series of PS-*b*-PFOA copolymers studied by SANS and SAXS form polydisperse spherical core–shell micelles, where PS forms the spherical core and PFOA forms the corona, or shell, swollen with CO_2 (Londono et al., 1997; McClain et al., 1996b). In most cases, the core radius was ~ 25 Å, the shell radius was ~ 85 Å, and the aggregation number was ~ 7. With an increase in CO_2 density implemented by decreasing the temperature, the polydispersity increased, likely from breaking a collection of aggregates of low polydispersity into a collection of smaller aggregates of higher polydispersity.

High-pressure static and dynamic light scattering were used to closely examine the behavior of block copolymers of poly(vinyl acetate) (PVAc) and poly(1,1,2,2-tetrahydroperfluoroalkyl acrylate (PTAN) as a function of CO_2 density (Buhler et al., 1998). The phase diagram for PVAc-*b*-TAN shows three distinct phases as a function of polymer concentration and CO_2 density at a fixed temperature of 45 °C (see Figure 9.1). The block copolymer forms a precipitated phase at low CO_2 densities, spherical micelles at intermediate CO_2 densities, and unimers, or free polymer chains in solution, at high densities. The micelles-to-unimer transition was found to be very

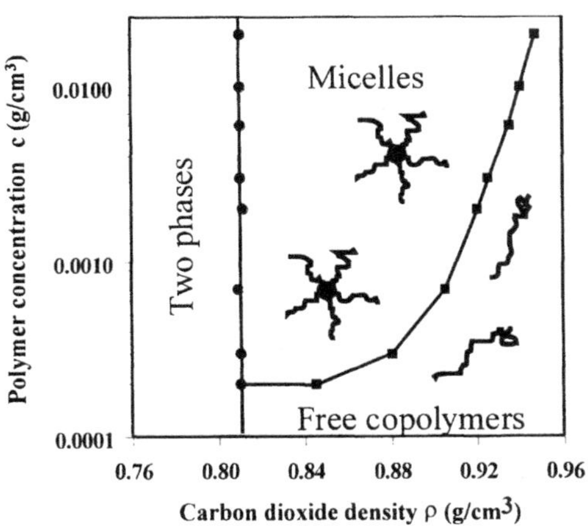

Figure 9.1. The phase diagram for PVAc-*b*-TAN determined by light scattering at 45 °C.

sharp. The aggregation number was dependent on CO_2 density or solvent quality, with a value as high as 120 at lower density, decreasing to 1, representing unimers. The hydrodynamic size of micelles, however, was independent of CO_2 density, because of a balance between the free energy of the corona and the interfacial free energy.

By in situ analysis of dispersion polymerizations in CO_2, Johnston's group has gained some insight into the mechanism of the polymerizations. The steric stabilization and flocculation of poly(2-ethylhexyl acrylate) stabilized by homopolymer or block copolymer surfactants was studied by turbidimetry and tensiometry (O'Neill et al., 1997) and by dynamic light scattering (Yates et al., 1997). The results confirmed that the critical flocculation density occurred at the theta-point density of the surfactant in CO_2. Dispersion polymerizations of vinyl acetate with homopolymer or block copolymer surfactants were studied by turbidimetry (Canelas et al., 1998). This method for measuring the polymer particle size is particularly useful since the particle shape of this low-T_g polymer is lost during the removal of CO_2. The dispersion polymerization of MMA stabilized by PDMS-monomethacrylate has been the subject of several investigations. In the study of the particle-formation mechanism by turbidimetry, coagulative nucleation and controlled coagulation were seen (O'Neill et al., 1998b). The particle growth stage was also studied (O'Neill et al., 1998a). The results of these two studies agreed with other experimental studies (Shaffer et al., 1996) and with the conventional dispersion polymerization model of Paine (1990).

Conclusions

Clearly, CO_2 is a suitable, inert solvent for many kinds of polymerization methods, ranging from chain-growth to step-growth polymerizations, from solution to heterogeneous polymerizations, and from fluoropolymers to hydrocarbon polymers. Depending on the type of polymerization, numerous environmental advantages can be realized with the use of CO_2. Despite the need for high-pressure equipment and the limited solubility of most polymers, CO_2 can be an important solvent alternative in the fight for pollution prevention and energy efficiency.

References

Beckman, E.; Porter, R. S. *J. Polym. Sci, Part B: Polym. Phys.* **1987**, *25*, 151.

Betts, D. E.; McClain, J. B.; DeSimone, J. M. The Importance of Surfactants for Polymerizations in Carbon Dioxide. In *High Pressure Chemical Engineering*, Process Technology Proceedings; von Rohr, P. R., Trepp, C., Eds.; Elsevier Science: New York, 1996; Vol. 12, pp. 23–30.

Betts, D. E.; Johnson, T.; LeRoux, D.; DeSimone, J. M. Controlled Radical Polymerization Methods for the Synthesis of Nonionic Surfactants for CO_2.

In *Controlled Radical Polymerization*; Matyjaszewski, K., Ed.; ACS, Washington, DC, 1998; Vol. 685, p. 418.

Buhler, E.; Dobrynin, A. V.; DeSimone, J. M.; Rubinstein, M. Light-Scattering Study of Diblock Copolymers in Supercritical Carbon Dioxide: CO_2 Density-Induced Micellization Transition. *Macromolecules* **1998**, *31*, 7347–7355

Bunyard, W. C.; Romack, T. J.; DeSimone, J. M. Perfluoropolyether Synthesis in Liquid Carbon Dioxide by Hexafluoropropylene Photooxidation. *Macromolecules* **1999**, *32*, 8224–8226.

Canelas, D. A.; DeSimone, J. M. Polymerizations in Liquid and Supercritical Carbon Dioxide. *Adv. Polym. Sci.* **1997b**, *133*, 103–140.

Canelas, D. A.; Betts, D. E.; DeSimone, J. M. Dispersion Polymerization of Styrene in Supercritical Carbon Dioxide: Importance of Effective Surfactants. *Macromolecules* **1996**, *29*, 2818–2821.

Canelas, D. A.; Betts, D. E.; DeSimone, J. M.; Yates, M. Z.; Johnston, K. P. Poly(vinyl acetate) and Poly(vinyl acetate-*co*-ethylene) Latexes via Dispersion Polymerizations in Carbon Dioxide. *Macromolecules* **1998**, *31*, 6794–6805.

Charpentier, P. A; Kennedy, K. A.; DeSimone, J. M.; Roberts, G. W. Continuous Polymerizations in Supercritical Carbon Dioxide: Chain-Growth Precipitation Polymerizations. *Macromolecules* **1999**, *32*, 5973–5975.

Chillura-Martino, D.; Triolo, R.; McClain, J. B.; Combes, J. R.; Betts, D. E.; Canelas, D. A.; Samulski, E. T.; DeSimone, J. M.; Cochran, H. D.; Londono, J. D.; Wignall, G. D. Neutron Scattering Characterization of Homopolymers and Graft-Copolymer Micelles in Supercritical Carbon Dioxide. *J. Mol. Struct.* **1996**, *383*, 3–10.

Combes, J. R.; Guan, Z.; DeSimone, J. M. Homogeneous Free-Radical Polymerizations in Carbon Dioxide. 3. Telomerization of 1,1-Difluoroethylene in Supercritical Carbon Dioxide. *Macromolecules* **1994**, *27*, 865–867.

Cooper, A. I.; Holmes, A. B. Synthesis of Molded Monolithic Porous Polymers Using Supercritical Carbon Dioxide as the Porogenic Solvent. *Adv Mater.* **1999**, *11*, 1270.

Cooper, A. I.; Hems, W. P.; Holmes, A. B. Synthesis of Highly Cross-Linked Polymers in Supercritical Carbon Dioxide by Heterogeneous Polymerization. *Macromolecules* **1999**, *32*, 2156–2166.

DeSimone, J. M.; Guan, Z.; Eisbernd, C. S. Synthesis of Fluoropolymers in Supercritical Carbon Dioxide. *Science* **1992**, *257*, 945–947.

DeSimone, J. M.; Maury, E. E.; Menceloglu, Y. Z.; McClain, J. B.; Romack, T. R.; Combes, J. R. Dispersion Polymerizations in Supercritical Carbon Dioxide. *Science* **1994**, *265*, 356–359.

DeYoung, J. P.; Romack, T. J.; DeSimone, J. M. Synthesis of Fluoropolymers in Liquid and Supercritical Carbon Dioxide Solvent Systems. In *Fluoropolymers 1: Synthesis*; Hougham, G. et al., Eds.; Plenum Press: New York, 1999; pp. 191–205.

Fulton, J. L.; Pfund, D. M.; McClain, J. B.; Romack, T. J.; Maury, E. E.; Combes, J. R.; Samulski, E. T.; DeSimone, J. M.; Capel, M. Aggregation of Amphiphilic Molecules in Supercritical Carbon Dioxide: A Small Angle X-ray Scattering Study. *Langmuir* **1995**, *11*, 4241–4249.

Gangal, S. V. In *Encyclopedia of Polymer Science and Engineering*, 2nd Ed.; Wiley-Interscience: New York, 1988; Vol. 16, p. 577.

Gross, S. M.; Givens, R. D.; Jikei, M.; Royer, J. R.; Khan, S.; DeSimone, J. M.; Odell, P. G.; Hamer, G. K. Synthesis and Swelling of Poly(bisphenol) A carbonate) Using Supercritical CO_2. *Macromolecules* **1998**, *31*, 9090–9092.

Gross, S. M.; Flowers, D.; Roberts, G.; Kiserow, D. J.; DeSimone, J. M. Solid-State Polymerization of Polycarbonates Using Supercritical CO_2. *Macromolecules* **1999**, *32*, 3167–3169.

Gross, S. M.; Roberts, G. W.; Kiserow, D. J.; DeSimone, J. M. Crystallization and Solid State Polymerization of Poly(bisphenol A carbonate) Facilitated by Supercritical CO_2. *Macromolecules* **2000**, *33*, 40–45.

Guan, Z.; DeSimone, J. M. Fluorocarbon-Based Heterophase Polymeric Materials. 1. Block Copolymer Surfactants for Carbon Dioxide Applications. *Macromolecules* **1994**, *27*, 5527–5532.

Hsiao, Y.-L.; Maury, E. E.; DeSimone, J. M.; Mawson, S. M.; Johnston, K. P. Dispersion Polymerization of Methyl Methacrylate Stabilized with Poly(1, 1-dihydroperfluorooctyl acrylate) in Supercritical Carbon Dioxide. *Macromolecules* **1995**, *28*, 8159–8166.

Kendall, J. L.; Canelas, D. A.; Young, J. L.; DeSimone, J. M. Polymerizations in Supercritical Carbon Dioxide. *Chem. Rev.* **1999**, *99*, 543–563.

Londono, J. D.; Dharmapurikar, R.; Cochran, H. D.; Wignall, G. D.; McClain, J. B.; Betts, D. E.; Canelas, D. A.; DeSimone, J. M.; Samulski, E. T.; Chillura-Martino, D.; Triolo, R. The Morphology of Block Copolymer Micelles in Supercritical Carbon Dioxide by Small-Angle Neutron and X-ray Scattering. *J. Appl. Crystallogr.* **1997**, *30*, 690–695.

McClain, J. B.; Londono, D.; Combes, J. R.; Romack, T. J.; Canelas, D. A.; Betts, D. E.; Wignall, G. D.; Samulski, E. T.; DeSimone, J. M. Solution Properties of a CO_2-Soluble Fluoropolymer via Small Angle Neutron Scattering. *J. Am. Chem. Soc.* **1996a**, *118*, 917–918.

McClain, J. B.; Betts, D. E.; Canelas, D. A.; Samulski, E. T.; DeSimone, J. M.; Londono, J. D.; Cochran, H. D.; Wignall, G. D.; Chillura-Martino, D.; Triolo, R. Design of Nonionic Surfactants for Supercritical Carbon Dioxide. *Science* **1996b**, *274*, 2049–2052.

Melnichenko, Y. B.; Kiran, E.; Wignall, G. D.; Heath, K. D.; Salaniwal, S.; Cochran, H. D.; Stamm, M. Pressure- and Temperature-Induced Transitions in Solutions of Poly(dimethylsiloxane) in Supercritical Carbon Dioxide. *Macromolecules* **1999**, *32*, 5344–5347.

O'Neill, M. L.; Yates, M. Z.; Harrison, K. L.; Johnston, K. P.; Canelas, D. A.; Betts, D. E.; DeSimone, J. M.; Wilkinson, S. P. Emulsion Stabilization and Flocculation in CO_2. 1. Turbidimetry and Tensiometry. *Macromolecules* **1997**, *30*, 5050–5059.

O'Neill, M. L.; Yates, M. Z.; Johnston, K. P.; Smith, C. D.; Wilkinson, S. P. Dispersion Polymerization in Supercritical CO_2 with a Siloxane-Based Macromonomer: The Particle Growth Regime. *Macromolecules* **1998a**, *31*, 2838–2847.

O'Neill, M. L.; Yates, M. Z.; Johnston, K. P.; Smith, C. D.; Wilkinson, S. P. Dispersion Polymerization in Supercritical CO_2 with Siloxane-Based Macromonomer. 2. The Particle Formation Regime. *Macromolecules* **1998b**, *31*, 2848–2856.

Paine, A. J. Dispersion Polymerization of Styrene in Polar Solvents. 7. A Simple Mechanistic Model to Predict Particle Size. *Macromolecules* **1990**, *23*, 3109–3117.

Romack, T. J.; DeSimone, J. M. Synthesis of Tetrafluoroethylene-Based, Non-aqueous Fluoropolymers in Supercritical Carbon Dioxide. *Macromolecules* **1995**, *28*, 8429–8431.

Romack, T. J.; Combes, J. R.; DeSimone, J. M. Free-Radical Telomerization of Tetrafluoroethylene in Supercritical Carbon Dioxide. *Macromolecules* **1995a**, *28*, 1724–1726.

Romack, T. J.; Kipp, B. E.; DeSimone, J. M. Polymerization of Tetrafluoroethylene in a Hybrid Carbon Dioxide/Aqueous Medium. *Macromolecules* **1995b**, *28*, 8432–8434.

Shaffer, K. A.; Jones, T. A.; Canelas, D. A.; DeSimone, J. M.; Wilkinson, S. P. Dispersion Polymerizations in Carbon Dioxide Using Siloxane-Based Stabilizers. *Macromolecules* **1996**, *29*, 2704–2706.

Shiho, H.; DeSimone, J. M. Preparation of Micron-Size Polystyrene Particles in Supercritical Carbon Dioxide. *J. Polym. Sci., Part A: Polym. Chem.* **1999**, *37*, 2429–2437.

Van Bramer, D. J.; Shiflett, M. B.; Yokozeki, A. U.S. Patent 5 345 013, 1994.

Xia, J. H.; Johnson, T.; Gaynor, S. G.; Matyjaszewski, K.; DeSimone, J. Atom Transfer Radical Polymerization in Supercritical Carbon Dioxide. *Macromolecules* **1999**, *32*, 4802–4805.

Yates, M. Z.; O'Neill, M. L.; Johnston, K. P.; Webber, S.; Canelas, D. A.; Betts, D. E.; DeSimone, J. M. Emulsion Stabilization and Flocculation in CO_2. 2. Dynamic Light Scattering. *Macromolecules* **1997**, *30*, 5060–5067.

Yates, M. Z.; Li, G.; Shim, J. J.; Maniar, S.; Johnston, K. P.; Lim, K. T.; Webber, S. Ambidextrous Surfactants for Water-Dispersible Polymer Powders from Dispersion Polymerization in Supercritical CO_2. *Macromolecules* **1999**, *32*, 1018–1026.

10

Preparation and Studies of Polymer/Polymer Composites Prepared Using Supercritical Carbon Dioxide

EDWARD KUNG
ALAN J. LESSER
THOMAS J. McCARTHY

Because of the recent emphasis on green chemistry, there has been interest in using supercritical carbon dioxide (sc CO_2) as a solvent or swelling agent to aid in polymer processing and polymer chemistry (Adamsky and Beckman, 1994; DeSimone et al., 1992; Hayes and McCarthy, 1998; Kung et al., 1998; Mistele et al., 1996; Romack et al., 1995; Watkins and McCarthy, 1995). Supercritical CO_2 is a very weak solvent for most polymers (some fluoropolymers and silicones are exceptions); however, it swells most polymers and dissolves many small molecules (Berens and Huvard, 1989). The density of a supercritical fluid (SCF), and thus its solvent strength, is continuously tunable as a function of temperature or pressure up to liquidlike values. This provides the ability to control the degree of swelling in a polymer as well as the partitioning of small-molecule penetrants between a swollen polymer phase and the fluid phase. The low viscosity and zero surface tension of SCFs allows for fast transfer of penetrants into swollen polymers. The lack of vapor/liquid coexistance in SCFs allows the sorption to proceed without the penetrant solution wetting the substrate surface. Since most of the common SCFs are gases at ambient conditions, the removal and recovery of the solvent from the final product is extremely facile. All of these factors aid in a new method we have developed for preparing polymer composites.

This method involves the absorption of a supercritical solution of a monomer, initiator, and CO_2 into a solid polymer substrate and subsequent thermal polymerization of the monomer to yield a composite system of the two polymers. We have focused on radical polymerization of styrene within various

Table 10.1. Polymer–Polystyrene Composites

Polymer	Polystyrene (wt %)
High-density polyethylene	5–150
Poly(chlorotrifluoroethylene)	5–75
Poly(4-methyl-1-pentene)	2–300
Nylon 66	1–30
Poly(oxymethylene)	2–35
Poly(tetrafluoroethylene)	1–10
Poly(tetrafluoroethylene-co-hexafluoropropylene)	3–40

solid semicrystalline polymer substrates (Hayes and McCarthy, 1998; Kung et al., 1998; Watkins and McCarthy, 1995). Table 10.1 lists a number of systems that we have studied to make polymer–polystyrene composites.

The method for preparing the polymer blends listed in Table 10.1 involves the soaking of the substrate polymer in a supercritical solution of styrene, a thermal radical initiator, and CO_2 at a temperature where the initiator decomposes very slowly (half-lives of hundreds of hours). The system is then heated to initiate polymerization and held at that temperature for various durations (in contact with the solution) and then vented. We emphasize several generalities of this process. The polymerizations are carried out below the melting transition of the polymers. Polystyrene is immiscible with these polymers and insoluble in sc CO_2; it precipitates upon formation. The SC solution penetrates only the amorphous regions of the polymers and polymerization is confined to these regions. The nascent polystyrene does not undergo large-scale phase separation; the crystalline domains frustrate this. Polystyrene contents can be much greater than the equilibrium solubility of styrene in the sc CO_2-swollen polymers; as polymerization occurs, styrene re-equilibrates into the substrate. Polymerization is largely confined to the substrate, but some polymerization occurs in the fluid phase. The polymer formed in the substrates is of much higher molecular weight than that formed in the fluid phase.

We focus here on the polystyrene/high-density polyethylene (HDPE) system. We have studied this system in greater detail than any other and describe here the phase behavior of this system, the blend synthesis, and some mechanical properties of the composites.

Experimental Section

Materials

Coleman-grade (99.99% pure) CO_2 was purified by flowing through columns packed with activated alumina and a copper catalyst (Engelhard Q-5) to remove water and oxygen, respectively.

The HDPE samples were compression-molded plaques (1.25 mm thick) prepared from Dow 04452N HDPE pellets. Styrene was vacuum distilled from calcium hydride, and ethylbenzene and *tert*-butyl perbenzoate (TBPB) were used as received (Aldrich).

Phase Behavior

We determined the phase behavior of the HDPE/styrene/CO_2 using the method described by Berens et al. (1992), modeling mass uptake data as Fickian diffusion into a planar sheet (Crank, 1975). Ethylbenzene was used as the penetrant to model styrene.

Blend Preparation

An HDPE sample together with aliquots of TBPB/styrene solution (0.3 mol % TBPB) were sealed in stainless-steel-high pressure vessels that were purged with CO_2 gas, tared, and heated to the soak temperature of 80 °C. A heated high-pressure CO_2 manifold was used to pressurize the vessel to 243 bar. The vessel was then weighed to determine the mass of CO_2 transferred and then heated for 5 h at 80 °C. The styrene in CO_2 concentration was 28 wt %. After this initial soaking period, the vessel was heated to 100 °C and maintained at this temperature for the desired reaction time. The vessel was then vented, refilled with nitrogen, and maintained at 100 °C for 6 h.

Characterization

The compositions of the composites were determined simply by weighing the tared solid sample. Melting endotherms and glass transition temperatures were determined using a TA Instruments 2910 modulated differential scanning calorimeter (DSC) operated with a 3 °C/min ramp rate, a 0.75 °C oscillation amplitude, and a 60-s oscillation period.

Phase morphology was determined by cryomicrotoming (at −120 °C) a sample to produce flat and featureless surfaces and then etching using a solution of potassium permanganate, sulfuric acid, and orthophosphoric acid. The etched samples were gold coated (~ 200 Å) and examined under a JEOL 35CF SEM in the secondary electron imaging (SEI) mode.

Molecular weights of the polystyrene were determined using a Polymer Laboratories gel permeation chromatograph (GPC) with tetrahydrofuran (THF) as the mobile phase; the values reported are relative to narrow molecular-weight polystyrene standards. The polystyrene and polyethylene were separated by dissolving the blends in xylenes at 120 °C and precipitating the polyethylene in a 2:1 (vol/vol) mixture of acetone and cyclohexane. After filtration, polystyrene was isolated from the resulting solution by rotary evaporation.

Mechanical tests were carried out with an Instron 1123 mechanical test machine operated at a crosshead speed of 2 mm/min. Moduli were determined using rectangular bar specimens that were pulled in tension using an extensometer to obtain accurate strain measurements. Initial slopes of the stress–strain curves represent the moduli. Strength measurements were made using ASTM Type V tensile bars (cut after the composites were produced) that were pulled in tension, and the maximum tensile stresses attained were taken as the strength values.

Results and Discussion

Absorption Kinetics/Phase Behavior

All experiments in this study were carried out under conditions where CO_2 and styrene are miscible. The solubilities of CO_2 and ethylbenzene (a model for styrene) in HDPE were determined at 80 °C and 243 bar. The HDPE samples were immersed in either pure CO_2 or a 36 wt % ethylbenzene/CO_2 solution within pressure vessels under these conditions for various times. Figure 10.1 shows results of a typical desorption experiment to determine the mass uptake of ethylbenzene for a given soak time; the equilibrium mass uptake was found to be $\sim$ 4% and this was reached after approximately 5 h. Figure 10.2 illustrates the mass uptakes as a function of soak time; the diffusivity of ethylbenzene in CO_2-swollen HDPE under these conditions was calculated by curve fitting to be 9.23×10^{-7} cm^2/s. Attempts to determine the equilibrium mass uptake of neat ethylbenzene in HDPE at 80 °C failed because ethylbenzene dissolves polyethylene under these conditions.

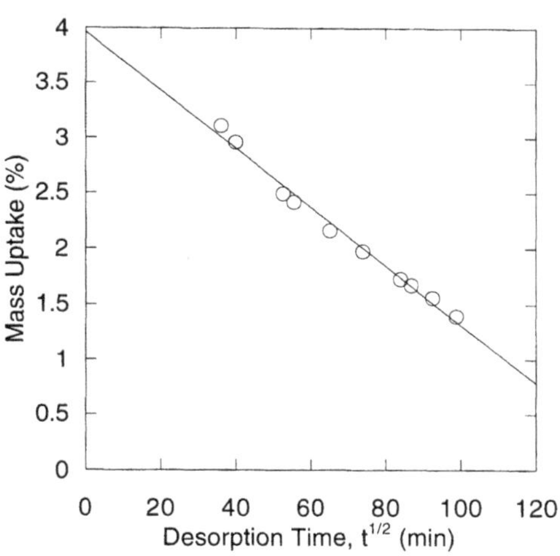

Figure 10.1. An example of an ethylbenzene desorption experiment used to determine the equilibrium mass uptake of ethylbenzene in HDPE. The specimen was soaked in a 36 wt % ethylbenzene/CO_2 solution for 5 h at 80 °C and 243 bar.

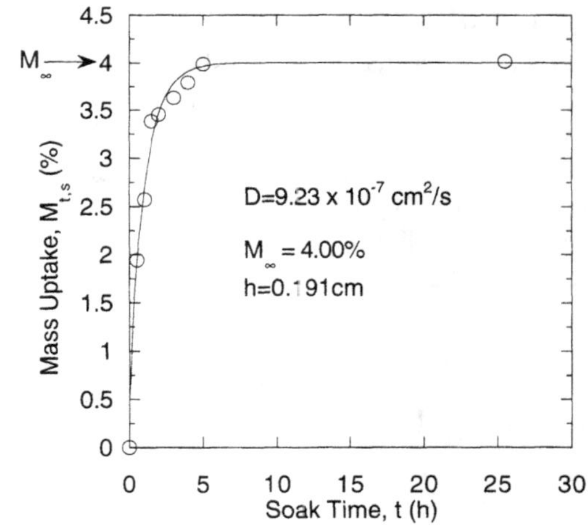

Figure 10.2. Absorption kinetics of ethylbenzene into HDPE for specimens soaked in a 36 wt % ethylbenzene/CO_2 solution at 80 °C and 243 bar.

Neither CO_2 nor ethylbenzene/CO_2 were found to dissolve HDPE or polystyrene under these conditions.

Composite Synthesis

The HDPE samples were soaked in styrene/TBPB/CO_2 solution for 5 h at 80 °C and 243 bar and placed in a 100 °C temperature bath for various times. The vessels were then vented, refilled with nitrogen, and maintained at 100 °C for various durations. All experiments were complete within two half-lives of the initiator. In Figure 10.3, polystyrene incorporation (as wt % of substrate) is plotted as a function of reaction time. Data for two sets of experiments are reported; the size and mass of the HDPE samples (and thus the HDPE:styrene ratio) were different in the two sets of experiments. The experiments run with the lower HDPE concentration yielded higher mass uptakes for a given reaction period. The nonzero intercepts at zero reaction time indicate that polymerization occurs during the soaking period. These reaction conditions permit very high polystyrene incorporation and can produce blends with more polystyrene than HDPE. We note that the mass uptake data reported in Figure 10.3 are significantly higher than the equilibrium uptake of styrene expected as a result of the ethylbenzene sorption experiments. The polymerization of styrene within the HDPE substrate is a heterogeneous process; polystyrene precipitates from the CO_2-swollen amorphous HDPE phase and depletes the matrix of styrene. Styrene in the fluid phase repartitions to make up for this depletion. The nascent solid polystyrene phase must affect this partitioning. The linearity of the data is likely fortuitous and is the result of a combination of the phase behavior and both polymerization and absorption kinetics. Because of the complexity of

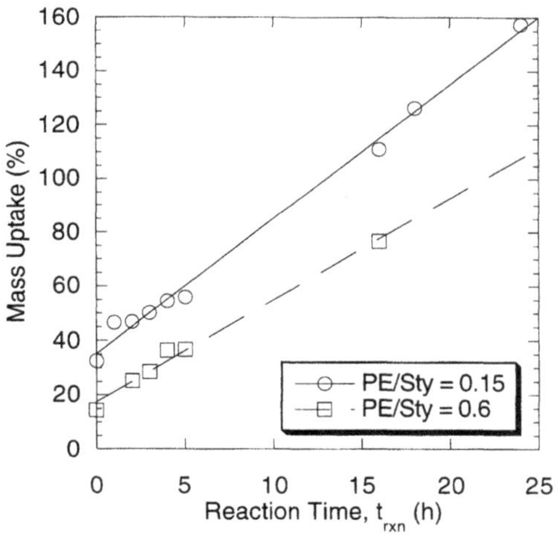

Figure 10.3. Mass uptake of polystyrene into HDPE as a function of reaction time at 100 °C for two concentrations of HDPE. The specimens were soaked in 28 wt % styrene/CO_2 for 5 h at 80 °C and 243 bar prior to the reaction period. The specimens were then postreacted under a nitrogen ambient for 6 h at 100 °C after the reaction period.

the system, how these various properties change over time cannot easily be deconvoluted.

We note that polymerization also occurs outside the substrate. This external polymerization also depletes the fluid phase of styrene and further complicates the system. The polymerization occurring outside the substrate is not nearly as effective as the polymerization occurring within the substrate, as indicated by the difference in the molecular weights of the polystyrene formed in the two types of polymerization. Molecular weight data is presented in the following section.

Characterization

Figure 10.4 shows DSC traces of virgin HDPE and a 43 wt % polystyrene/HDPE blend. The melting endotherms indicate that the crystalline regions of the HDPE substrate are not affected during the blend synthesis. Using a heat of fusion for polyethylene of 292.65 J/g, the unmodified HDPE substrates are 71% crystalline and the 43 wt % polystyrene/HDPE samples are 41% crystalline. The reduced crystallinity is due entirely to dilution by the addition of amorphous polystyrene. The melting temperature of the HDPE also remains unchanged. These results indicate that the styrene polymerization occurs solely within the amorphous regions of the polymer. As the temperatures used in the processes are below the melting point of HDPE, it is unlikely that penetrants (CO_2, monomer, or initiator) enter the crystalline regions. The polystyrene glass transition at 100 °C is faint but discernible in the DSC traces. Wide-angle X-ray diffraction confirms the DSC data; both the positions and breadths of the diffraction peaks for the (110) and (200) planes of crystalline polyethylene are identical for a virgin HDPE specimen and for a 39 wt % polystyrene/HDPE blend specimen.

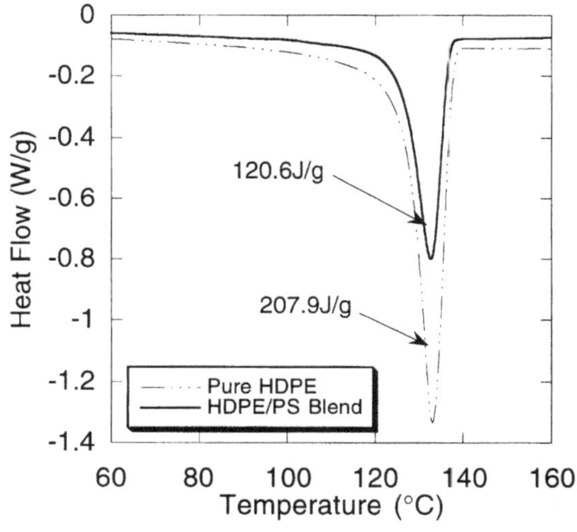

Figure 10.4. Differential scanning colorimeter traces showing the melting endotherms for a pure HDPE sample and a 43 wt % polystyrene/HDPE sample. Integration of the endotherms provides the heat of fusion of the samples.

Figure 10.5 shows scanning electron micrographs of blend samples that were prepared as described in the "Experimental Section". The etchant preferentially attacks polyethylene, producing a topography in which the polystyrene-rich domains are raised above the polyethylene domains. The interlamellar amorphous material provides a location for styrene to penetrate and polymerize. A considerable amount of polystyrene is present in the center of the spherulites. This is due either to amorphous polyethylene that is present in these locations or to voids that develop during crystallization

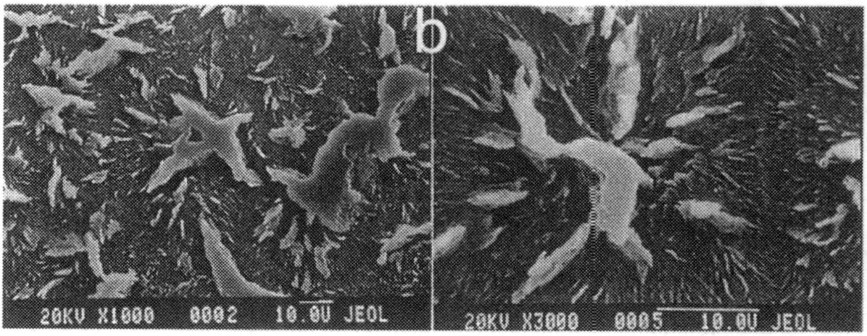

Figure 10.5. Secondary electron micrographs at two magnifications (note the 10-μm bars) of a 43 wt % polystyrene/HDPE specimen etched in potassium permanganate/acid solution. The raised domains are polystyrene-rich.

and/or annealing. As polystyrene domains form, the styrene preferentially partitions there, enhancing growth of these domains. This growth of polystyrene domains appears to break up the spherulites to some degree. A second, finer level morphology may also exist. We also note that the micrographs show no evidence of polystyrene at the spherulite boundaries. This suggests that the polyethylene spherulites in this substrate do not possess distinct boundaries. Similar blends prepared with isotactic polypropylene substrates exhibit polystyrene at the spherulite boundaries.

Some grafting between polystyrene and polyethylene may occur, but we think not. Substantial amounts of polystyrene (but not all) have been extracted from the blend samples by soaking the specimens in refluxing THF for several days. We suspect that if grafting does occur, it is not a significant contributor to polystyrene mass uptake. All the polystyrene could be extracted from a 50 wt % polystyrene/poly(4-methyl-1-pentene) (PMP) blend that was prepared by essentially the same procedure The backbone of PMP (with two tertiary C−H bonds per repeat unit) is likely more susceptible to radical grafting than HDPE.

Gel permeation chromatography of polystyrene separated from blend samples indicates that high-molecular-weight polymer is formed. Polystyrene removed from a 43 wt % polystyrene/HDPE specimen exhibited $M_w = 628$ K with polydispersity index, PDI = 2.7. The corresponding polystyrene formed outside this specimen exhibited $M_w = 48$ K with PDI = 3.1.

Mechanical Tests

Mechanical tests indicate that these blends do not behave like conventional blends and suggest that the polystyrene phase is continuous in the substrate. The moduli of the blends as a function of blend composition is plotted in Figure 10.6. The Voigt and Reuss models are provided for comparison (Nielsen, 1978) These are the theoretical upper and lower bounds, respectively, on composite modulus behavior; our data follows the Voigt model, suggesting that both the polystyrene and polyethylene phases are continuous. In most conventional composites of polystyrene and HDPE, the moduli fall below the Voigt prediction indicating that the phases are discontinuous and dispersed (Barentsen and Heikens, 1973; Wycisk et al., 1990).

Figure 10.7 shows that the tensile strength is improved as polystyrene is incorporated. Data for conventional melt-blended samples (Fayt et al., 1989) are provided for comparison. We note that the ductile-to-brittle transition for our system is shifted toward much higher polystyrene content. Fayt and others have shown that conventionally prepared polyethylene/polystyrene blends are relatively poor materials (Barentsen and Heikens, 1973; Wycisk et al., 1990). Blends of most compositions are weaker than polystyrene or polyethylene homopolymers because of the poor interfacial adhesion between the two immiscible polymers. The electron micrographs and the mechanical data for the blends described here indicate that poly-

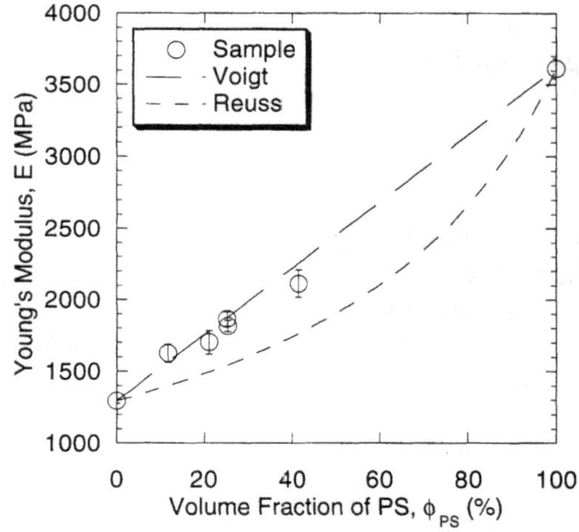

Figure 10.6. Polystyrene/HDPE composite tensile modulus as a function of polystyrene content.

styrene forms a reinforcing scaffold throughout the spherulites that strengthens the entire structure. The adhesion between the phases is high enough to support load transfer.

Conclusions

Supercritical CO_2 aids the infusion of styrene and a radical initiator into HDPE substrates and kinetically trapped (non-equilibrium) composites are prepared by polymerization of the styrene in the substrate. The amount of polystyrene that is incorporated into the HDPE substrates can be controlled

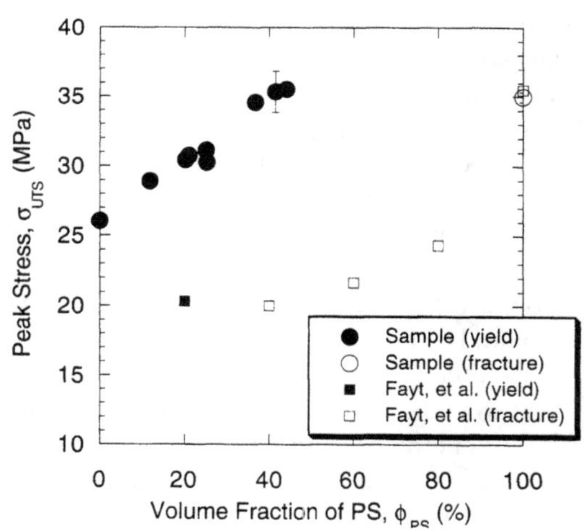

Figure 10.7. Polystyrene/HDPE composite tensile strength as a function of polystyrene content.

with reaction time to values far in excess of the equilibrium uptake of styrene. This process does not affect the crystalline regions of the HDPE. The styrene absorbs into and polymerizes within the amorphous domains of the substrate. Polystyrene also exists in the spherulite centers, implying a certain lack of crystallinity there. The polystyrene thus forms a scaffold that reinforces the polyethylene spherulites. This reinforcement effects efficient modulus enhancement and strength improvement.

Note Address all correspondence to T. J. McCarthy.

References

Adamsky, F. A.; Beckman, E. J. Inverse Emulsion Polymerization of Acrylamide in Supercritical Carbon Dioxide. *Macromolecules* **1994**, *27*, 312.

Barensten, W. M.; Heikens, D. Mechanical Properties of Polystyrene/Low Density Polyethylene Blends. *Polymer* **1973**, *14*, 579.

Berens, A. R.; Huvard, G. S. In *Supercritical Fluid Science and Technology*; Johnston, K. P., Penniger, J. M. L., Eds.; American Chemical Society: New York, 1989; pp. 207–223.

Berens, A. R.; Huvard, G. S.; Korsmeyer, R. W.; Kunig, F. W. Application of Compressed Carbon Dioxide in the Incorporation of Additives into Polymers. *J. Appl. Polym. Sci.* **1992**, *46*, 231.

Crank, J. *The Mathematics of Diffusion*; 2nd ed.; Clarendon Press: Oxford, U.K., 1975.

DeSimone, J. M.; Guan, Z.; Elsbernd, C. S. Synthesis of Fluoropolymers in Supercritical Carbon Dioxide. *Science* **1992**, *257*, 945.

Fayt, R.; Jerome, R.; Teyssie, P. Molecular Design of Multicomponent Polymer Systems. XIV. Control of the Mechanical Properties of Polyethylene–Polystyrene Blends by Block Copolymers. *J. Polym. Sci., Part B: Polym. Phys.* **1989**, *27*, 775.

Hayes, H. J.; McCarthy, T. J. Maleation of Poly(4-Methyl-1-Pentene) Using Supercritical Carbon Dioxide. *Macromolecules* **1998**, *31*, 4813.

Kung, E.; Lesser, A. J.; McCarthy, T. J. Morphology and Mechanical Performance of Polystyrene/Polyethylene Composites Prepared in Supercritical Carbon Dioxide. *Macromolecules* **1998**, *31*, 4160.

Mistele, C. D.; Thorp, H. H.; DeSimone, J. M. Ring-Opening Metathesis Polymerizations in Carbon Dioxide. *J. Macromol. Sci., Pure Appl. Chem.* **1996**, *A33*, 953.

Nielsen, L. E. *Predicting the Properties of Mixtures*; Marcel Dekker: New York, 1978.

Romack, T. J.; Maury, E. E.; DeSimone, J. M. Precipitation Polymerization of Acrylic Acid in Supercritical Carbon Dioxide. *Macromolecules* **1995**, *28*, 912.

Watkins, J. J.; McCarthy, T. J. Polymerization of Styrene in Supercritical CO_2-Swollen Poly(chlorotrifluoroethylene). *Macromolecules* **1995**, *28*, 4067.

Wycisk, R.; Trochimczuk, W. M.; Matys, J. Polyethylene–Polystyrene Blends. *Eur. Polym. J.* **1990**, *26*, 535.

11

Rheological Properties of Polymers Modified with Carbon Dioxide

CHARLES W. MANKE
ESIN GULARI

Use of supercritical fluids (SCFs), particularly supercritical carbon dioxide, as alternative solvents in polymer synthesis and processing is a rapidly growing research area with successful industrial applications (McCoy, 1999). In some cases, the need for alternative solvents is based on environmental concerns, with regulations mandating replacement solvents. An environmentally mandated example is the 1995 ban of the use of chlorofluorocarbons (CFCs) as physical blowing agents in the manufacture of polymeric foams after CFCs were classified as class-I-ozone-depleting substances (ODPs). Among the alternative blowing agents are gases like CO_2 and N_2 and refrigerants such as 1,1-difluoroethane (R152a) and 1,1,1,2-tetrafluoroethane (R134a). Under the foaming conditions, at temperatures above the glass transition temperature of a polymer, and at pressures required for flow of highly viscous polymer melts, these alternative blowing agents are frequently supercritical.

When polymers are formed into final products by various melt-processing techniques, such as extrusion, injection molding, blow molding, foaming, and spin-coating, extremely high melt viscosity presents a major difficulty. A common method to moderate the processing conditions is to add a liquid solvent or plasticizer to the melt. Solvents and plasticizers lower the glass transition temperature, T_g, of the polymer so that the polymer can be made to flow at lower pressures and temperatures. Replacing liquid solvents with SCFs presents unique processing advantages. Higher diffusivity and lower viscosity of SCFs, compared with liquid solvents, increase rates of dissolution and mixing. The properties of polymer–SCF solutions are tunable via pressure or temperature changes, thus allowing efficient downstream separations. Most importantly, dissolution of an SCF produces very large reductions in melt viscosity compared with a liquid solvent dissolved in the melt.

Whether the interest in using SCFs in polymer synthesis and processing is driven by environmental concerns or processing advantages, it is important to understand the rheological behavior of polymer–SCF mixtures. In this chapter, we describe rheological measurements of polymer melts containing dissolved gases for two polymers, polydimethylsiloxane (PDMS) swollen with CO_2 at 50 °C and 80 °C and polystyrene (PS) swollen with CO_2, R152a, and R134a at 150 °C and 175 °C. We illustrate that classical viscoelastic scaling methods can be used to superimpose the concentration-dependent viscosity curves onto the viscosity curve of pure polymer. We predict the viscoelastic scaling factor or the Newtonian viscosity reduction from a modified free-volume theory where the fractional free volumes of pure polymer and polymer–SCF mixtures are determined from thermodynamic data and equation-of-state models.

Experimental Measurements

Rheological measurements for polymeric liquids containing dissolved gases entail all of the usual requirements for polymer-melt rheometry (Dealy and Wissbrun, 1990). These requirements include the establishment of a fully developed viscometric flow, the accurate control of temperature, the accurate measurement of stress or pressure differential, and the characterization of effects, such as non-Newtonian velocity profiles and viscoelastic entry. When dissolved gas or volatile solvents are present, rheological measurements must be conducted under conditions that ensure that the dissolved gas is homogeneously mixed and equilibrated with the polymer melt, and that the liquid remains single-phase during the measurement. Since high pressure is usually required to hold dissolved gas in solution in a polymer melt, the effects of pressure on viscosity must also be considered in interpreting the experimental viscosity measurements on such systems.

High-pressure capillary rheometers are widely used to evaluate viscosity as a function of strain rate for conventional polymer melts (without gas). These instruments establish a viscometric shear flow in a small-diameter capillary, and the shear rate and viscosity are determined from the volumetric flow rate and the pressure drop over the length of the capillary. The use of capillaries with large L/D ratios assures that the pressure drop is predominately due to the shear stress generated by fully developed shear flow, although capillary entry pressures can also be significant for viscoelastic materials. For capillary flow, the effects of non-Newtonian velocity profiles and viscoelastic entry pressure can be characterized by the well-known Rabinowitch and Bagley corrections (Dealy and Wissbrun, 1990). In his studies of the rheology of polystyrene melts containing ethylbenzene, Mendelson (1979, 1980) extended the capillary rheometer method to the measurement of polymer melts that dissolved volatile solvents. The basic modification needed for systems containing volatile components is that the capillary rheometer must be operated as a closed, sealed system, where a well-mixed, equilibrated sample is loaded in the rheometer reservoir prior to the measurement.

Han and Ma (1983a, 1983b) developed another approach to viscosity measurements for polymer melts with dissolved volatile solvents. In their investigations of polymer melts containing volatile blowing agents, they employed an extruder to mix the volatile component with the polymer melt and to provide the pressure force driving flow. Molten material from the extruder is forced through a slit die with an array of pressure transducers along its length. The viscosity and shear rate are determined from the volumetric flow rate and the measured pressure gradient. While this method provides a convenient in-line method for viscosity measurement using industrial processing equipment, two possible drawbacks of this approach are that (1) flow from the slit is discharged to atmospheric pressure, allowing degassing and foaming of the volatile component during flow in the slit, and (2) the blending of the volatile material into the melt in the extruder may not produce a completely homogeneous sample.

The measurements reported here employ a sealed, high-pressure capillary rheometer, shown schematically in Figure 11.1, which is similar to the instrument developed by Mendelson (1979, 1980). Full details of the instrument are given in Gerhardt (1994), Gerhardt et al. (1997), Kwag (1998), and Kwag et al. (1999). Briefly, a standard Instron Capillary Rheometer, Model 3210 (Instron, Canton, MA), is modified with a back-pressure chamber that seals against the outflow end of the capillary. The O-ring piston seals of the standard instrument are also supplemented by a Teflon plug and a viscous liquid seal to prevent gas leakage from the rheometer sample reservoir. With these modifications, the reservoir, capillary, and back-pressure chamber form a completely sealed system that can be pressurized to prevent gas evolution from the sample. Constant volumetric flow in the capillary is achieved by driving the piston at constant displacement speeds, thereby achieving constant shear rate flow, as in a conventional capillary rheometer. The pressure drop across the capillary is determined by the difference between the pressure at the capillary entrance, measured by the force applied to the piston, and the pressure at the capillary exit, measured by a pressure transducer mounted in the back-pressure chamber.

The polymer–gas solution to be tested is prepared separately, and is loaded into the rheometer barrel as a well-mixed, equilibrated sample. The loading method depends on whether the polymer itself is a liquid or a solid at room temperature. For liquid polymers such as PDMS, Gerhard (1994) developed a sealed module that attached to the top of the rheometer barrel, with a side valve though which an equilibrated liquid polymer–gas sample is injected. For solid polymers such as polystyrene, Kwag (1998) found that the diffusion of gas from solid pellets at room temperature was sufficiently slow to allow direct loading of the sample as solid pellets pre-

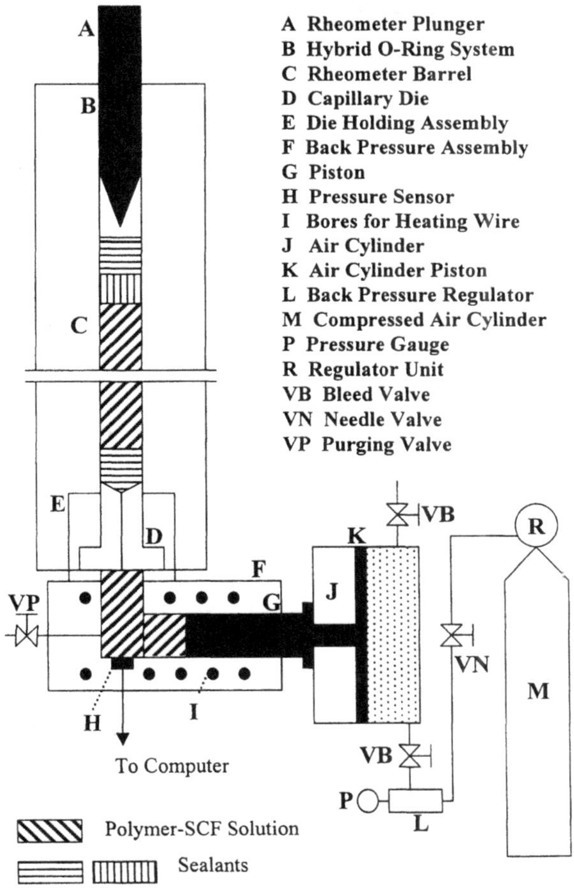

Figure 11.1. Schematic diagram of sealed, high-pressure capillary rheometer with attached back-pressure chamber.

viously equilibrated with gas. Data for the rates of diffusion of various gases from polystyrene pellets at room temperature are shown in Figure 11.2. With certain experimental precautions, such as chilling both the sample and the rheometer barrel, the gas-loaded pellets can be weighed to determine composition and then loaded and sealed in the rheometer without appreciable loss of gas. After loading, the rheometer is heated to the temperature of the viscosity measurement, melting the polymer pellets, and mechanical equilibrium with the back-pressure chamber is established. Flow through the capillary is then initiated to commence the viscosity measurements.

The raw data from the capillary rheometer measurements are subject to four important corrections to obtain true viscosity and shear rate values. These corrections, which are fully described in Kwag (1998) and Kwag et al.

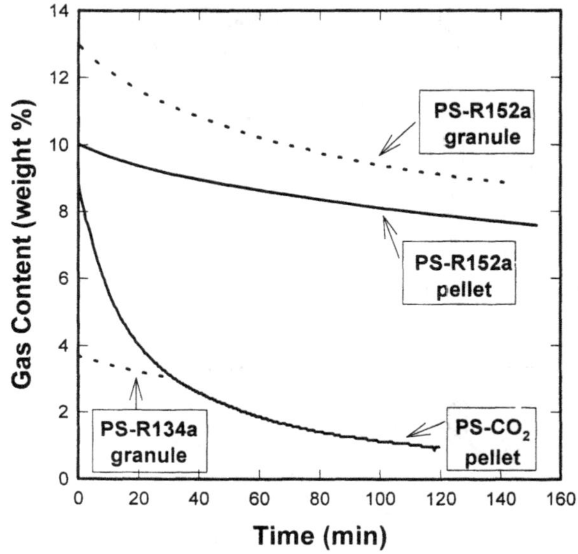

Figure 11.2. Rates of desorption of gases from polystyrene pellets and granules. Samples of solid polystyrene were saturated with gas at high pressure prior to measurement. The sample weights are measured continuously as gas desorbs from the sample at room temperature and atmospheric pressure. The label "pellets" refers to the 1/8 in × 1/8 in diameter polystyrene pellets received from the manufacturer, while "granules" refers to pellets ground to approximately #14 mesh size. Data from Kwag (1998).

(1999), are (1) correction of the load on the rheometer piston to account for friction; (2) Bagley correction of the measured pressure differential to account for the effect of viscoelastic entry pressure; (3) correction for the effect of non-Newtonian velocity profiles, by means of either the Rabinowitsch or Schümmer correction; and (4) correction for the effect of pressure on viscosity. The latter is accomplished by use of an exponential term $\exp\{b(P^* - P)\}$, where b is an empirical constant, to scale viscosity values taken at individual measurement pressures P^* to a single reference pressure P, thereby producing an isobaric set of data.

The systems selected for evaluation are the PDMS–CO_2 system studied by Gerhardt et al. (1997, 1998) and PS–gas systems studied by Kwag et al. (1999). Properties for these systems are listed in Table 11.1. The variation in physical properties between these systems provides a very broad basis for evaluating the rheological properties of polymer–gas systems. The PDMS–CO_2 system exhibits a favorable thermodynamic affinity between the polymer and dissolved gas, and provides the opportunity to evaluate the rheology of melts with very high dissolved gas content (up to 21 wt %). Carbon dioxide is much less soluble in polystyrene than in PDMS, so the PS–CO_2

Table 11.1. Properties of Polymer–Gas Systems

(a) Polymers

Polymer	M_w	T_g (°C)	T of Experiments (°C)
PS	132,000	100	150–175
PDMS	308,000	−128	50–80

(b) Critical Properties of Gases

Gas	T_c (°C)	P_c (MPa)
CO_2	31.1	7.38
R152a	113.5	4.49
R134a	101.1	4.06

system represents the contrasting case of relatively poor thermodynamic affinity between the polymer and gas. Polystyrene and PDMS also exhibit markedly different glass transition temperatures. Thus, the PDMS–CO_2 system represents rheological behavior of a polymer–gas system at temperatures well above T_g of the pure polymer, whereas the measurements on the polystyrene systems were conducted relatively close to T_g.

Rheological Measurements and Viscoelastic Scaling

PDMS–CO_2 System

Capillary viscometer measurements for the PDMS–CO_2 system, performed by Gerhardt et al. (1997), are shown in Figure 11.3a. The viscosity of pure PDMS is shown as a function of shear rate in the top curve (solid triangles). These data represent a typical viscosity curve for a high-molecular-weight linear polymer. Substantial shear-thinning of viscosity is observed as the shear rate increases, and a power-law region of constant slope is approached at high shear rates. At low shear rates, a constant Newtonian viscosity is approached, which is more evident in Figure 11.3b, where supplementary viscosity measurements for pure PDMS taken with a cone-and-plate viscometer are shown as the solid circles. The viscosity curves for PDMS with dissolved CO_2 are similar in shape to the pure PDMS viscosity curve, but overall viscosity values are markedly reduced as CO_2 content increases. Careful inspection of the data in Figure 11.3a also reveals that the positions

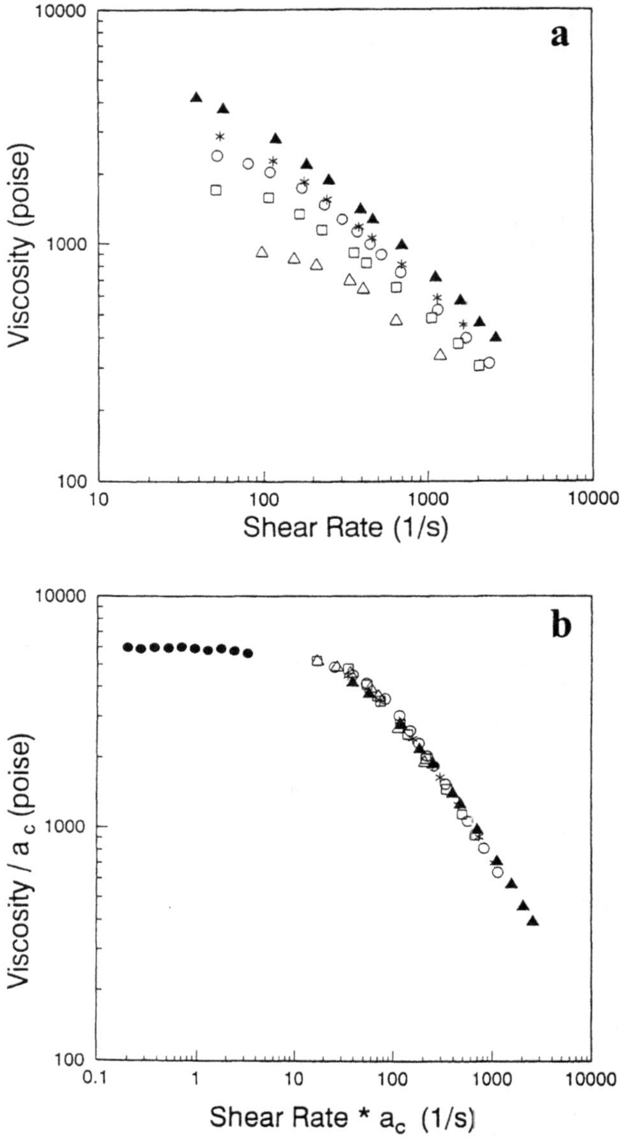

Figure 11.3. (a) Viscosity of PDMS melts swollen with carbon dioxide measured as a function of shear rate and carbon dioxide content at 50 °C. (b) Master viscosity curve produced by applying concentration-dependent scaling factors (see text). ▲ Pure PDMS, * 4.84 wt % CO_2, ○ 9.03 wt % CO_2, □ 14.4 wt % CO_2; △ 20.7 wt % CO_2. Data denoted by ● are Newtonian viscosity measurement for pure PDMS. Data from Gerhardt (1994).

of the viscosity curves appear to be shifted progressively to higher shear rates, relative to pure PDMS, with increasing CO_2 concentration. In particular, we notice the presence of a Newtonian plateau at about 100 s^{-1} for the PDMS melt containing 20.7 wt % CO_2, whereas the comparable Newtonian plateau region occurs below 10 s^{-1} for pure PDMS.

The data displayed in Figure 11.3a are similar in appearance to viscosity curves for a polymer melt measured at different temperatures, with increasing gas content causing an effect very similar to the effect of increasing temperature. This suggests that classical viscoelastic scaling methods, analogous to the principle of time–temperature superposition (Ferry, 1980), may be employed to reduce the concentration-dependent viscosity curves to a single master curve. Gerhardt et al. (1997) have shown that a single concentration-dependent viscoelastic scaling factor, a_c, can be used to produce a master curve of scaled viscosity $\eta(c, \gamma)/a_c$ as a function of scaled shear rate, $a_c\gamma$, as shown in Figure 11.3b. This master viscosity curve is identical to the viscosity curve for the pure polymer. The viscoelastic scaling factor a_c can also be interpreted as the ratio of the low shear rate (Newtonian) viscosity of the polymer–gas mixture, $\eta_0(c)$, to the Newtonian viscosity, $\eta_{p,o}$, of the pure polymer: $a_c = \eta_0(c)/\eta_{p,o}$. Thus, a_c is a measure of the viscosity reduction produced by the dissolved gas.

Polystyrene–Gas Systems

Kwag et al. (1999) have performed capillary viscosity measurements on polystyrene melts with dissolved CO_2, R152a, and R134a. Their data for the PS–CO_2 system at 150 °C are shown in Figure 11.4. The viscosity data displayed in this figure cover several logarithmic decades, thus the reductions in viscosity with gas content seen in this figure are much greater than those observed for the PDMS–CO_2 system, even though the gas content is less (5 wt % for PS–CO_2 vs. 21 wt % for PDMS–CO_2). As in Figure 11.3a, we see that the viscosity curves for PS–CO_2 mixtures exhibit the same shape as the viscosity curve for the pure polymer. Viscosity data for the PS-R152a system at 150 °C follow a similar trend, shown in Figure 11.5, but here even higher viscosity reductions are observed because of the high solubility (up to 10 wt %) of R152a in polystyrene. Data taken at 175 °C, and data for the PS-R134a system exhibit behavior similar to data shown in Figures 11.4 and 11.5. These data are not displayed here, but they are available in Kwag (1998) and Kwag et al. (1999). In all the viscosity measurements on polystyrene systems reported here and in Kwag et al. (1999), the effect of pressure on viscosity is very significant, and the pressure-correction procedure developed by Kwag et al. (1999) is needed to obtain true isobaric viscosity curves from the capillary viscosity data. This correction has been implemented to obtain the curves displayed in Figures 11.4 and 11.5.

The PS–gas systems studied by Kwag (1998) follow the same viscoelastic scaling principle as the PDMS–CO_2 system. Figure 11.6 shows a master

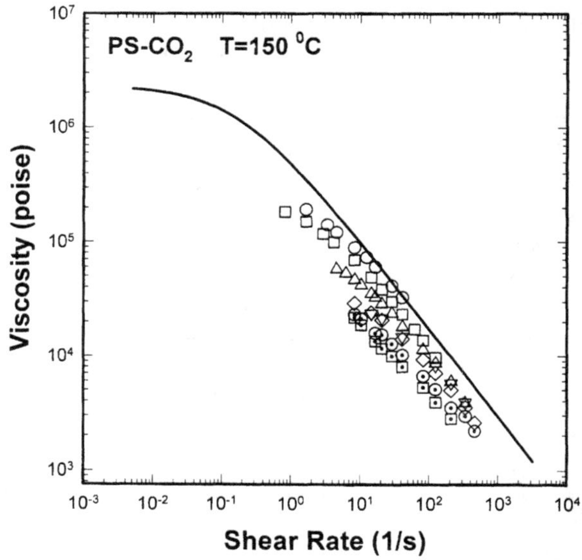

Figure 11.4. Viscosity of polystyrene with dissolved carbon dioxide at 150 °C measured as a function of shear rate. The Schümmer correction and a pressure correction have been applied, as described in the text, to obtain isobaric viscosity curves corresponding to the back-pressures applied during the measurements. Carbon dioxide composition and back-pressure values are ○ 1.0 wt % CO_2, 9.79 MPa; □2.0 wt %, 11.37 MPa; △ 3.0 wt%, 11.98 MPa; ▽3.5 wt %, 8.36 MPa; ◇ 4.5 wt %, 14.84 MPa; ⊙ 5.0 wt %, 12.98 MPa; ⊡5.2 wt %, 12.18 MPa. The solid curve is the viscosity curve for pure polystyrene at 150 °C and 1 atm pressure. Data from Kwag (1998).

curve constructed by scaling viscosity data of Kwag (1998) for all three PS–gas systems at 150 °C and 175 °C. In addition to the compositional scaling factor a_c, a pressure scaling factor a_p and a temperature scaling factor a_T have been used to scale the data to reference conditions of 150 °C and 1 atm pressure. Ideal viscoelastic scaling behavior is observed for all three PS–gas systems, producing a master curve identical to the viscosity curve for the pure polystyrene melt.

Variation of Viscoelastic Scaling Factor a_c with Composition

We have seen in the previous sections that viscoelastic scaling, employing the scaling factor a_c, produces master viscosity curves for polymer–gas solutions that are identical to the master curve for the pure polymer. This means that the effect of dissolved gas on the rheology of polymer melts can be described entirely by the variation of a_c with gas content. We have not, of course, demonstrated that all polymer–gas systems follow this scaling beha-

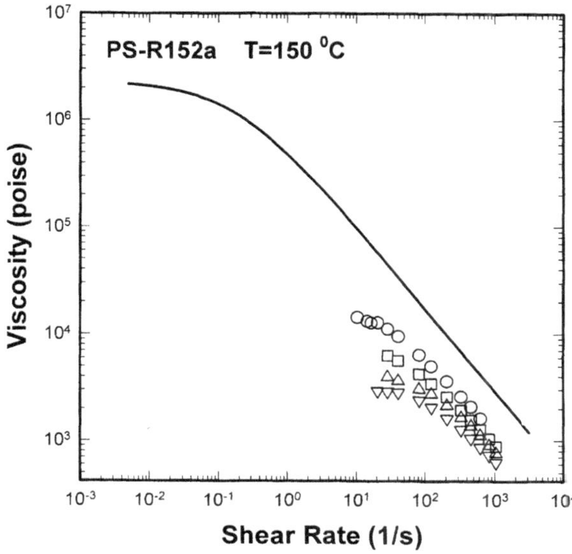

Figure 11.5. Viscosity of polystyrene with dissolved R-152a at 150 °C measured as a function of shear rate. The Schümmer correction and a pressure correction have been applied, as described in the text, to obtain isobaric viscosity curves corresponding to the back-pressures applied during the measurements. The R-152a composition and back-pressure values are: ○ 5.6 wt% R-152a, 12.06 MPa; □ 7.0 wt%, 12.48 MPa; △ 8.3 wt %, 17.18 MPa; ▽ 10.4 wt%, 16.41 MPa. The solid curve is the viscosity curve for pure polystyrene at 150 °C and 1 atm pressure. Data from Kwag (1998).

vior. However, the wide range of gas content, temperatures relative to T_g, and polymer–gas thermodynamic affinity encompassed by the systems examined here (see Table 11.1) suggest that the observed scaling behavior is general.

The viscoelastic scaling factor a_c is displayed as a function of CO_2 content for PDMS-CO_2 systems and PS–CO_2 systems in Figure 11.7. Clearly, a_c varies much more sharply with CO_2 content for the PS–CO_2 systems, which are 45–70 °C above T_g of pure PS, than for the PDMS–CO_2 systems, which are 178–208 °C above T_g of pure PDMS. Moreover, the fine details of the data show that the slope of the a_c versus w_{CO_2} curves decrease with increasing temperature for both the PDMS–CO_2 and PS–CO_2 data. Thus, temperature exerts a strong influence on a_c. At temperatures within 75 °C of T_g of the pure polymer, a few percent dissolved gas dramatically reduces the viscosity of the melt, reflected by the a_c values that are on the order of 10^{-2} to 10^{-3}.

The scaling of rheological behavior with dissolved gas content via the scaling factor a_c is closely related to the reduction of the glass transition temperature by dissolved gas, a phenomena that has been investigated by

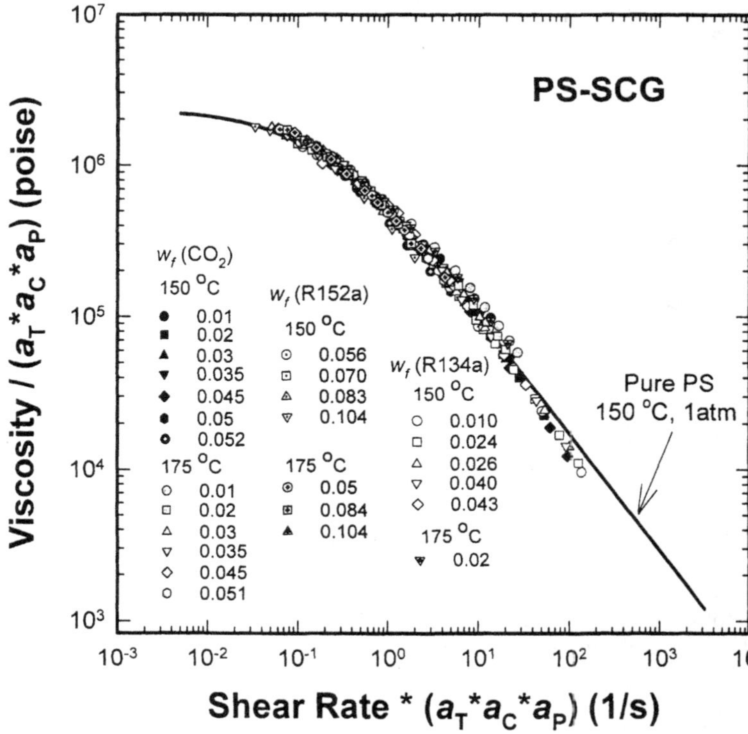

Figure 11.6. Master viscosity curve produced by superposing all data for all systems. Viscosity data taken at 175 °C have been shifted to 150 °C by employing the temperature scaling factor a_T for pure polystyrene. The master viscosity curve is identical to the viscosity curve for pure polystyrene at 1 atm and 150 °C, which is displayed as the solid line. Data from Kwag (1998).

Chow (1980), Condo et al. (1994), Wissinger and Paulaitis (1991), and others. Condo et al. (1994) and Wissinger and Paulaitis (1991) and have shown that the T_g of polystyrene can be reduced to values as low as 35 °C by the addition of about 10 wt % CO_2. Rudimentary calculations employing the Williams–Landel–Ferry (WLF) equation show that the scaling factors presented in Figure 11.7 are consistent with the T_g measurements cited earlier for the PS–CO_2 system.

Prediction of Viscoelastic Scaling Factor from Free-Volume Theory

Simple free-volume theories such as Doolittle's equation (Doolittle, 1951) suggest that the viscosity of liquids varies with the exponential of the fractional free volume. Viscoelastic scaling theories based on the free-volume

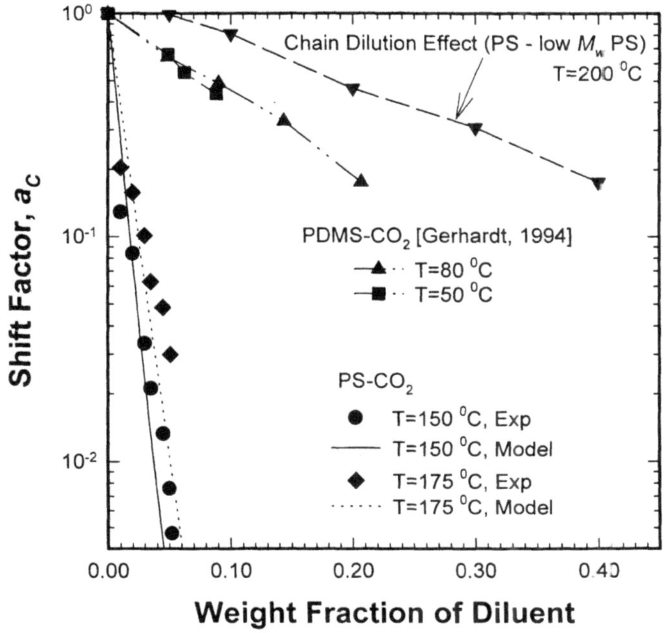

Figure 11.7. Variation of viscoelastic scaling factors with gas content for PS–CO_2 and PDMS–CO_2 systems. Lower scaling factor values for PS–CO_2 system, compared with PDMS–CO_2 system, are due to the closer proximity of the experimental temperatures to T_g of the pure polymer. The top curve displaying results for iso-free volume dilution of high-M_w polystyrene by low-M polystyrene represents the effect on viscosity of volumetric dilution of high-M_w chains. Viscosity reductions for polymer–gas systems are significantly lower than the iso-free volume dilution curve, indicating that viscosity reduction is primarily due to free volume contributed by dissolved gas.

concept, such as the WLF equation, have proven to be very successful in correlating the dependence of rheological properties of polymer melts on temperature and pressure (see Ferry, 1980). Gerhardt et al. (1998) developed a method for prediction of the viscoelastic scaling factor a_c by employing equation-of-state (EOS) predictions for specific volume V in the following free-volume-based expression for a_c:

$$a_c = (1 - w_c)^n \left[\frac{V_p}{V_m}\right]^n \exp\left(\frac{1}{f_m} - \frac{1}{f_p}\right) \tag{11.1}$$

where w_c is the mass fraction of dissolved gas, n is a constant exponent near 3.4, and V_p and V_m are the specific volumes of the pure polymer and polymer–gas mixture, respectively. The fractional free volumes $f = (V - V_o)/V_o$ of the pure polymer and polymer–gas mixture, f_p and f_m, are calculated from the EOS-predicted specific volume V and the occupied volume V_o that is inaccessible to polymer chain motion (Gerhardt et al., 1998). The a_c values predicted by this method, shown in Figure 11.8, are in excellent agreement with experimental values for the PDMS–CO_2 system. Preliminary comparisons (Kwag et al., 1999) of a_c values predicted for the PS–CO_2 and PS-R152a systems to experimental data are also very good, although the final predictions for these systems are still in progress.

The success of the dual-axis viscoelastic scaling method, the success of eq. 11.1 in predicting the scaling factors, and the known effects of dissolved gas in depressing T_g, offer strong evidence that the reduction in viscosity observed in polymer–gas mixtures is primarily due to free volume in addition to swelling of the melt with dissolved gas. Gerhardt et al. (1998) have performed thermodynamic calculations with the Sanchez–Lacombe and Panayioutou–Vera equations of state showing that the fractional free volume of PDMS increases substantially as CO_2 is added. The dependence of a_c on temperature relative to T_g can be readily understood in this context. As temperature approaches T_g, as in the PS–CO_2 system, the difference between V and V_o becomes very small, leading to very large values of $1/f$. A relatively small increase in $(V - V_o)$ due to the presence of added gas produces a disproportionate reduction in $1/f$, which is further amplified

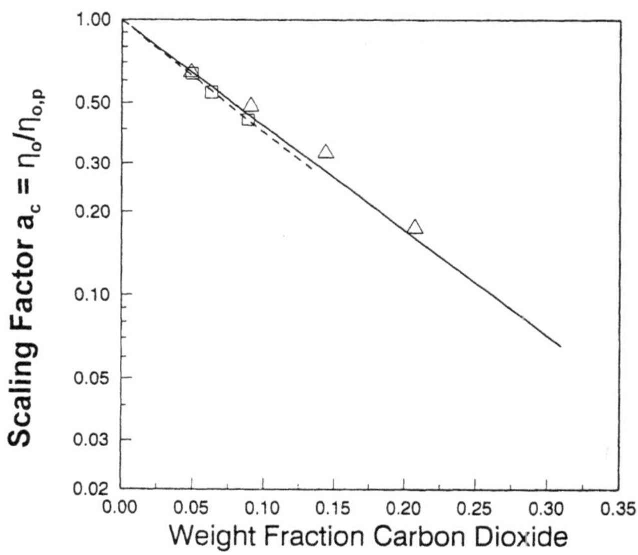

Figure 11.8. Viscoelastic scaling factor for the PDMS–CO_2 system predicted by eq. 11.1, with the Sanchez–Lacombe equation of state employed to evaluate specific volume, as calculated by Gerhardt et al. (1998). The solid line is the prediction at 50 °C, and the broken line is the prediction for 80 °C. The predicted curves are compared to the data of Gerhardt et al. (1997); △ data for 50 °C; □ data for 80 °C.

through the exponential dependence of a_c on $1/f$. Thus a_c decreases very sharply with dissolved gas content when the temperature is near T_g of the pure polymer, and very large reductions in viscosity can be produced by dissolved gas. Conversely when T is far away from T_g, as in the PDMS–CO_2 system, $1/f$ assumes more moderate values and small increases in $(V - V_o)$ do not produce large changes in $1/f$, and consequently in a_c.

Extrusion Experiments

Several processing studies on polymer/carbon dioxide systems have been performed with laboratory-scale extrusion equipment, where the gas is mixed into the polymer melt in an extruder and a viscosity measurement is performed in an attached capillary or slit die. This method is similar to the viscosity measurements of Han and Ma (1983a, 1983b) discussed earlier. Lee et al. (1999) measured the viscosity of polystyrene with 0–4 wt % dissolved carbon dioxide in a foaming extruder equipped with a capillary die at 220 °C. They also applied viscoelastic scaling to produce master viscosity curves similar to that in Figure 11.6. Elkovitch et al. (1999, 2000) have investigated the effects of dissolved carbon dioxide on the viscosity and morphology of blends of polymethylmethacrylate (PMMA) and polystyrene. Their experiments show that dissolved carbon dioxide can rectify viscosity contrasts between immiscible polymer melt phases, thereby promoting smaller domain sizes for the dispersed phase. Morphology control in polymer blends is a very important processing problem, and this work suggests that supercritical fluids may have very important applications as processing agents for such systems.

Conclusions

The effects of dissolved gases on the viscosity of polymer melts are investigated. A high-pressure capillary rheometer developed for measuring the viscosity of polymer melts with dissolved gases is described. Sample preparation along with the sealing and loading of the rheometer is outlined. Viscosity measurements for two polymers, PDMS containing up to 20 wt % CO_2 and polystyrene containing up to 5 wt % CO_2, up to 10 wt % R152a, and up to 3 wt % R134a, are reported. The results show that dissolved gases reduce viscosity of polymer melts by orders of magnitude. For example, viscosity of polystyrene swollen with 10 wt % R152a, is three orders of magnitude lower than the viscosity of pure polymer at 150 °C.

All polymer–gas systems studied here exhibit ideal viscoelastic scaling, whereby viscosity measurements taken at different gas compositions can be unified to a master curve of reduced viscosity $\eta(c, \gamma)/a$ versus reduced shear rate $a\gamma$ by a viscoelastic scaling factor a that depends on composition and pressure. This viscosity master curve is identical to the viscosity curve for pure polymer at the same temperature, meaning that the rheological effect of dissolved gas can be represented completely through the scaling factor a.

An important difference between the PS–gas systems (Kwag et al., 1999) and the PDMS–CO_2 system (Gerhardt et al., 1997) is that the viscosity measurements of the PS–gas systems are conducted at temperatures within 75 °C of T_g of PS, whereas the PDMS–CO_2 measurements were performed nearly 200 °C above T_g of PDMS. The difference between these two thermal regimes leads to several differences in the observed rheological behavior. The viscosity reductions relative to the pure polymer are much greater for PS–gas systems than for PDMS–CO_2 systems at similar dissolved gas compositions, and the dependence of a_c on temperature is much more pronounced for the PS–gas systems. These trends are consistent with the observations of Gerhardt et al. (1997, 1998) that the effect of dissolved gas on polymer melt viscosity occurs primarily through a free-volume mechanism.

A modified version of the free-volume theory is used to calculate the viscoelastic scaling factor or the Newtonian viscosity reduction where the fractional free volumes of pure polymer and polymer–SCF mixtures are determined from thermodynamic data and equation-of-state models. The significance of the combined EOS and free-volume theory is that the viscoelastic scaling factor can be predicted accurately without requiring any mixture rheological data.

References

Chow, T. S. *Macromolecules* **1980**, *13*, 362.

Condo, P. D.; Paul, D. R.; Johnston, K. P. *Macromolecules* **1994**, *27*, 365.

Dealy, J. M.; Wissbrun, K. F. *Melt Rheology and Its Role in Plastics Processing*; Chapman & Hall: London, 1990.

Doolittle, A. K. *J. Appl. Phys.* **1951**, *31*, 1003.

Elkovitch, M. D.; Lee, L. J.; Tomasko, D. L. *Polym. Eng. Sci.* **1999**, *39*, 2075.

Elkovitch, M. D.; Lee, L. J.; Tomasko, D. L. *Polym. Engi. Sci.*, **2000**, *40*, 1850–1861.

Ferry, J. D. *Viscoelastic Properties of Polymers*, 3rd ed; John Wiley and Sons: New York, 1980.

Gerhardt, L. J. Ph.D. Dissertation, Wayne State University, Detroit, MI, 1994.

Gerhardt, L. J.; Manke, C. W.; Gulari, E. *J. Polym. Sci. B: Polym. Phys.* **1997**, *35*, 523–534.

Gerhardt, L. J.; Manke, C. W.; Gulari, E. *J. Polym. Sci. B: Polym. Phys.* **1998**, *36*, 1911–1918.

Han, C. D.; Ma, C.-Y. *J. Appl. Polym. Sci.* **1983a**, *28*, 831–850.

Han, C. D.; Ma, C.-Y. *J. Appl. Polym. Sci.* **1983b**, *28*, 851–860.

Kwag, C. Ph.D. Dissertation, Wayne State University, Detroit, MI, 1998.

Kwag, C.; Manke, C. W.; Gulari, E. *J. Polym. Sci. B: Polym. Phys.* **1999**, *37*, 2771–2781.
Lee, M.; Park, C. B.; Tzoganakis, C. *Polym. Eng. Sci.* **1999**, *39*, 99.
McCoy, M. *Chem. Eng. News* **1999**, June 14, 11–14.
Mendelson, R. A. *J. Rheol.* **1979**, *23*, 545–556.
Mendelson, R. A. *J. Rheol.* **1980**, *24*, 765–781.
Wissinger, R. G.; Paulaitis, M.E. *J. Polym. Sci. B: Polym. Phys.* **1991**, *29*, 631–633.

PART III

INDUSTRIAL PROCESSES AND APPLICATIONS UTILIZING CARBON DIOXIDE

12

Coatings from Liquid and Supercritical Carbon Dioxide

YURY CHERNYAK
FLORENCE HENON
ERIK HOGGAN
BRIAN NOVICK
JOSEPH M. DeSIMONE
RUBEN CARBONELL

This chapter describes several aspects of the use of carbon dioxide as a solvent or cosolvent in coating applications. The primary impetus for using carbon dioxide for this purpose has been the alleviation of volatile emissions and liquid solvent wastes. However, the special physical properties of liquid and supercritical carbon dioxide may offer some processing advantages over conventional organic or aqueous solvents. Liquid carbon dioxide is quite compressible, and a reduction in temperature results not only in a reduction in the operating pressure, but also in a significant increase in the liquid density to values of approximately 0.9 g/cm^3. At these high liquid densities, carbon dioxide exhibits improved solvent performance, but with much lower viscosities and interfacial tensions than aqueous or organic liquid solvents. Under supercritical conditions, carbon dioxide also exhibits high densities, low viscosities, and improved solvent power. Low viscosities and interfacial tensions tend to facilitate the transport of the solvents into any crevices or imperfections on the surface to be covered, and this might prove advantageous in the coating of patterned or etched surfaces. Since carbon dioxide dissolves and diffuses easily into many different polymers and organic liquids, it can also be used to reduce the viscosity of coating solutions. Whether in the liquid or the supercritical state, the temperature and pressure of the mixture can be used to control its physical properties in ways that are impossible to achieve with traditional solvents. These distinguishing features have raised the level of industrial interest in carbon dioxide as a solvent for coating applications, beyond those based solely on environmental concerns.

In this chapter, we will discuss current applications and research on the use of CO_2 as a solvent for coatings. The first section deals with spray coating from supercritical CO_2. Subsequent sections deal with the use of liquid coatings, such as spin and free meniscus coatings, and impregnation coatings.

Spray Coatings with Supercritical CO_2

Since the start of the 20th century (ca. 1907), atomization has been the basis for conventional spray coating applications (Muirhead, 1974). Typically, atomization is caused by high shear of the coating fluid in air (Figure 12.1A), leading to droplet or particle formation. The energy for atomization is supplied by the kinetic energy of the high-velocity stream that is generated by the large pressure drop at the spray orifice. A number of different volatile organic compounds (VOCs) have been used to dilute coating materials. This dilution reduces the mixture viscosity, which is crucial for improving flow properties of coating formulations and leveling them on the surface to form

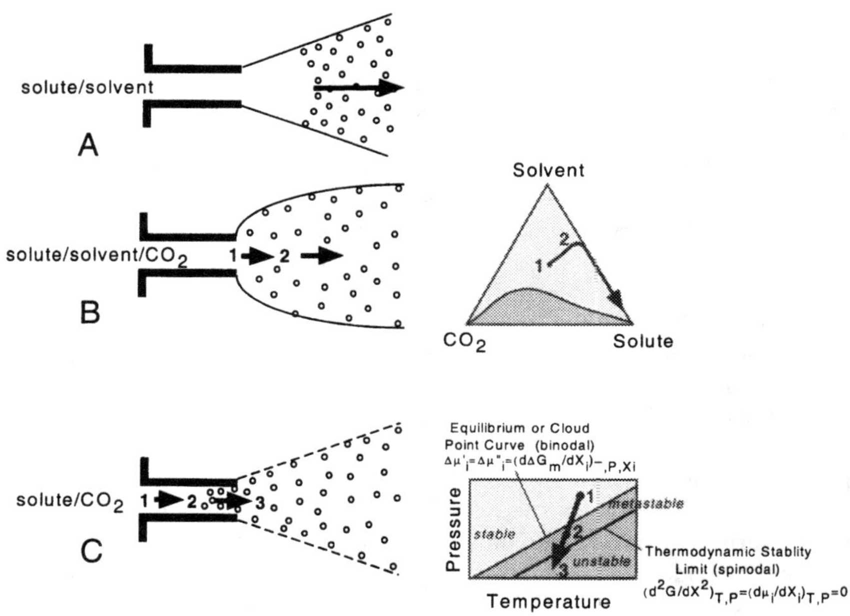

Figure 12.1. The process of spray atomization. (A) Conventional atomization: break up of a liquid film from the shear with surrounding air; (B) UNICARB® vigorous decompressive atomization produced by the expansive forces of the compressed carbon dioxide; (C) RESS: solute precipitation due to solution supersaturation followed by nucleation and spinodal decomposition.

a uniform film. For many years VOC-based air and airless spraying technologies have provided high-quality coatings that have good appearance and physical properties. However, with increased environmental concerns and regulation of VOCs since the late 1980s (Montreal Protocol, 1987, and Clean Air Act amendments, 1990), considerable effort has been put into developing new coating systems that utilize little or no organic solvents. Examples include waterborne, high-solids, ultraviolet, and powder coatings (Donohue et al., 1995). Although these systems have had some success, their inherent drawbacks have limited their acceptance (Donohue et al., 1995; Lewis et al., 1997; Woods and Busby, 1995). Because of the great need to eliminate VOC emissions, and the interest in technologies that would match the general utility and acceptance of conventional solvent-borne coatings, research efforts have focused on developing effective alternative coating methods utilizing environmentally benign solvents. This research has reinforced the belief that supercritical carbon dioxide can be used as the diluent for conventional solvent formulations (Donohue et al., 1995; Henon et al., 1999; Hutchenson and Foster, 1995; Johnston, 1989; Nielsen et al., 1991b; Schut, 1991; Tom and Debenedetti, 1991).

In 1990, the supercritical CO_2 fluid spray process, now known as the UNICARB process, was commercialized by Union Carbide to reduce the concentration of organic solvents in coating formulations (Figure 12.1B) (Donohue et al., 1995; Lewis et al., 1997; Schut, 1991). The UNICARB spray solution usually consists of 10–50 wt % dissolved CO_2 in the coating material. The amount used in any given application depends upon the CO_2 solubility, the viscosity, the solids level, the pigment loading of the coating formulation, and the spray pressure and temperature (Hoy and Donohue, 1990; Nielsen et al., 1991a, 1991b). Carbon dioxide not only functions as a good viscosity reducer, but also, more importantly, its presence results in a novel mechanism for atomization. As the dissolved CO_2 in the spray solution leaves the nozzle, it undergoes a rapid decompression due to the sudden pressure drop. This pressure drop creates a large driving force for nucleation, coagulation, and expansion of the dissolved carbon dioxide, resulting in the rapid formation of small liquid droplets in the expansion zone. The formation of liquid droplets occurs within a short distance from the spray orifice instead of downstream from the nozzle as in normal spray coating (Busby et al., 1991). It has been shown that the UNICARB process can produce fine droplets within the same range as conventional spray systems, but with a narrower size distribution that improves the appearance of the coating (Donohue et al., 1995). The UNICARB process has been used to apply a variety of high-quality clear, pigmented, and metallic coatings. The UNICARB process has undergone considerable technical development since it was introduced. Recent results obtained at Union Carbide support the notion that organic solvents can be eliminated to produce zero-VOC coatings (Lewis et al., 1997; Woods and Busby, 1995).

Another approach to using carbon dioxide as the only solvent in coating formulations arose from research on particle formation by the expansion of supercritical solutions (Tom and Debenedetti, 1991; Perrut, 1997). These efforts began with a comprehensive study on powder formation from the decompression of supercritical CO_2 solutions published in 1984 (Krukonis, 1984). In order for coating materials to dissolve in CO_2, it is usually necessary to pressurize the mixture to supercritical conditions. Its high diffusivity, low viscosity, and accessible critical temperature and pressure make carbon dioxide almost an ideal solvent for this purpose. The new spraying technique utilizing CO_2 was named the rapid expansion of supercritical solution (RESS) process. In this process, a dilute solution of a solute in a supercritical fluid is expanded through a nozzle from a high upstream pressure to a low downstream pressure (Figure 12.1C). This expansion causes the precipitation of the solute from solution due to the low solubility of the materials at the low solvent densities. The decompression of the solution across a capillary or pinhole nozzle provokes a mechanical perturbation that travels at the speed of sound, thus favoring a rapid attainment of uniform conditions within the expanding fluid. This rapid perturbation leads to supersaturation of the solute with characteristic times for phase separations on the order of 10^{-5}–10^{-6} s (Lele and Shine, 1994; Matson et al., 1986a; Mohamed et al., 1989a). This produces very small and monodisperse precipitates consisting of particles and droplets, depending on the nature of the solute and the operating conditions. Unlike conventional spraying methods and UNICARB, in RESS the carbon dioxide does not function as a viscosity reducer or as a generator of the atomization process. The density of CO_2 is directly related to its solvent power and it can be adjusted to dissolve the solute and then to precipitate it by expanding the solution with changes in pressure. Thus, the precipitation of solute in RESS from the solution is driven by other physical forces, namely, by supersaturation-induced nucleation and diffusive- and shear-driven coagulation.

The RESS process has been extensively studied since the late 1980s (Chang and Randolf, 1989; Debenedetti, 1990; Larson and King, 1986; Matson et al., 1986b) and continues to be an active area of investigation (Domingo et al., 1997; Kim et al., 1996; Liu and Nagahama, 1997; Shim et al., 1999). The major application of RESS has been in the processing of inorganic and organic solids sensitive to heat, oxidizing environments, or mechanical shocks, and in the formation of small particles for coating applications. The particle size reduction has generally been better than that obtained by conventional methods such as milling and grinding, or the airless spray used in paint application. Tom and Debenedetti (1991) reviewed all the studies on RESS published prior to 1991. Mawson et al. (1995) summarized the solid polymeric and inorganic materials involved in earlier RESS investigations, and Subramaniam et al. (1997) reported different pharmaceutical compounds precipitated by the rapid expansion of supercritical CO_2, together with the precipitate characteristics.

The RESS process attracted researchers hoping to overcome many of the issues facing the modern coatings industry. Initially, reduction of VOC emissions was the goal, but as the process developed, a number of other benefits were recognized. These benefits include the ability of the RESS process to accommodate various coating systems, the applicability to different substrates, the high-quality appearance of coatings owing to the uniform size distribution of precipitate, different morphologies of precipitated coatings from submicron powders to supermicron fibers, and many other technological and economic benefits such as improved transfer efficiency compared with conventional applications, cost effectiveness, and so on. Because of these advantages, the potential uses of the RESS process spread beyond areas of conventional spraying into areas such as optical coatings, semiconductor applications, and thin corrosion-protection barriers (Matson et al., 1986a, 1986b; Smith, 1986a, 1986b). The RESS process is limited only by the compatibility of supercritical carbon dioxide or other supercritical solvent with the coating material of interest. The possibility of controlling product characteristics by varying RESS operating conditions produces a wide variety of unique products.

Effect of RESS Process Variables

The pre-expansion temperature is the temperature of the solution prior to entering the nozzle region in which the expansion occurs. The difference between the pre-expansion and equilibrium temperature at the same pressure and composition of the solution provides a relative scale for the driving force for phase separation. When the pre-expansion temperature is higher than the equilibrium temperature, thermally induced separation occurs and the solution enters the nozzle as a two-phase mixture. This gives more time for the solute to grow and to form larger precipitates upon expansion (Tom and Debenedetti, 1991). On the other hand, precipitate sizes of smaller diameter are produced when pre-expansion temperatures below the equilibrium temperature cause phase separation to occur from a homogeneous solution (Mawson et al., 1995).

It has been shown that small changes in the pre-expansion pressure at constant pre-expansion temperature and solute concentration do not affect the precipitate characteristics (Mohamed et al., 1989a). On the other hand, for changes in the pre-expansion pressure of 69 bar, the observed precipitate morphology varied from micron-size spherical to fiber shape (Lele and Shine, 1994). However, Domingo et al. (1997) reported that changes in the pre-expansion pressure of their system proved to be inconclusive. Interestingly, Ksibi (1995) noticed a decrease in particle size with an increase in the pre-expansion pressure of about 41 bar. This was accompanied by a narrower particle size distribution.

Among the various factors influencing precipitate morphology, nozzle design is believed to be of major importance. In the vast majority of

RESS experiments, the supercritical solution flows through a calibrated orifice (Mawson et al., 1995; Mohamed et al., 1989b) or a capillary tube (Ksibi and Subra, 1996; Ksibi et al., 1995; Lele and Shine, 1990) but flow though a frit nozzle has also been investigated (Domingo et al., 1997). Some of the expansion configurations tested have been capillary nozzles of 5-mm length and 60-μm i.d. (inner diameter) (Matson et al., 1987), 40-mm length and 200-μm i.d. (Lele and Shine, 1992), 25.4-mm length and 50-μm i.d. (Mawson et al., 1995), or calibrated orifices of 15–30 μm i.d. (Mawson et al., 1995; Mohamed et al., 1989b). Pore diameters were varied from 0.5–10 μm for the frit nozzle design (Domingo et al., 1997). In this last study, it was found that an increase in the pore diameter led to an increase in the average precipitate size. The general trend for the capillary and the orifice nozzle seems to indicate that the precipitate morphology depends more on the length-to-diameter ratio of the nozzle than on its diameter alone (Ksibi et al., 1995). Low L/D ratios, like pinhole orifices, are found to produce spherical submicron precipitate shapes. A large ratio of L/D, as found in the capillary nozzle, seems to produce fibers or elongated precipitates.

Although some authors (Matson et al., 1987; Mawson et al., 1995) have reported that a decrease in solute concentration brings a decrease in particle size, Liu and Nagahama (1997) found a decrease in the size of particles with an increase of solute concentration in supercritical CO_2. Similar conclusions were also drawn by Mohamed and co-workers a few years earlier (Mohamed, et al., 1989b).

Thermodynamic and Rate Processes in RESS

The conclusions drawn from the different experiments with respect to the effect of RESS process parameters on precipitate size and morphology reflect the importance of specific solute/solvent properties on the process path. This probably results in the apparent contradictions found in experimental results. To date, the theoretical work aimed at gaining a fundamental understanding of the RESS expansion phenomenon and physical processes relevant to droplet, particle or film formation has been quite limited (Debenedetti, 1990; Debenedetti et al., 1993; Kwauk and Debenedetti, 1993; Lele and Shine, 1994).

From a thermodynamic viewpoint, the expansion of the solution in the nozzle is the process of transferring the solute/CO_2 system from a single- to a two-phase state. An example of a typical RESS thermodynamic process path of the solution representing low critical solution temperature (LCST) phase behavior is illustrated schematically in Figure 12.1C. Point 1 on the P–T diagram corresponds to the thermodynamic parameters at the nozzle entrance. Because of pressure and temperature drop in the nozzle, the process path crosses the equilibrium or cloud-point curve entering the two-phase metastable range (point 2). Further decrease of pressure and temperature leads to the conditions corresponding to point 3 at the nozzle exit, which could be either in the two-phase metastable or, more likely, in the two-phase unstable region. The criteria of phase equilibrium (binodal) and stability limit (spinodal) can be obtained from the First and Second laws of thermodynamics in terms of the Gibbs free energy of mixing (Figure 12.1C). The thermodynamic parameters corresponding to those criteria can be predicted for a given mixture from its equation of state (Skripov, 1974). Crossing the equilibrium curve into the two-phase metastable range, the solution becomes supersaturated, initiating nuclei growth. In general, crossing the binodal from a single- to a two-phase region is a necessary condition for nucleation to begin, whereas crossing the spinodal is a sufficient condition for the irreversible and spontaneous phase decomposition. However, in a real process the transition from nucleation to spinodal decomposition does not necessarily occur at the spinodal curve. The kinetic nucleation theory (Debenedetti, 1996) predicts the rate at which nuclei exceed threshold size and grow spontaneously. It also establishes the dependence between the supersaturation of the solution (thermodynamic parameter) and τ_{life}, the characteristic time for the formation of a new phase (kinetic parameter).

Fluid dynamics also play an important role in the RESS process. By setting the RESS operating parameters and nozzle geometry, the process path and characteristic process time τ_{RESS} (the residence time of the solution in the nozzle) will be predetermined for the solution at given thermodynamic and physical conditions. The particle-formation criteria in the nozzle can be established (Figure 12.2) by combining the fluid dynamic analysis of the solution with the thermodynamic/kinetic model. The case when spinodal decomposition happens prior to entering the nozzle, and the case when the solution flows rapidly through the nozzle in the homogeneous state ($\tau_{RESS} \ll \tau_{life}$) will lead to larger precipitates and to submicron powder, respectively. In both of these cases, the average size of the precipitate becomes quite uncontrollable. Kwauk and Debenedetti (1993) suggested that practical benefits may result from maximizing the amount of particle formation and growth occurring inside the nozzle to provide a more controlled environment than in the free-jet area. Adjusting the RESS operating conditions and nozzle geometry to satisfy $\tau_{RESS} > \tau_{life}$ should favor a controlled particle-formation process and should allow the recovery of most of

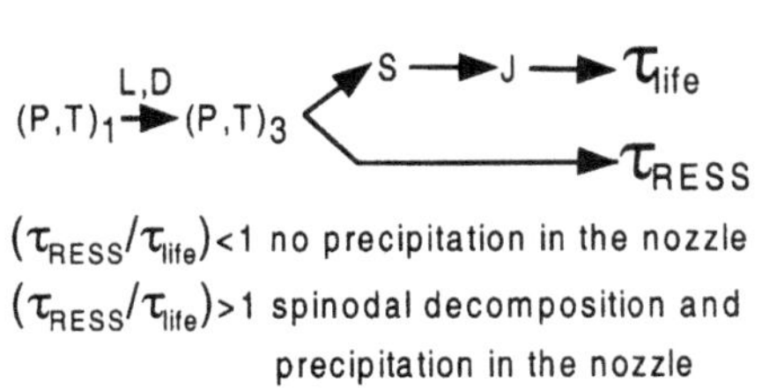

Figure 12.2. Particle-formation criteria; L is the nozzle length D is the nozzle internal diameter, S is the supersaturation, J is the nucleation rate, τ_{RESS} is the characteristic time of the process (the residence time of the solution in the nozzle), and τ_{life} is the lifetime of the supersaturated solution in a homogeneous state.

the solute from the solution at the nozzle exit. Mathematical modeling of particle formation in RESS was done by Debenedetti and co-workers to uncover the basic trends linking particle size and process variables (Debenedetti et al., 1993; Kwauk and Debenedetti, 1993).

Liquid Coating Systems

Many important coating processes utilize a liquid–gas interface during the deposition process. These include spin coating, free meniscus coating, flow coating, and metered coating techniques (blade, roll, etc.) (Brinker et al., 1997; Coeling, 1992; Cohen Edward et al., 1990). By definition, the supercritical state of carbon dioxide or any compressed gas lacks the necessary liquid/gas interface. However, this does not make these techniques impossible in CO_2. Coating can be performed either with liquid carbon dioxide as the solvent or from the supercritical state with a cophase such as nitrogen.

There are many advantages to using liquid carbon dioxide. It has a variable density similar to that of the supercritical state, but at a lower pressure. Unlike most conventional solvents, the density of liquid carbon dioxide is not a linear function of temperature at ambient conditions (see Figure 12.3). This is important for coating processes as it provides a method to drive and control deposition. Just as in the supercritical state, a change in density in

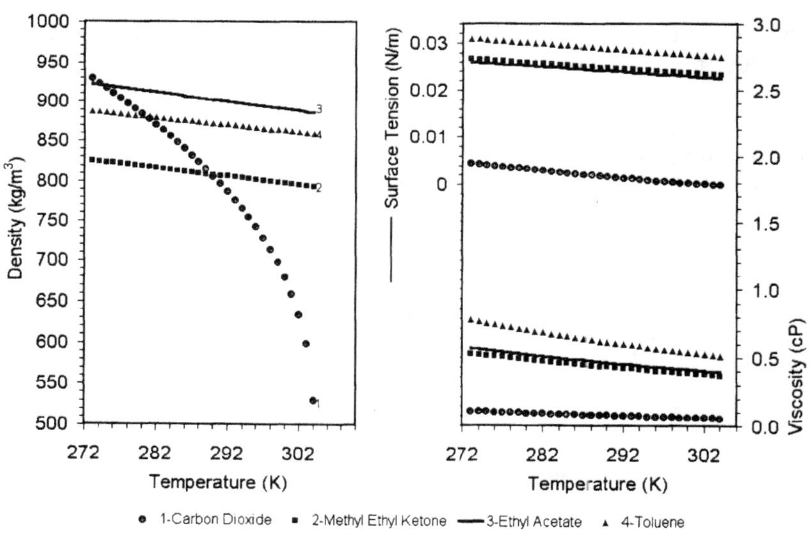

Figure 12.3. Properties of liquid carbon dioxide. Adapted from Daubert, T. E.; Danner, R. P. *Data Compilation Tables of Properties of Pure Compounds*. Design Institute for Physical Property Data, American Institute of Chemical Engineers: New York, 1985.

the liquid state can change solubility. As mentioned previously, the surface tension and viscosity of liquid CO_2 are also much lower than those of conventional solvents (Figure 12.3). This can lead to improved transport phenomena and improved penetration into porous materials. These possible benefits have led researchers in our group to begin investigating the use of coatings from liquid-state carbon dioxide (Hoggan et al., 1999a, 1999b; Kendall et al., 1999; Novick et al., 1999).

Spin Coating

Motivation

Recognizing the potential of CO_2, several groups have used supercritical CO_2 (sc CO_2) in the development of photoresist films (Gabor et al., 1997; Gallagher-Wetmore et al., 1995; Ober et al., 1997). Development is the process in which a solvent is used to selectively remove portions of a polymer film from silicon wafers, subsequently creating a pattern in the silicon (McGillis, 1983). Ober, Allen, and others have shown that in some cases it is possible to replace the traditional aqueous developers with sc CO_2 (Gabor et al., 1997; Gallagher-Wetmore et al., 1995; Ober et al., 1997). However, there are no examples in the literature of efforts to deposit photoresist films using CO_2. Realizing that films of photoresists could be deposited with CO_2 as the only solvent, we have investigated spin coating polymer resists using liquid CO_2.

Traditionally, spin coating of photoresists has used hazardous organic solvents such as 2-ethoxy ethyl acetate and diglyme (Moreau, 1988). While the amount of solvent used has decreased in recent years due to tighter controls and improved equipment, a typical fab line still consumes about 1 million gallons of photoresists and organic solvents per year at a cost of $400 million (Moreau et al., 1997). Not included in this amount are the costs associated with ventilation equipment and other safety measures that must be used when handling these hazardous solvents. The benefits of using CO_2 for such a process are clearly apparent.

Background

Before describing spin coating from CO_2, it is necessary to give a brief introduction to conventional spin coating. Spin coating is a process in which a thin liquid film is spread by centrifugal force onto a rotating substrate. Spin coating is used to form very thin (1–10 μm) films with nonuniformities no greater than 1% (Moreau, 1988). When the spin coated liquid contains polymers or colloids (up to 25 wt % solute), solid films may be formed on the substrate (Bornside et al., 1991; Rehg and Higgins, 1989). The most common application for spun films is in the microelectronics industry, but the technique is also used in the manufacture of compact disks, flat panel displays, and highly reflecting mirrors (Parodi et al., 1996).

Applicability to CO_2

Several detailed analyses of the spin coating process have been conducted (Bornside et al., 1989; Emslie et al., 1958; Meyerhofer, 1978; Yonkoski and Soane, 1992). The analyses conducted by these authors have been shown to hold for a variety of volatile organic solvents (Birnie, 1997; Paul et al., 1996). However, this model may not apply well to spin coating in CO_2 because coatings from CO_2 are formed under drastically different conditions. This difficulty in relating the conventional process to CO_2-based spin coating is illustrative of the hurdles encountered when developing new processes in CO_2.

Spin coating requires a liquid–vapor interface. This requirement necessitates operation in a temperature and pressure range under which CO_2 is a liquid. In addition, there are other more subtle restrictions that must be considered. Being restricted to liquid CO_2 limits the solvent strength of the CO_2, thus limiting the variety of polymers that may be used in the coating. The polymer used must have a fairly strong affinity for CO_2 under liquid conditions. Also, because the process is carried out under high pressure, the evaporation rate (a variable that traditionally depends only upon the organic solvent used) must be carefully regulated. This is accomplished by controlling the partial pressure of CO_2 in the coating chamber, either by altering the total system pressure or by adding inert gases. If the spin coating is carried out at a pressure that is too low, the CO_2 will evaporate immediately, leaving a highly viscous polymer that cannot be spread uniformly. Likewise, if the pressure is too high, the CO_2 vapor may condense on the wafer, thus destroying film quality.

Films Formed in CO_2

As stated previously, being restricted to subcritical pressures limits the solvent strength of the CO_2 and thus the solubility of the polymer in the fluid. However, CO_2 is highly tunable by changing the temperature as well as the pressure. By simply cooling the spin coating system from 25 °C to 8 °C, it is possible to increase the solvent density by nearly 50% (Figure 12.3). Therefore, spin coating can be conducted in a cold cabinet to maintain high CO_2 densities, allowing higher concentration solutions.

Solutions of 10–25 wt % poly(1*H*,1*H*-perfluorooctyl methacrylate) (PFOMA) in CO_2 were spun on 125-mm Si wafers. Because of the difficulty in controlling the evaporation of CO_2 during the spinning process, initial work focused on coating with no evaporation at all. The chamber pressure was simply maintained at the vapor pressure of the polymer/CO_2 solution. After spinning was completed, gas was slowly vented from the chamber, causing the CO_2 in the film to evaporate. This method was successful in producing films with thicknesses of 0.5–1.5 μm (Figure 12.4) (Kendall et al., 1999). Through various refinements, films with 3–6% uniformity have been

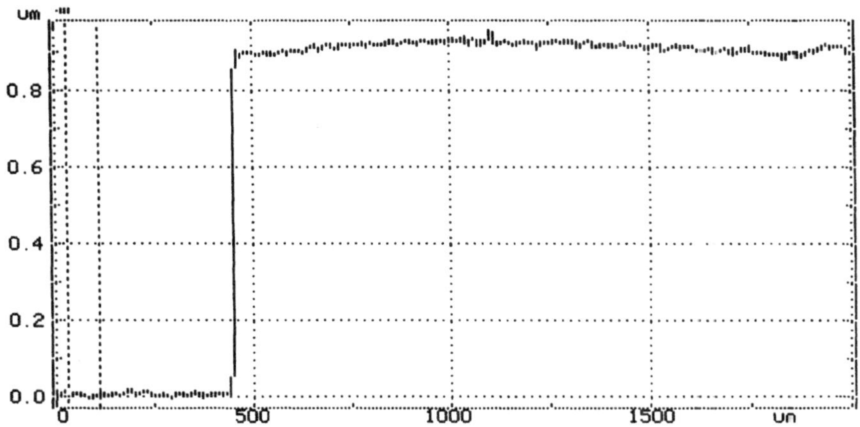

Figure 12.4. Profile of PFOMA film cast from liquid CO_2.

achieved. As work in this area is in its early stages, results should improve significantly in the future.

Free Meniscus Coating

Another important coating method is known as free meniscus coating (FMC). This technique utilizes the shape of a meniscus and process variables to determine the final coating properties. There are many forms of "free meniscus coating" including dip coating, slot coating, and "true" roll coating (Brinker et al., 1997; Coeling, 1992; Cohen Edward et al., 1990). The use of carbon dioxide as a solvent for these techniques is not described in the literature. The closest references involve the impregnation of textiles using a dipping method (see later section "Impregnation Coatings"). The following section compares theoretical predictions of liquid carbon dioxide with organic solvents for dip coating processes and presents a preliminary result for the deposition of thin lubricant films relevant to the hard disk industry.

Dip coating can be accomplished using withdrawal, drainage, or continuous coating processes. In withdrawal a substrate is removed from a bath, thereby entraining solution. During the entrainment process there is simultaneous evaporation of the solvent, resulting in deposition of a coating. Drainage is based on the same principles as withdrawal coating but the substrate is held stationary while the fluid level is allowed to decrease. Continuous processes involve moving an object (such as wire) through a liquid–gas interface so that the coating process takes place under steady-state conditions (Brinker et al., 1997; Coeling, 1992; Denson, 1970; Deryagin and Levi, 1964; Scriven and Suszynski, 1990; Tallmadge and Gutfinger, 1967).

Applications

Research involving free meniscus processes is plentiful. Haaland reported that in 1994 nearly 47% of the 141,517 U.S. coating patents and over 114,000 technical publications refer to dipping processes (Haaland et al., 1995). These techniques can be used to coat objects of irregular or discrete shape. Hard disk lubricant coatings, protective coatings, optical coatings, sol–gel coatings, and langmuir blodgett coatings are just a few of the films formed by a dipping method (Baudry et al., 1990; Brinker et al., 1991; Chung Ching et al., 1980; Cohen Edward et al., 1990; Deryagin and Levi, 1964; Gao et al., 1995; Gesang et al., 1997; Jansen et al., 1994; Khedkar Sachin and Radhakrishnan, 1997; Koestlin et al., 1997; Kusano et al., 1997; Schelle et al., 1997; Takahashi et al., 1997; Vroon and Spee, 1997).

Free Meniscus Theory

The theory used to predict film thickness in free meniscus processes is well developed, having first been investigated with the work of Jeffreys (1930). Since then, authors including Landau, Levich, Tallmadge, Deryagin, Brinker, Hurd, and Scriven have continuously refined FMC theory, making it an accurate method of predicting wet film thickness (Brinker and Hurd, 1994; Deryagin and Levi, 1964; Groenveld, 1971; Guglielmi and Zenezini, 1990; Landau and Levich, 1942; Riley and Carbonell, 1993; Scriven and Higgins, 1979; Tallmadge, 1970, 1971a, 1971b, 1971c; Tallmadge and Gutfinger, 1967; Tallmadge and Lang, 1971; van de Ven, 1998). Many versions of these theories exist for the prediction of film thickness. One of these theories is Tallmadge's four force inertial theory (FFIT), which can be used to predict entrained film thickness for low capillary numbers (Riley and Carbonell, 1993; Tallmadge and Gutfinger, 1967; Tallmadge and Lee, 1974).

Tallmadge's theory and other modern FMC theories predict that three distinct coating regions should exist. The "meniscus region" and "nonconstant-film-thickness region" exist at the top and bottom of the entrained film. However, the most important of the regions is the "constant-film-thickness region" (CFTR). In dip coating, it is desirable to cover the majority of the substrate with the CFTR to ensure uniformity of the film. The shape and thickness of each coating region is dependent on physical properties and process variables such as viscosity, surface tension, density, withdrawal velocity, and drainage time (Brinker and Hurd, 1994; Deryagin and Levi, 1964; Groenveld, 1971; Guglielmi and Zenezini, 1990; Landau and Levich, 1942; Riley and Carbonell, 1993; Scriven and Higgins, 1979; Tallmadge, 1970, 1971a, 1971b, 1971c; Tallmadge and Gutfinger, 1967; Tallmadge and Lang, 1971; van de Ven, 1998).

Theoretical Results

Figure 12.5 compares the results of the four force inertial theory for various solvents including carbon dioxide. The film thickness and slope of withdrawal velocity versus entrained film thickness is smallest for carbon dioxide. This may result in the ability to produce thin films at higher production rates. The small slope may also result in a decrease in nonuniformities because of small changes in withdrawal velocity.

Advantages

There are other advantages for using carbon dioxide for FMC processes. The low viscosity and surface tension will lead to increased transport of material within the entrained film as well as into porous materials (Levien et al., 1998). This should help decrease nonuniformities that develop during the evaporation process because of concentration differences. Also, the rate of evaporation can be controlled utilizing pressure gradients so that the likelihood of concentration gradients is decreased further. It may also be a convenient solvent because of the solubility of low-surface-energy materials (fluorocarbons and siloxanes) that are commonly used for lubricant and protective coatings. Lastly, there should be no need to dry coatings created with liquid carbon dioxide.

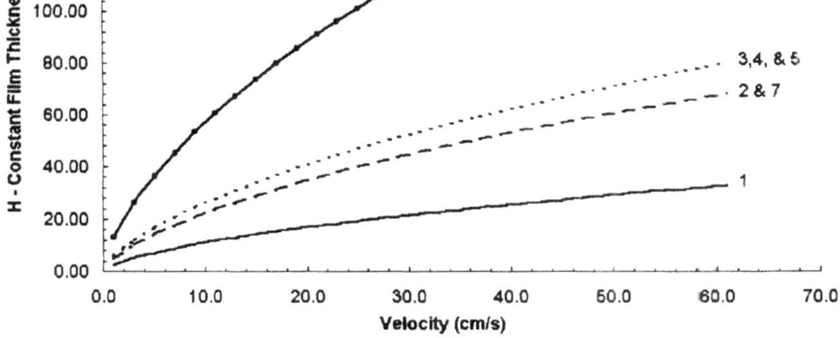

Figure 12.5. Theoretical constant film thickness immediately after withdrawal of a 5-cm plate during dip coating at 293 K and 39 cm/s withdrawal velocity.

Liquid Carbon Dioxide FMC Results

Despite these advantages there are no known references to a free meniscus coater that utilizes liquid carbon dioxide as the solvent. Because of this, we have begun investigating the development of such an apparatus based on the drainage configuration (Novick et al., 1999). Our simple design has allowed us to create films of soluble perfluoropolyethers, such as Fomblin ZDol, used for lubrication of magnetic media (Figure 12.6). The film thickness can be varied from about 2.5 to 30 nm, although uniformity of the film has not been measured. These polymers are highly soluble in compressed carbon dioxide and are an ideal case study because of the large collection of literature (Bhushan, 1996; Gao et al., 1995; Gellman, 1998). A future apparatus should have the ability to cast films from both homogeneous and colloidal solutions.

Impregnation Coatings

The coating methods presented earlier are all used to create films of materials upon surfaces. There is another class of coatings from CO_2 that has not yet been discussed. Using both liquid and supercritical CO_2, it is possible to coat or impregnate porous and woven materials with a desired polymer or chemical agent. Three specific applications will be discussed in turn.

Impregnation of Wood

To expand the usefulness of wood products, it is often desirable to increase the physical strength, improve chemical resistance, and slow degradation. One method of effecting these changes is to impregnate the porous wood

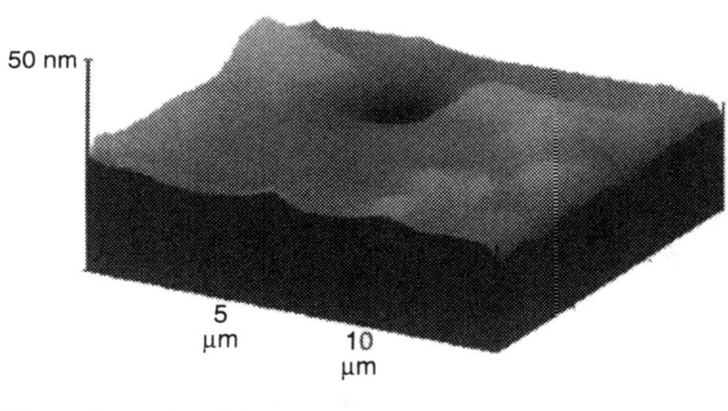

Figure 12.6. Three-dimensional AFM plot of Fomblin ZDol on a glass slide after free meniscus coating from liquid carbon dioxide. The hole in the center of the slide is a reference to the bare substrate.

material with polymers or other chemical agents (Kiran, 1995). However, because polymer solutions of organic liquids often possess high viscosities, it is difficult to achieve good penetration. In an attempt to alleviate these problems, conventional solvents may be replaced with supercritical fluids including CO_2.

The use of supercritical fluids in such coatings was first suggested by Ito et al. (1984). Kayihan (1992) used sc CO_2 in the deposition of a dye within small wood dowels. Sahle-Demessie et al. (1995) have built upon this work and conducted an extensive study of the impregnation of wood with biocides using sc CO_2. They successfully impregnated TCMTB [2-(thiocyanomethylthio)benzothiazole] and PCP (pentachlorophenol) into wood samples. Their experiments were carried out by dissolving the biocide in the sc CO_2, often with the aid of a cosolvent. The solution was then added to a second chamber containing the wood sample. The wood remained in contact with the sc CO_2 solution for a specified treatment time, after which the solution was vented into a third chamber. As the solution was vented, the pressure dropped and the biocide fell out of solution. In this manner, a large portion of the biocide remained in the porous surface while the CO_2 was removed.

These efforts to impregnate wood with the two biocides were successful. As expected, the distribution of the material throughout the wood was highly dependent on the treatment time, operating pressure, and the dimensions of the wood sample. Not surprisingly, the investigators also found that the addition of the cosolvent increased the solubility of the biocide and resulted in higher concentration and more uniform impregnation.

Coating of Fibrous Polymers Using Supercritical Carbon Dioxide

Just as wood may be impregnated with additives to improve physical properties, a variety of porous polymers also benefit from polymer impregnation. As more and more specialized applications arise, it is necessary to produce polymeric materials with very specific surface and bulk properties. The correct combination of properties may not be obtainable using a single polymer or copolymer. However, polymeric materials may be impregnated with an additive in order to modify the surface energy, bondability, hydrophilicity, and other important properties.

The first work impregnating polymers with additives was done by Sand in 1985 (Sand, 1986). In the early 1990s, several groups began work in the same area (Berens et al., 1992; Howdle et al., 1994; Perman and Riechert, 1993). Building upon that work, Ma and Tomasko (1997) investigated the coating and impregnation of fibrous polyethylene with a nonionic surfactant using sc CO_2.

The method employed by Ma and Tomasko differs from the coatings discussed previously in that they did not employ a pressure drop to precipitate the additive. Instead, they rely on the affinity of the additive for the polymer substrate. In their experiments, they flowed a solution of the nonionic surfactant (*N*,*N*-dimethyldodecylamine *N*-oxide) in sc CO_2 through a high-pressure vessel containing a sample of the fibrous high-density polyethylene (HDPE). As the solution flowed through the polymer, the surfactant preferentially adhered to the surface. By simply monitoring the effluent, they were able to determine when the uptake was completed and then purge the vessel, leaving behind the polymer coated with additive.

Ma and Tomasko found that the structural integrity and strength of the polymer were unchanged. They also found that the contact angle and the wicking rate were indeed altered by the procedure. Their method has an advantage over previous work in that it is less disruptive of the polymer network and therefore the surfactant–polymer interactions can be more precisely investigated. However, such a method is only applicable to a limited number of materials because of the required solubility behavior.

Efforts in this area are ongoing. Work by Bayraktar and Kiran (1999) has explored the use of CO_2 to impregnate polydimethylsiloxane (PDMS) into polystyrene and polyethylene bulk polymers.

Application of Dyes in Supercritical Carbon Dioxide

The conventional dyeing of textiles consumes large quantities of water. While much of this water may be recovered, wastewater treatment imposes significant costs. In addition, the hydrophobicity of many synthetic fabrics necessitates the use of dispersing agents and surfactants in the dyeing liquor (Saus et al., 1993). One possible solution to these problems is the use of sc CO_2 as the solvent.

Saus et al. (1993) have conducted work using sc CO_2 in the dyeing of synthetic fibers. They were successful in dyeing samples of poly(ethylene terephthalate) PET with a variety of dyestuffs. The procedure used for dyeing was a combination of the two impregnation methods discussed previously. They placed the synthetic fabric in contact with the dye/CO_2 solution for a period of 0–60 mins. At the end of the dyeing process, the chamber was vented and the sample removed. Thus, they rely on the preferential absorption of the dye onto the PET, as well as utilizing the precipitation impregnation discussed earlier.

They found sc CO_2 to be an excellent solvent for the dye. By controlling the temperature they were able to manipulate the diffusivity of the solution and to obtain dyeing uptakes comparable to those of conventional methods. In addition, their method allowed for easier recovery of residual dyestuff and eliminated the need for after-treatments such as reductive washing.

Conclusion

Research up to now has certainly shown that it is possible to produce uniform and effective coatings from both supercritical and liquid CO_2 for a wide variety of applications. A great deal remains to be learned about the exact mechanisms that control the ultimate particle or droplet sizes, and the resulting film thickness and uniformity of the coatings. Once these mechanisms are well understood, the design of CO_2-based coating processes offer great promise as effective environmentally responsible replacements for coating operations based on water or organic liquid solvents.

References

Baudry, P.; Rodrigues, A. C. M.; Aegerter, M. A.; Bulhoes, L. O. Dip-Coated TiO Sub 2-CeO Sub 2 Films as Transparent Counter-Electrode for Transmissive Electrochromic Devices. *J. Non-Cryst. Solids* **1990**, *121*, 319–322.

Bayraktar, Z.; Kiran, E. Presented at AIChE National Meeting, Dallas, TX, 1999.

Berens, A. R.; Huvard, G. S.; Korsmeyer, R. W.; Kunig, F. W. Application of Compressed Carbon Dioxide in the Incorporation of Additives into Polymers. *J. Appl. Polym. Sci.* **1992**, *46* (2), 231–242.

Bhushan, B. *Tribology and Mechanics of Magnetic Storage Devices*; Springer-Verlag: New York, 1996.

Birnie, D. P. Combined Flow and Evaporation During Spin Coating of Complex Solutions. *J. Non-Cryst. Solids* **1997**, *2118*, 174–178.

Bornside, D. E.; Macosko, C. W.; Scriven, L. E. Spin Coating: One-Dimensional Model. *J. Appl. Phys.* **1989**, *66*, 5185–5193.

Bornside, D. E.; Macosko, C. W.; Scriven, L. E. Spin Coating of a PMMA/Chlorobenze Solution. *J. Electrochem. Soc.* **1991**, *138*.

Brinker, C. J.; Hurd, A. J. In *Piezo Pyro Ferroelectric Materials and Their Applications: Bulk Materials and Thin Films. 8–9 June*; Applied-Physics: Limoges, France, 1994, Vol. 4, Journal-de-Physique-III, pp. 1231–1242.

Brinker, C. J.; Hurd, A. J.; Schunk, P. R. In *Liquid Film Coating*; Kistler, S. F., Schweizer, P. M., Eds.; Chapman & Hall: London, 1997; pp. 673–708.

Brinker, C. J.; Hurd, A. J.; Frye, G. C.; Schunk, P. R.; Ashley, C. S. Sol–gel thin film formation. *J. Ceram. Soc. Jpn., Int. Ed.* **1991**, *99*, 862–877.

Busby, D. C.; Glancy, C. W.; Hoy, K. L.; Kuo, A. C.; Lee, C.; Nielsen, K. A. Supercritical fluid spray application technology: a pollution-prevention technology for the future. *Dev. Sci. Surf. Coat., Conf. Pap.* **1991**, 23–29.

Chang, C. J.; Randolf, A. D. Precipitation of Microsize Organic Particles from Supercritical Fluids. *AIChE J.* **1989**, *35*, 1876–1882.

Coeling, K. J. In *Encyclopedia of Chemical Technology*; Kirk, R. E., Othmer, D. F., Kroschwitz, J. I., Howe-Grant, M., Eds.; Wiley: New York, 1992, Vol. 6, pp. 606–669.

Cohen, E. D.; Lightfoot, E. J.; Gutoff, E. B. A Primer on Forming Coatings. *Chem. Eng. Prog.* **1990**, *86*, 30–36.

Debenedetti, P. G. Homogeneous Nucleation in Supercritical Fluids. *AIChE J.* **1990**, *36*, 1289–1298.

Debenedetti, P. G. *Metastable Liquids: Concepts and Principles*; Princeton University Press; Princeton, NJ, 1996.

Debenedetti, P. G.; Tom, J. W.; Kwauk, X.; Yeo, S.-D. Rapid Expansion of Supercritical Solutions (RESS): Fundamentals and Applications. *Fluid Phase Equilib.* **1993**, *82*, 311–321.

Denson, C. D. The Drainage of Newtonian Liquids Entrained on a Vertical Surface. *Ind. Eng. Chem. Fundam.* **1970**, *9*, 443–448.

Deryagin, B. M.; Levi, S. M. *Film Coating Theory*; The Focal Press: New York, 1964.

Domingo, C.; Berends, E.; Van Rosmalen, G. M. Precipitation of Ultrafine Organic Crystals from the Rapid Expansion of Supercritical Solutions over a Capillary and a Frit Nozzle. *J. Supercrit. Fluids*, **1997**, *10*, 35–55.

Donohue, M. D.; Geiger, J. L.; Kiamos, A. A.; Nielsen, K. A. In *Green Chemistry*; ACS: Washington, DC, 1995, pp. 152–166.

Emslie, A. G.; Bonner, F. T.; Peck, L. G. (1958). Flow of a Viscous Liquid on a Rotating Disk. *J. Appl. Phys.* **1958**, *29*, 858–862.

Gabor, A. H.; Allen, R. D.; Gallagher-Wetmore, P.; Ober, C. K. In *Proceedings SPIE, Advances in Resist Technology and Processing XIII*; Kunz, R. R., Ed., 1997; Vol. 2724, pp. 410–417.

Gallagher-Wetmore, P.; Wallraff, G. M.; Allen, R. D. In *Proceedings SPIE, Advances in Resist Technology and Processing XII*; Allen, R. D., Ed., 1995; Vol. 2438, pp. 694–708.

Gao, C.; Lee, Y. C.; Chao, J.; Russak, M. Dip-Coating of Ultra-Thin Liquid Lubricant and Its Control for Thin-Film Magnetic Hard Disks. *IEEE Trans. Magn.* **1995**, *31*, 2982–2984.

Gellman, A. J. Lubricants and Overcoats for Magnetic Storage Media. *Curr. Opin. Colloid Interface Sci.* **1998**, *3*, 368–372.

Gesang, T.; Hoeper, R.; Possart, W., Petermann, J.; Hennemann, O. D. Adsorption and Growth of Dip-Coating Prepolymer Films on Silicon Wafers. An Atomic Force Microscope Study. *Appl. Surf. Sci.* **1997**, *115*, 10–22.

Groenveld, P. Drainage and Withdrawal of Liquid Films. *AIChE J.* **1971**, *17*, 489–490.

Guglielmi, M.; Zenezini, S. The Thickness of Sol–Gel Silica Coatings Obtained by Dipping. *J. Non-Cryst. Solids* **1990**, *121*, 303–309.

Haaland, P.; McKibben, J.; Parodi, M. The Art and Science of Thin-Film Coating: A Progress Report. *Solid State Technol.* **1995**, *38*, 83–89.

Henon, F. E.; Carbonell, R. G.; Camaiti, M.; Piacenti, F.; Burks, A.; DeSimone, J. M. Supercritical CO_2 as a Solvent for Polymeric Stone Protective Materials. *J. Supercrit. Fluids* **1999**, *15*, 173–179.

Hoggan, E. N.; Carbonell, R. G.; Flowers, D.; DeSimone, J. M. Presented at AIChE National Meeting, Dallas, TX, 1999a.

Hoggan, E. N.; Flowers, D.; Carbonell, R. G.; DeSimone, J. M. Presented at ACS National Meeting, New Orleans, LA, 1999b.

Howdle, S. M.; Ramsay, J. M.; Cooper, A. I. (1994). Spectroscopic Analysis and in situ Monitoring of Impregnation and Extraction of Polymer Films and Powders Using Supercritical Fluids. *J. Polym. Sci., Part B: Polym Phys.* **1994**, *32*.

Hoy, K. L.; Donohue, M. D. In *Polymer Preprints—Division of Polymer Chemistry*; Culbertson, B. M., Ed.; ACS, Vol. 31 (1); New York, 1990; pp. 679–680.

Hutchenson, K. W.; Foster, N. R. In *Supercritical Fluid Science and Technology*; Hutchenson, K. W., Foster, N. R., Eds.; American Chemical Society: Washington, DC, 1995, pp. 1–31.

Ito, N. T.; Someya, T.; Taniguchi, M.; Inamura, H. Japanese Patent 59-101311, 1984.

Jansen, H. V.; Gardeniers, J. G. E.; Elders, J.; Tilmans, H. A. C.; Elwenspoek, M. Applications of Fluorocarbon Polymers in Micromechanics and Micromachining. *Sens. Actuators, A: Physical* **1994**, *41*, 136–140.

Jeffreys, H. Draining of a Vertical Plate. *Proc. Cambridge Philos. Soc.* **1930**, *26*, 204–205.

Johnston, K. P. In *Supercritical Science and Technology*; Johnston, K. P., Penninger, J. M. L., Eds.; American Chemical Society: Washington, DC, 1989; pp. 1–12.

Kayihan, F. U.S. Patent 4,552,786, 1992.

Kendall, J. L.; DeSimone, J. M.; Carbonell, R. G.; Hoggan, E. N.; Askew, K. L.; Haynie, M. L. In Fluorine in Coatings III; Presented at Grenelefe, FL, 1999.

Khedkar Sachin, P.; Radhakrishnan, S. Application of Dip-Coating Process for Depositing Conducting Polypyrrole Films. *Thin Solid Films* **1997**, *303*, 167–172.

Kim, J.-H.; Paxton, T. E.; Tomasko, D. L. Microencapsulation of Naxopren Using Rapid Expansion of Supercritical Solutions. *Biotechnol. Prog.* **1996**, *12*, 650–661.

Kiran, E. In *Innovations in Supercritical Fluids: Science and Technology*; Hutchenson, K. W.; Foster, N. R., Eds.; American Chemical Society: Washington, DC, 1995.

Koestlin, H.; Frank, G.; Hebbinghaus, G.; Auding, H.; Denissen, K. Optical Filters on Linear Halogen-Lamps Prepared by Dip-Coating. *J. Non-Cryst. Solids* **1997**, *218*, 347–353.

Krukonis, V. Presented at AIChE Annual Meeting, San Francisco, paper 140 f, 1984.

Ksibi, H. Chemical Engineering, Ph.D. Thesis, Paris XIII, Paris, France, 1995.

Ksibi, H.; Subra, P. Influence of Nozzle Design on the Nucleation Conditions in the RESS Process. *Chem. Biochem. Eng. Q.* **1996**, *10*, 69–73.

Ksibi, H.; Subra, P; Garrabagos, Y. Formation of Fine Powders of Caffeine by RESS. *Advanced Powder Technol.* **1995**, *6*, 25–33.

Kusano, H.; Kimura Shin, I.; Kitagawa, M.; Kobayashi, H. Application of Cellulose Langmuir–Blodgett Films as Humidity Sensors, and Characteristics of the Sorption of Water Molecules into Polymer Monolayers. *Thin Solid Films* **1997**, *295*, 53–59.

Kwauk, X.; Debenedetti, P. G. Mathematical Modeling of Aerosol Formation by Rapid Expansion of Supercritical Solutions in a Converging Nozzle. *J. Aerosol Sci.* **1993**, *24*, 445–469.

Landau, L. D.; Levich, V. G. Dragging of a Liquid by a Moving Plate. *Acta Physicochimica U.R.S.S.* *17*, 42–54.

Larson, K. A.; King, M. L. Evaluation of Supercritical Fluid Extraction in the Pharmaceutical Industry. *Biotechnol. Prog.* **1986**, *2* (2) 73–82.

Lele, A. K.; Shine, A. D. Dissolution and Precipitation of Polymers Using a Supercritical Solvent. *Polymer Prepr.* **1990**, *31*, 677–678.

Lele, A. K.; Shine, A. D. Morphology of Polymers Precipitated from a Supercritical Solvent. *AIChE J.* **1992**, *38*, 742–752.

Lele, A. K.; Shine, A. D. Effect on RESS Dynamics on Polymer Morphology. *Ind. Eng. Chem. Res.* **1994**, *33*, 1476–1485.

Levien, K. L.; Morrell, J. J.; Sahle-Demessie, E. Impregnating Porous Solids Using Supercritical CO_2. CHEMTECH *March*, **1998**, 12–18.

Lewis, J.; Argyropoulos, J. N.; Nielson, K. A. Supercritical Carbon Dioxide Spray Systems. *Metal Finishing* **1997**, *95* (4), 33–41.

Liu, G. T.; Nagahama, K. Solubility and RESS Experiments of Solid Solution in Supercritical Carbon Dioxide. *J. Chem. Eng. Jpn.* **1997**, *30*, 293–301.

Ma, X.; Tomasko, D. L. Coating and Impregnation of a Nonwoven Fibrous Polyethylene Material with a Nonionic Surfactant Using Supercritical Carbon Dioxide. *Ind. Eng. Chem. Res.* **1997**, 736, 1586–1597.

Matson, D. W.; Petersen, R. C.; Smith, R. D. Formation of Silica Powders from the Rapid Expansion of Supercritical Solutions. *Adv. Ceram. Mater.* **1986a**, *1*, 242–246.

Matson, D. W.; Petersen, R. C.; Smith, R. D. The Preparation of Polycarbosilane Powders and Fibers during Rapid Expansion of Supercritical Fluid Solutions. *Mater. Lett.* **1986b**, *4*, 429–432.

Matson, D. W.; Fulton, J. L.; Petersen, R. C.; Smith, R. D. Rapid Expansion of Supercritical Fluid Solution: Solute Formation of Powders, Thin Films and Fibers. *Ind. Eng. Chem. Res.* **1987**, *26*, 2298–2306.

Mawson, S.; Johnston, K. P.; Combes, J. R.; DeSimone, J. M. Formation of Poly(1,1,2,2-tetrahydroperfluorodecyl acrylate) Submicron Fibers and Particles from Supercritical Carbon Dioxide Solutions. *Macromolecules* **1995**, *28*, 3182–3191.

McGillis, D. A. In *VLSI Technology*; Sze, S. M., Ed.; McGraw Hill: New York, 1983; p. 267.

Meyerhofer, D. Characteristics of Resist Films Produced by Spinning. *J. Appl. Phys.* **1978**, *49*, 3993–3997.

Mohamed, R. S.; Halverson, D. S.; Debenedetti, P. G.; Prud'homme, R. K. In *Supercritical Fluid Science and Technology*; Johnston, K. P., Penninger, M. L., Eds.; American Chemical Society: Washington, DC, 1989a; pp. 355–378.

Mohamed, S. R.; Debenedetti, P. G.; Prud'homme, R. K. Effects of Process Conditions on Crystals Obtained from Supercritical Mixtures. *AIChE J.* **1989b**, *35*, 325–328.

Moreau, W. M. *Semi-Conductor Lithography*; Plenum Press: New York, 1988.

Moreau, W.; Cornett, K.; Fahey, J.; Linehan, L.; Montgomery, W.; Plat, M.; Smith, R.; Wood, R. In *Proceedings SPIE*; Fuller, G. E., Ed.; The International Society for Optical Engineering: 1997; Vol. 2438.

Muirhead, J. In *Science and Technology of Surface Coating*; Chapman, B. N., Anderson, J. C., Eds.; Academic Press: London and New York, 1974.

Nielsen, K. A.; Busby, D. C.; Glancy, C. W.; Hoy, K. L.; Kuo, A. C.; Lee, C.; Perry, K. M. Spray Application of Low-VOC Coatings Using Supercritical Fluids. *Soc. Automot. Eng. Trans.* **1991a**, *100*, 9–16.

Nielsen, K. A.; Glancy, C. W.; Hoy, K. M.; Perry, K. M. In *Proceedings of the Fifth International Conference on Liquid Atomization and Spray Systems*; Hratch, G. S., Ed.; Gaithersburg, MD, 1991b; NIST SP813, pp. 367–374.

Novick, B. J.; Carbonell, R. G.; DeSimone, J. M. Presented at AIChE National Meeting, Dallas, TX, 1999.

Ober, C. K.; Gabor, A. H.; Gallagher-Wetmore, P.; Allen, R. D. Imaging Polymers with Supercritical Carbon Dioxide. *Adv. Mater.* **1997**, *9*, 1039–1043.

Parodi, M.; Batchelder, T.; Haaland, P.; McKibben, J. Spin Coating and Alternative Techniques for Flat Panel Displays. *Semicond. Int.* **January 1996**, 101–105.

Paul, S.; Halle, O.; Einsiedel, H.; Menges, B.; Mullen, K.; Knoll, W.; Mittler-Neher, S. An Anthracene-Containing PMMA Derivative for Photoresist and Channel Waveguide Applications. *Thin Solid Films* **1996**, *288*, 150–154.

Perman, C. A.; Bartkus, J. M.; Choi, H. H.; Riechert, M. E.; U.S. Patent 5,340,614, 1993.

Perrut, M. Presented at Future Trends for Non Conventional Use of Supercritical Fluids, Fourth Italian Conference on Supercritical Fluids and their Applications, Capri, Italy, September 7–10, 1997.

Rehg, T. J.; Higgins, B. G. Ph.D. Thesis, University of California, Davis, 1989.

Riley, D. J.; Carbonell, R. G. Mechanisms of Particle Deposition from Ultrapure Chemicals onto Semiconductor Wafers: Deposition from a Thin Film of Drying Rinse Water. *J. Colloid Interface Sci.* **1993**, *158*, 274–288.

Sahle-Demessie, E.; Levien, K. L.; Morrell, J. J. In *Innovations in Supercritical Fluids: Science and Technology*; Hutchenson, K. W., Foster, N. R., Eds.; American Chemical Society: Washington, DC, 1995.

Sand, M. L. U.S. Patent 4,598,006, 1986.

Saus, W.; Knittel, D.; Schollmeyer, E. Dyeing of Textiles in Supercritical Carbon Dioxide. *Textile Res. J.* **1993**, *63*, 135–142.

Schelle, C.; Mennig, M.; Krug, H.; Jonschker, G.; Schmidt, H. One Step Antiglare Sol–Gel Coating for Screens by Sol–Gel Techniques. *J. Non-Cryst. Solids* **218**, 163–168.

Schut, J. H. Novel Low-VOC Paint Technology. *Plast. Technol.* **1991**, *37*, 29–37.

Scriven, L. E.; Higgins, B. G. Interfacial Shape and Evolution Equations for Liquid Films and Other Viscocapillary Flows. *Ind. Eng. Chem. Fundam.* **1979**, *18*, 208–215.

Scriven, L. E.; Suszynski, W. J. Take a Closer Look at Coating Problems. *Chem. Eng. Prog.* **1990**, *86*, 24–29.

Shim, J.-J.; Yates, M. Z.; Johnston, K. P. Polymer Coatings by Rapid Expansion of Suspensions in Supercritical Carbon Dioxide. *Ind. Eng. Chem. Res.* **1999**, *38*, 3655–3662.

Skripov, V. P. *Metastable Liquids*; John Wiley & Sons: New York, 1974.

Smith, R. D. U. S. Patent 4,582,731, 1986a.

Smith, R. D.; Fulton, J. L.; Peterson, R. C.; Kopriva, A. J.; Wright, B. W. Performances of Capillary Restrictors in Supercritical Fluid Chromatography. *Anal. Chem.* **1986b**, *58* (9), 2055–2084.

Subramaniam, B.; Rajewski, R. A.; Snavely, W. K. Pharmaceutical Processing with Supercritical Carbon Dioxide. *J. Pharm. Sci.* **1997**, *86*, 885–890.

Takahashi, Y.; Okada, S.; Tahar Radhouane Bel, H.; Nakano, K.; Ban, T.; Ohya, Y. Dip-Coating of ITO Films. *J. Non-Cryst. Solids* **1997**, *218*, 129–134.

Tallmadge, J. A. Withdrawal of Flat Plates from Power Law Fluids. *AIChE J.* **1970**, *16*, 925–930.

Tallmadge, J. A. Draining Films in Unsteady Withdrawal. *AIChE J.* **1971a**. *17*, 760–761.

Tallmadge, J. A. On the Liquid Film which Occurs in a Draining Vessel. *J. Phys. Chem.* **1971b**, *75*, 583–585.

Tallmadge, J. A. A Theory of Entrainment for Angular Withdrawal of Flat Supports. *AIChE J.* **1971c**, *17*, 243–245.

Tallmadge, J. A.; Gutfinger, C. Entrainment of Liquid Films Drainage, Withdrawal, and Removal. *Ind. Eng. Chem.* **1967**, *59*, 19–35.

Tallmadge, J. A.; Lang, K. C. A Postwithdrawal Expression for Drainage on Flat Plates. *Ind. Eng. Chem. Fundam.* **1971**, *10*, 648–650.

Tallmadge, J. A.; Lee, C. Y. Dynamic Meniscus Profiles in Free Coating III. Predictions Based on Two Dimensional Flow Fields. *AIChE J.* **1974**, *20*, 1079–1086.

Tom, J. W.; Debenedetti, P. G. Particle Formation with a Supercritical Fluids—A Review. *J. Aerosol Sci.* **1991**, *22*, 555–584.

van de Ven, T. G. M. The Capture of Colloidal Particles on Surfaces and in Porous Material: Basic Principles. *Colloids Surf.* **1998**, *138*, 207–216.

Vroon, Z.; Spee, C. Sol–Gel Coatings on Large Area Glass Sheets for Electrochromic Devices. *J. Non-Cryst. Solids* **1997**, *218*, 189–195.

Woods, R. G.; Busby, D. C. Cost-Effective Coating with Supercritical Fluid Technology. *Mater. Proc.* **1995**, *34* (11), 45–58.

Yang, C. C.; Josefowicz, J. Y.; Alexandru, L. Deposition of Ultrathin Films by a Withdrawal Method. *Thin Solid Films* **1980**, *74*, 117–27.

Yonkoski, R. K.; Soane, D. S. Model for Spin Coating in Microelectronic Applications. *J. Appl. Phys.* **1992**, *72* (2), 725–740.

13

Dry Cleaning with Liquid Carbon Dioxide

GINA STEWART

The process of cleaning one item invariably involves making something else dirty. Whether that something else is an organic or halogenated solvent, soapy water, or a rag, we seldom address the dirtying that accompanies any cleaning process. If we are to achieve environmentally benign cleaning, we must look at the life cycle of solvents employed for cleaning, including the potential for recycling, reuse, or release into the environment. Truly "green" cleaning processes not only minimize the amount of waste generated; but also they prevent the dispersal of that waste into large amounts of solvent, water, soil, or air.

Dense-phase carbon dioxide is a great cleaning solvent from a pollution-prevention viewpoint. By-product CO_2 generated by other industrial processes can be captured, so it is not necessary to generate CO_2 specifically for cleaning. Spills of CO_2 will not contaminate groundwater or create a need for soil remediation. Carbon dioxide even has advantages for the work environment, since no chronic, harmful effects are known from repeated inhalation of low concentrations of CO_2.

The barriers to using CO_2 as a cleaning solvent have centered around two issues: the expense of high-pressure equipment and the poor solubility of many contaminants in CO_2. Micell Technologies, Inc., based in Raleigh, NC, has addressed the equipment issue by using liquid CO_2 just below ambient temperature (~18–22 °C) and vapor pressure (~50 bar). The equipment needed to contain this pressure is considerably less expensive than that needed for supercritical CO_2 processes. As for the second barrier, Micell has surfactant packages that enhance the ability of CO_2 to dissolve many contaminants commonly found on clothes or on metal parts.

Micell is in the process of designing and bringing to market integrated CO_2 solutions, including equipment and appropriate chemistries, to replace the organic solvents or water traditionally used in garment dry cleaning, metal degreasing, and textile processing.

Conventional Garment Dry Cleaning

Dry cleaning is a bit of a misnomer, in that clothes are cleaned in a liquid solvent. "Dry" simply means that exposure of a garment, such as a wool suit or silk blouse, to water is minimized to prevent damage to hydrophilic fibers. Some visible soils are treated with soil-specific cleaning chemistries; this is termed pre- or post-spotting, depending on whether the treatment takes place before or after processing in the bulk dry cleaning solvent.

Dry cleaning establishments (plants) are typically located in neighborhoods as close to the consumer as possible. In this decentralized industry, a typical plant processes 24,000 kg of clothing per year (USEPA, 1998), although volumes vary widely. "Mom and pop" shops prevail, with many plants being run by owner/operators and a few family members. Although this may at first glance seem an unlikely arena in which to introduce a sophisticated new technology, environmental pressure on the industry has reached epic proportions. Those who want to remain in the business are looking for green alternatives.

Perchloroethylene

Since the 1960s, the prevailing solvent at the 36,000 dry cleaning establishments in the United States has been perchloroethylene (also called perc or PCE) (USEPA, 1998). Process improvements implemented in the dry cleaning industry have greatly reduced perc usage in recent years, but annual consumption is approximately 53 million kg (Mannsville, 1997).

Despite the widespread use, there are many health, safety, and environmental concerns associated with the use of perc. Perchloroethylene has been described as a probable human carcinogen based on both laboratory animal studies and human epidemiological studies (IARC, 1995). According to the USEPA (1998), there "is a reasonable basis to conclude that there can be a health risk for cancer and some non-cancer effects to workers from the relatively high PCE exposures observed on the average in the dry cleaning industry." Risks also exist for apartment residents colocated with a perc dry cleaning facility.

Perc spills have led to groundwater and soil contamination. Of 1190 National Priorities List sites in the United States, nearly two-thirds (771) have perchloroethylene contamination (ATSDR, 1995). Even filters and lint from perc machines must be disposed of as hazardous waste. In addition, ever-increasing taxes and regulations on the use of perchlorothylene reflect the growing public concern over continued use of this solvent.

Other Solvents

Hydrocarbon (or petroleum) solvents are also used for dry cleaning. From the original Stoddard solvent to Exxon's latest DF 2000 offering, the main drawback has been the fuel potential of hydrocarbon solvents. It was the flammability of Stoddard solvent that turned the industry away from petroleum solvents to the nonflammable halogenated solvents such as perc, in the first place. Other barriers to the widespread use of petroleum arise over zoning restrictions, taxation, and inevitable regulation related to its use.

Professional wet cleaning—while environmentally safe when effluent is treated—can not clean the range of "dry clean only" garments that CO_2 can. Water exposure is not appropriate for many fabric types, and drying and pressing garments after water exposure is often time- and labor-intensive. Furthermore, water is an increasingly scarce resource, and purifying water and drying garments are both energy-intensive processes.

Carbon Dioxide

The introduction of environmentally friendly CO_2 as a dry cleaning solvent is the most significant innovation in dry cleaning since the 1950s. As of this writing, CO_2 dry cleaning has moved from a developmental product to commercial reality. Over 100,000 kg of customer clothing have been processed in liquid CO_2, with both dry cleaning operators and consumers pleased with the results.

The use of CO_2 circumvents the environmental concerns and the worker and consumer health issues associated with perc use. Carbon dioxide is nonflammable and thus does not have the site restrictions often encountered with petroleum. The controlled level of moisture in a CO_2 system does not damage hydrophilic fibers. And best of all, CO_2 cleaning leads to a high-quality finished product that is safer for clothes, workers, consumers, and the environment.

The Micare System

The Micare system includes everything needed to dry clean clothes in liquid carbon dioxide, including the machine, detergents, spotters, pretreatment, and liquid carbon dioxide. The components of this integrated package—machine, detergents, and process—have been designed to work in harmony. Each is essential to the working of the Micare system and is not interchangeable with other suppliers' offerings.

Design for the Environment, a cooperative project between the U.S. Environmental Protection Agency and the garment and textile care industry, recently published a case study of liquid CO_2 cleaning based on the Micare system (USEPA, 1999).

The Micare system utilizes a specially designed, 25-kg-capacity $MICO_2$ machine (Figure 13.1) that can hold liquid CO_2. A detergent system improves the cleaning ability of the liquid CO_2, allowing it to remove the dirt from the garments. After the cleaning cycle, the machine pulls the wash fluid away from the clothes, and then cleans and recycles it. Clothes are not heated in the Micare process, which is very gentle to fabric.

The Machine

The $MICO_2$ machine operates in a manner similar to conventional dry cleaning machines and front-loading washers. Garments are placed in the large horizontally mounted rotating basket in the wash tank (or "wheel", in dry cleaning industry jargon). The wash tank is not completely filled with wash fluid—in fact, the tank is only about one-third full to establish a liquid–vapor interface. The basket can be rotated clockwise or counterclockwise, so that the mechanical action of clothes tumbling through the liquid–vapor interface enhances cleaning. The direction of rotation is reversed periodically to prevent garment entanglement. The machine is programmed with several different cycles optimized for different cleaning needs, such as

Figure 13.1. The $MICO_2$ machine.

gentle wash, power wash, and acetate cycle. Salient features of the machine are summarized in Table 13.1.

The Chemistry

The first available detergent formulation is analogous to a conventional "charged" dry cleaning system, with makeup detergent added based on the rate of distillation. An injection detergent is currently under development as well. Micare detergent formulations are proprietary and are designed to improve the cleaning capacity of the carbon dioxide wash fluid. Micare products are nontoxic and contain no chlorinated chemicals.

Carbon Dioxide Delivery and On-Site Storage

$NuCO_2$, a CO_2-delivery company based in Stuart, FL, supplies liquid CO_2 to Micell's dry cleaning customers. The Micare System uses the same beverage-grade bulk carbon dioxide that $NuCO_2$ delivers to more than 50,000 restaurants, stadiums, bars, convenience stores, and other fountain beverage dispensers in the United States. $NuCO_2$ leases and installs bulk CO_2 systems and has a 24-h refilling service.

Table 13.1. Salient Features of the $MICO_2$ Dry Cleaning Machine

Made in the USA	35–45 min cycle
	25 kg capacity
	Large rotating basket
Ultrafine filtration system	5–10 μm lint filter
	32 kg granular carbon canister
Automated detergent addition	Maintains correct concentration
Superior mechanical action	Traditional rotating basket
	High fluid turnover
	Liquid–gas interface
	Low rpm spin extract
	Tumbling during vapor recovery
Automated still sweep	Automated still bottom collection
	Ensures clean solvent
	Residue recycled by Micell
Efficient recovery of CO_2	No heated drying cycle
	Gentle tumbling throughout
	98% CO_2 recovered

The Process

The marriage of machine, chemistry, and process leads to effective cleaning. The process is best explained with the use of a diagram of the machine components as shown in Figure 13.2.

Garments (up to 27 kg) are placed inside the basket of the $MICO_2$ machine, the door is closed, and the system is sealed. A vacuum pump removes the majority of the air in the system, then CO_2 gas is added from bulk CO_2 storage to prepressurize the wash tank.

Carbon dioxide gas is then added from the working tank until the pressures in the working tank and wash tank are equal. Once the pressure has equilibrated, the wash fluid, which consists of liquid CO_2 and the Micare detergent package, is transferred from the working tank to the wash tank. The basket inside the wash tank is rotated, and the clothes are agitated for a desired time period and with a selected speed depending on the nature of the garments (e.g., delicate cycle, heavy-duty cycle). The wash fluid is circulated in a loop comprising a lint filter to capture loose fibers and lint; a carbon filter to remove odorants, soils, and dyes; and the wash tank. The basket continues to rotate as the fluid is circulated.

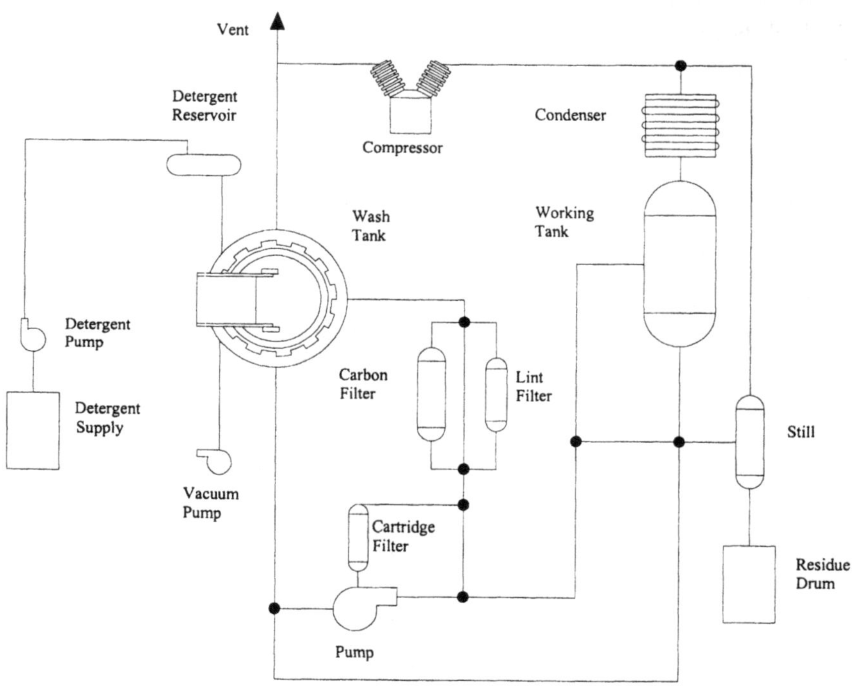

Figure 13.2. Simplified diagram of the components of the $MICO_2$ machine.

After a predetermined time, the wash fluid is pumped out of the wash tank to the working tank. Excess liquid that clings to the garments is removed by a high-speed spin extract cycle. A portion of the wash fluid is sent to the still to be cleaned up through a distillation process. Carbon dioxide gas is removed from the wash tank using a compressor, liquified in a condenser, and returned to the working tank for reuse. Most (98%) of the CO_2 is recycled, while a nominal amount (about 5 kg) of CO_2 gas is vented outside the plant. The cleaned garments are then removed from the wash tank after a cycle time of 35–45 min. Garments are cool to the touch and easy to finish.

Some of the wash fluid is distilled each cycle to remove dirt and contaminated detergent. The distillation residue, which contains detergent and dirt removed from the clothes, is automatically eliminated from the machine and collected for recycling. The $MICO_2$ machine does have the capability to distill on demand for a dye bleed or similar situation, although fewer dye bleeds have been seen with the Micare system compared with conventional dry cleaning solvents.

Performance Advantages

In addition to the distinct advantages to the environment and to the health of both dry cleaning workers and consumers because of the benign nature of the solvent, there are many performance advantages inherent to CO_2 cleaning.

A Gentler System for Fabrics

Lint accumulates in any cleaning system. The lint is generated by fiber loss from the garments—it literally comes from decomposition of the clothes! The Micare system generates less lint per kilogram of garment cleaned than other professional fabricare methods, which is evidence that damage to clothing is minimized. The combination of heat and mechanical action is damaging to fibers, and the Micare process eliminates exposure of the garments to hot drying cycles. Therefore, garments processed in the Micare system last longer.

Another important indicator of gentleness is found in whether or not a cleaning method causes color loss or change in garments. Color change is less pronounced in the Micare system than in other professional fabricare processes for a wide variety of color and fabric combinations. For example, scientists from the College of Textiles at North Carolina State University working with Micell scientists have compared the colorfastness associated with dry cleaning three pairs of black 100% cotton men's chino pants and three red ladies sweater vests composed of 90% silk and 10% cotton (Table 13.2).

Table 13.2. Color Fastness for Garments Cleaning in Perchloroethylene and the Micare Liquid CO_2 Process

	ΔC	ΔE
Black pants cleaned in perc		
Average daylight	0.38	3.15
Incandescent	0.43	3.12
Fluorescent	0.46	3.11
Black pants cleaned in Micare		
Average daylight	0.12	0.40
Incandescent	0.13	0.40
Fluorescent	0.15	0.39
Red vest cleaned in perc		
Average daylight	−4.39	4.39
Incandescent	−5.08	5.12
Fluorescent	−2.87	2.96
Red vest cleaned in Micare		
Average daylight	−1.02	1.07
Incandescent	−1.32	1.39
Fluorescent	−0.48	0.50

One of each garment was dry cleaned 20 times in perchloroethylene, one was dry cleaned 20 times in the Micare liquid carbon dioxide system, and one was used as a standard control sample and was not cleaned. The color change in the cleaned garments was measured using a high-resolution color reflectometer, and the changes were scaled versus the noncleaned identical garment.

Two types of color changes were followed: change in intensity of reflected color and change in energy. The intensity is seen as color depth or color fading, while a change in energy appears as a color change, for example from red to orangey-red. The smaller the Δ, the less change caused by the dry cleaning process. The change in the intensity of the total chroma in a given color composition against a standard is quantified by ΔC. Positive numbers reflect increases in intensity while negative numbers reflect decreases in intensity.

The overall change in energy of reflected light is quantified by ΔE and usually is considered the best measurement for describing color loss. The nonlinearity of absorption and reflection of different wavelengths of light makes it difficult to compare color loss from one garment to another. Although it varies from one color to another, a ΔE of less than 1 typically cannot be distinguished from garment to garment by the human eye. In other words, the black pants and red vest cleaned 20 times in the liquid CO_2 Micare System are virtually indistinguishable from the identical garments never cleaned. However, an identical red vest and black pants cleaned 20 times in perchloroethylene show noticeable color loss and fading. Qualitatively similar trends have been observed for many other colors. It is clear from this example that colorfastness is superior in the Micare system.

A Gentler System for Decorations

Many buttons, sequins, and similar garment decorations can dissolve, deform, or suffer other damage or discoloration when exposed to perc. Traditionally, dry cleaners have removed suspect decorations before cleaning in perc and have had to resew the garment after processing. These decorations often can be left on for processing in CO_2 since it is a far less aggressive solvent than perchloroethylene for most substrates, resulting in substantial time savings for the dry cleaner.

It is somewhat surprising that the large variety of buttons and decorations survive the CO_2 cleaning, given the well-documented suppression of glass transition temperature for many polymers in CO_2. In early experiments, buttons were turned to "popcorn" by too-rapid outgassing of CO_2, especially after exposure to supercritical conditions followed by rapid depressurization. The operating pressures and temperatures of the Micare system, combined with controlled depressurization, prevent problems due to outgassing.

Ambient Temperature Advantages

In conventional systems, heating drying cycles are used to evaporate residual solvent from clothing. Clearly, this is not necessary with a liquified gas such as CO_2. Garments are kept at or slightly below room temperature during the entire cleaning process, including vapor recovery, which is analogous to a conventional drying step. The cooler temperatures result in less fiber damage and shrinkage. One study at Micell has shown that no shrinkage occurs upon processing in the Micare system for the following fabrics: grey cotton/polyester (65%/35%), purple silk garbadine, grey worsted wool, light blue rayon, blue cotton chintz, black polyester, beige linen, blue polyester/acetate blend, green nylon, and red silk. Minor shrinkage was noted for a navy acetate satin (1.75%) and for tan lycra/acetate (0.5%). In comparison, the light blue rayon, green nylon, and tan lycra/acetate shrank by 1.5%, 0.5%, and 2.5%, respectively, when processed in perc, while no shrinkage occurred in the other fabrics.

In addition, any stains not removed either by pretreatment or during the wash cycle can be heat-set in conventional systems through accelerated oxidation or denaturing of stain components. Some heat-set stains can never be removed. By avoiding an elevated-temperature drying step, the Micare process eliminates the heat-setting of stains, and minimizes time spent on post-spotting procedures.

For approximately 1400 kg of garments that were commercially cleaned at the Williams Hangers™ Cleaners in Wilmington, NC, in one and a half weeks during January 1999, about 12% of the garments required post-spotting. Over two-thirds of the garments received with visible soils did not require any post-spotting after cleaning in the Micare system. The average load was 19 kg (not machine limited) and contained 35–40 garments.

Garment Processing Advantages

The operating temperatures of the Micare system lead to another advantage—cool, damp garments are easier to press than garments dried at high temperatures. Since usually garments are placed individually on presses by workers, and since labor costs can amount to 40% of the typical dry cleaning plant operating costs, any reduction in pressing requirements results in substantial cost savings. In fact, one facility using the Micare system reports that in many cases garments can be finished in a steam tunnel instead of on presses, further reducing the amount of labor involved in finishing the garments (Ustanik, 1999).

Less Odor

Dry cleaning customers and employees of dry cleaning plants often complain about the "chemical" odor of perchloroethylene. More recently, with the recent process changes to reduce perc consumption and emission to the atmosphere, solvent turnover is minimized and complaints have arisen about a "stale, stagnant and unpleasant odor that the customer relates to as clothes being cleaned in a dirty solvent" (ISFA. 1999). Carbon dioxide dry cleaning clearly has an advantage here in that CO_2 is odorless.

Challenges

As with any cleaning process, CO_2 dry cleaning has benefits and limitations. Fiber plasticization in the presence of a good carrier for dyes (i.e., some detergent formulations) can result in dye loss, leading to color change over time. In general, plasticization is only problematic with triacetate and, to a lesser extent, acetate fabrics. The success of acetate cleaning in CO_2 systems is highly dependent on the detergent used. Micell has worked carefully to formulate a detergent that is safe for the majority of acetate fabrics.

The shrinkage of ordinary acetate fabrics, such as suit linings, is not a problem in the Micare system. Since susceptibility to shrinkage (and dye loss) increases with the degree of acetylation of the acetate fiber, the exposure of triacetate fabric to CO_2 should be avoided when possible. Triacetates, which are rarely encountered, can be successfully wet-cleaned. Most other acetates can be successfully processed in the Micare system, including acetate suit lining.

Costs

Capital Costs

The amount of capital required to purchase machinery is a primary concern for dry cleaners shopping for alternatives to perc cleaning systems. The up-front investment for the liquid CO_2 system is more than for conventional technologies, and will be more attractive to larger dry cleaning businesses with high volumes than to "mom and pop" operations. The $MICO_2$ machine, which can process up to 27 kg of clothing in a 35–45 min cycle time, allows for a much higher throughput than conventional machines. In some cases, one $MICO_2$ machine can replace two conventional machines. The $MICO_2$ machine retailed for approximately $150,000 in 1999. Low operating costs combined with high productivity ensure that the Micare system gives an adequate return on investment. In addition, $MICO_2$ machine lifetime is expected to be at least twice as long as conventional dry cleaning machines because the machine is manufactured using mostly stainless steel. As a comparison, a 50-lb perc machine can range from $40,000–$65,000 if it is totally up-fitted, while a petroleum machine with fire suppression systems and oxygen sensors can range from $75,000–$110,000.

Operating Costs

Compared with conventional dry cleaning technologies, the variable costs per kilogram of clothes cleaned in the Micare system is competitive with current technologies. The results of a detailed analysis of operating costs from three dry cleaning establishments that use perchloroethylene, one that uses Stoddard solvent, and three that use Exxon's DF-2000 petroleum have been published (USEPA, 1999). The analysis (which included cleaning-cycle time detergent costs, solvent cost and usage, labor and chemical costs associated with pre- and post-spotting, labor costs for garment finishing, filter cartridges, waste disposal costs, and equipment maintenance costs) showed the variable costs for the Micare system to be $1.34 per kilogram of clothes cleaned. Conventional technology variable costs ranged from $1 10–$2.27 per kilogram of clothes cleaned.

DryWash™

Another CO_2 dry cleaning system has been developed by Raytheon Environmental Systems, El Segundo, CA, and Los Alamos National Laboratory, NM, and is being licensed and marketed by Global Technologies of El Segundo (McCoy, 1999). The DryWash system utilizes jets of liquid CO_2 to agitate clothing. DryWash machines, with average 10-kg loads, have a smaller capacity than the $MICO_2$ machine, and are made by a number of manufacturers, including Alliance Laundry Systems LLC, Comeco2, Sailstar, and Wascator (Global, 1999). The machines use a proprietary CO_2-based solvent called DryWash Fluid that is delivered, premixed, by a local dry cleaning distributor (Global, 1999).

Conclusion

The Micare process is a commercial reality for garment dry cleaning. As acceptance of this new technology spreads, we at Micell anticipate that CO_2 will become the preferred solvent for dry cleaning, bringing environmental advantages inherent in CO_2 solvent technologies to the neighborhood cleaners. By virtue of its proximity to the marketplace and to the consumer, garment dry cleaning may very well become the breakthrough technology that motivates the ordinary citizen to demand environmentally responsible processes wherever possible, including the widespread adoption of CO_2 technology for other applications.

Acknowledgments The information presented in this chapter reflects the contributions of many of my colleagues at Micell Technologies. Special thanks are due to James McClain for the summary of machine attributes in Table 13.1, to Joseph Voelker for the $MICO_2$ diagram, to Stacy Geurin and to Dr. James DeYoung for supplying the dimensional stability and colorfastness data, to Roger Mustian for the variable costs financial analysis, and to Chris Harbinson, David Cauble, and James DeYoung for the data compiled at Williams Hangers.

References

ATSDR (Agency for Toxic Substances and Disease Registry). Toxicological Profile for Tetrachloroethylene Draft for Public Comment; Public Health Service: Atlanta, GA, 1995.

Global. Global Technologies Promotional Literature from Clean '99 Dry Cleaning Show, Orlando, FL, June 1999.

IARC (International Agency for Research on Cancer). Tetrachloroethylene. In *Dry Cleaning, Some Chlorinated Solvents and Other Industrial Chemicals*; IARC Monographs on the Evaluation of Carcinogenic Risks to Humans; IARC: Lyon, France, 1995, Vol. 63, pp. 159–221.

ISFA (Illinois State Fabricare Association). Odor in Clothes: A Rising Customer Complaint. *ISFA Newsletter* **1999**, *March/April*, p. 5.

Mannsville. Mannsville Perchloroethylene Chemical Products Synopsis. Mannsville Chemical Products Corporation: Asbury Park, NJ, 1997.

McCoy, M. Industry Intrigued by CO_2 as Solvent. *Chem. Eng. News* **1999**, *77* (24), 11–13.

USEPA (United States Environmental Protection Agency, Office of Pollution Prevention and Toxics). *Cleaner Technologies Substitutes Assessment for Professional Fabricare Processes*; EPA 744-B-98-001. USEPA: Washington, DC, 1998.

USEPA (United States Environmental Protection Agency, Design for the Environment). *Case Study: Liquid Carbon Dioxide (CO_2) Surfactant System for Garment Care*; EPA 744-F-99-002; USEPA: Washington, DC, May 1999.

Ustanik, T. Personal communication from Tom Ustanik of Lansing Cleaners to Joseph DeSimone, May 1999.

Note Added in Proof

Micell exited the dry cleaning business in 2001; the technology was acquired by Cool Clean Technologies, a spin-out of Chart. See http://www.co2olclean.com, and *Chemical and Engineering News*, September 2, 2002, Vol. *80* (35), p. 12, "ICI Enters CO_2 Dry Cleaning."

The Dry Wash system participants have changed since Chapter 13 was written. As of November 12, 2002, the Dry Wash web site (www.drywash.com) lists the following sublicensees:

Alliance Laundry Systems, LLC—Largest commercial laundry company in North America.

Chart Applied Technologies—Division of world's largest cryogenic equipment manufacturer.

Electrolux Wascator—World's largest commercial laundry company.

Sail Star USA (Charlotte, NC)—U.S. subsidiary of one of Asia's largest dry cleaning equipment manufacturers.

Nuevo Comeco, Spa—Europe's leading developer of pressure systems for gas technology.

Caled Chemical—Leading U.S. maker of detergents, additives, and filters for dry cleaners.

Laidlaw Corp.—Leading U.S. maker of hangers, detergents, and additives for dry cleaners.

Alex Reid, Ltd.—U.K.'s largest provider of chemistry and consumables for dry cleaners.

AGA/Linde—Recently merged company is now the world's second largest industrial gas company.

14

Selective and Complete Hydrogenation of Vegetable Oils and Free Fatty Acids in Supercritical Fluids

THOMAS TACKE
STEFAN WIELAND
PETER PANSTER

As described in other chapters of this book and elsewhere (Jessop, 1999), a wide range of catalytic reactions can be carried out in supercritical fluids, such as Fischer–Tropsch synthesis, isomerization, hydroformylation, CO_2 hydrogenation, synthesis of fine chemicals, hydrogenation of fats and oils, biocatalysis, and polymerization. In this chapter, we describe experiments aimed at addressing the potential of using supercritical carbon dioxide (and carbon dioxide/propane mixtures) for applications in the hydrogenation of vegetable oils and free fatty acids. Supercritical fluids, particularly carbon dioxide, offer a number of potential advantages for chemical processing including (1) continuously tunable density, (2) high solubilities for many solids and liquids, (3) complete miscibility with gases (e.g., hydrogen, oxygen), (4) excellent heat and mass transfer, and (5) the ease of separation of product and solvent. The low viscosity and excellent thermal and mass transport properties of supercritical fluids are particularly attractive for continuous catalytic reactions (Harrod and Moller, 1996; Hutchenson and Foster, 1995; Kiran and Levelt Sengers, 1994; Perrut and Brunner, 1994; Tacke et al., 1998).

There are a number of reports on hydrogenation reactions in supercritical fluids using homogenous and heterogeneous catalysts (Baiker, 1999; Harrod and Moller, 1996; Hitzler and Poliakoff, 1997; Hitzler et al., 1998; Jessop et al., 1999; Meehan et al., 2000; van den Hark et al., 1999). We have investigated the selective hydrogenation of vegetable oils and the complete hydrogenation of free fatty acids for oleochemical applications, since there are some disadvantages associated with the current industrial process and the currently used supported nickel catalyst. The hydrogenation of fats and oils is a very old technology (Veldsink et al., 1997). It was invented in 1901, by Normann, in order to increase the melting point and the oxidation stability of fats and oils through selective hydrogenation. Since the melting point increases during the hydrogenation, the reaction is also referred to as hardening. The melting behavior of the hydrogenated product is determined by the reaction conditions (temperature, hydrogen pressure, agitation, hydrogen uptake). Vegetable oils (edible oils) are hydrogenated selectively for application in the food industry; whereas free fatty acids are completely hydrogenated for oleochemical applications (e.g., detergents).

The reaction mechanism for the selective hydrogenation of edible oils is very complex. Figure 14.1 illustrates a reaction scheme for linoleic acid. In this scheme, $(n:m)$ is used to represent an oil with n carbon atoms and m double bonds. There are several parallel, consecutive, and side reactions. Oleic acid (*cis* 18:1) is the desired product when the reaction starts with linolenic (all-*cis* 18:3) or linoleic acid (*cis*, *cis* 18:2). In the hydrogenation of linolenic and linoleic acid, elaidic acid (*trans* 18:1) is formed in a *cis*/*trans* isomerization reaction. From the viewpoint of dietics, elaidic acid is an undesirable product; however, its presence increases the melting point of the product in a desirable way. Stearic acid (18:0) is formed in a consecutive reaction, but direct formation from linoleic acid is also possible.

The currently practiced commercial process using nickel catalysts, either supported on kieselguhr or silica, has some disadvantages:

- Discontinuous operation (i.e., batch or semibatch)
- Low space–time yields
- Undesirable by-products as a result of strong hydrogen mass-transfer control
- High variable costs (e.g., labor, energy, filtration)

The use of the supported nickel catalyst also leads to additional problems including undesirable by-products (*trans*-fatty acids) with an impact on

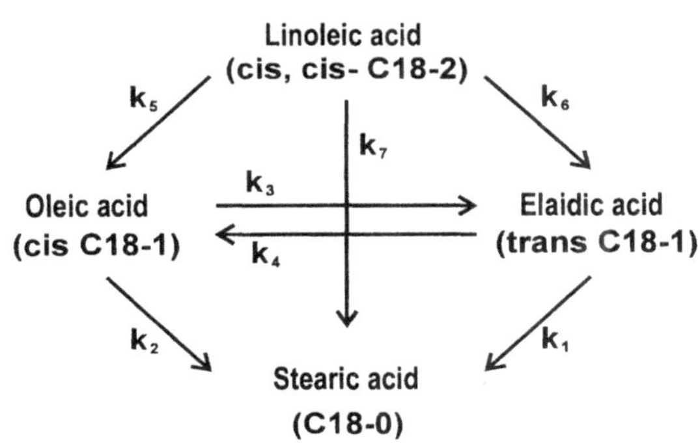

Figure 14.1. Reaction mechanism for the selective hydrogenation of vegetable oils (here, linoleic acid).

health (high cholesterol and lipid level in blood), catalyst deactivation through formation of nickel soaps in free fatty acid hydrogenation, high product-purification costs (e.g., distillation of free fatty acids), and disposal of nickel residues.

In order to overcome the existing problems with the state-of-the-art technology in vegetable oil and free fatty acid hardening, we have investigated hydrogenation reactions in liquid, near-critical, and supercritical CO_2, as well as CO_2/propane mixtures with precious-metal fixed-bed catalysts supported on acid-resistant supports.

Experimental Considerations

Heterogeneously catalyzed hydrogenation reactions can be run in batch, semibatch, or continous reactors. Our catalytic studies, which were carried out in liquid, near-critical, or supercritical CO_2 and/or propane mixtures, were run continuously in oil-heated (200 °C, 20.0 MPa) or electrically heated flow reactors (400 °C, 40.0 MPa) using supported precious-metal fixed-bed catalysts. The laboratory-scale apparatus for catalytic reactions in supercritical fluids is shown in Figure 14.2. This laboratory-scale apparatus can perform in situ countercurrent extraction prior to the hydrogenation step in order to purify the raw materials employed in our experiments. Typically, the following reaction conditions were used in our supercritical fluid hydrogenation experiments: catalyst volume, 2–30 mL; total pressure, 2.5–20.0 MPa; reactor temperature, 40–190 °C; carbon dioxide flow, 50–200 L/h;

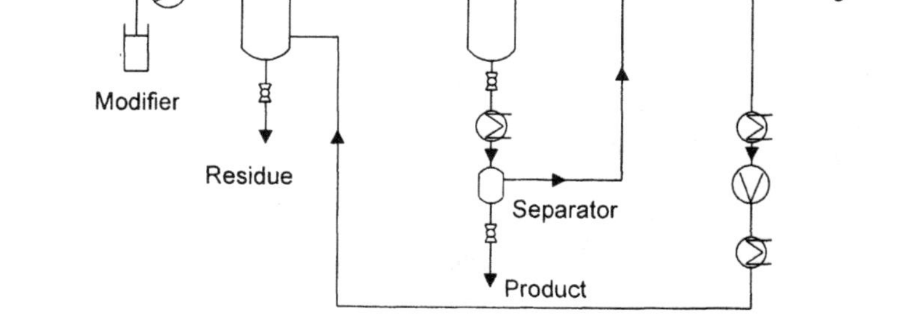

Figure 14.2. Laboratory-scale apparatus for catalytic reactions in supercritical fluids in combination with countercurrent extraction.

propane flow, 50–150 L/h; hydrogen flow, 5–100 L/h, and liquid hourly space velocities (LHSV) between 5 and 240 L/h.

The effect of reaction conditions (temperature, pressure, H_2 flow, CO_2 and/or propane flow, LHSV) and catalyst design on reaction rates and selectivites were determined. Comparative studies were performed either continuously with precious-metal fixed-bed catalysts in a trickle-bed reactor, or batchwise in stirred-tank reactors with supported nickel or precious metal on activated carbon catalysts. Reaction products were analyzed by capillary gas chromatography with regard to product composition, by titration to determine iodine and acid value, and by elemental analysis.

Catalyst Screening Test Results

A large number of heterogeneous catalysts have been tested under screening conditions (reaction parameters: 60 °C, linoleic acid ethyl ester at an LHSV of 30 L/h, and a fixed carbon dioxide and hydrogen flow) to identify a suitable fixed-bed catalyst. We investigated a number of catalyst parameters such as palladium and platinum as precious metal (both in the form of supported metal and as immobilized metal complex catalysts), precious-metal content, precious-metal distribution (egg shell vs. uniform distribution), catalyst particle size, and different supports (activated carbon, alumina, Deloxan®, silica, and titania). We found that Deloxan-supported precious-metal catalysts are at least two times more active than traditional supported precious-metal fixed-bed catalysts at a comparable particle size and precious-metal content. Experimental results are shown in Table 14.1 for supported palladium catalysts. The Deloxan-supported catalysts also led to superior linoleate selectivity and a lower *cis/trans* isomerization rate was found. The explanation for the superior behavior of Deloxan-supported precious-metal catalysts can be found in their unique chemical and physical properties—for example, high pore volume and specific surface area in combination with a meso- and macro-pore-size distribution, which is especially attractive for catalytic reactions (Wieland and Panster, 1995). The majority of our work has therefore focused on Deloxan-supported precious-metal catalysts.

Vegetable Oil Hardening in Supercritical Fluids

We have demonstrated that vegetable oils and fatty acid esters can be selectively hardened in liquid, near-critical, or supercritical CO_2 or propane and in mixtures thereof at temperatures between 60 °C and 120 °C and at a total pressure up to 20.0 MPa. Table 14.2 summarizes the results for the selective hydrogenation of vegetable oils in supercritical CO_2 in comparison with hydrogenation reactions performed in a discontinuous (i.e., batch or semibatch) stirred-tank reactor and in a continuous trickle-bed reactor.

Table 14.1. Catalyst Screening Test Results for the Hydrogenation of Fatty Acid Esters in Supercritical CO_2

Catalyst	Precious Metal	Precious-Metal Content (wt %)	Catalyst Support	Particle Size (mm)	Hydrogenation Activity (mole H_2/h $\times$ g active metal)
A	Palladium	1.0	Deloxan APII (spheres)	0.8–1.8	28.5
B	Palladium	1.0	Activated carbon (extrudates)	2.4 (diameter)	8.4
C	Palladium	0.5	Activated carbon (granules)	2.4–4.8	6.6
D	Palladium	2.0	Alumina (spheres)	0.5–2.0	9.5
E	Palladium	0.5	Alumina (spheres)	2.0–4.0	6.0
F	Palladium	0.5	Silica (tablets)	5.0 $\times$ 6.5	2.6
G	Palladium	0.5	Titania (extrudates)	4.5 (diameter)	1.1
H	Palladium complex	1.0	Deloxan HKI (spheres)	0.1–0.4	18.3
I	Platinum	2.0	Deloxan APII (spheres)	0.4–0.85	3.1

Table 14.2. Selective Hardening of Vegetable Oils in Liquid Phase (Slurry, Trickle Bed) and in Supercritical CO_2

Catalyst	Process	Temperature (°C)	H_2 pressure (MPa)	Space-Time Yield (m^3 oil/h × m^3 reactor vol.)	Hydrogenation Activity (mole H_2/h × g active metal)	Linoleate selectivity (–)	Max. *trans*-Fatty Acid Content by GC (area %)
25% Ni-silica (powder)	Batch/slurry/ mass-transfer controlled	120	0.3	< 1	4.1	10.8	40.0
5% Pd / C (powder)	Batch/slurry/ mass-transfer controlled	120	0.3	< 1	65.4	72.5	60.0
5% Pt / C (powder)	Batch/slurry/ mass-transfer controlled	120	0.3	< 1	43.1	6.2	37.0
1% Pd / Deloxan AP II	Continuous trickle bed	60	0.5	5	6.3	1.9	20.9
1% Pt / Deloxan AP II	Continuous trickle bed	60	2.0	30	28.5	2.1	20.6
1% Pd / Deloxan AP II	New continuous fixed-bed process	60	10.0 (H_2 + CO_2)	60	52.3	3.0	19.0
1% Pd / Deloxan HKI	New continuous fixed-bed process	60	10.0 (H_2 + CO_2)	30	13.3	13.1	15.5
2% Pt / Deloxan AP II	New continuous fixed-bed process	60	10.0 (H_2 + CO_2)	30	3.1	7.8	7.5

Space–time yields in batch hydrogenation reactions with supported nickel and activated carbon supported precious-metal catalysts are below $1 \times m^3$ oil/($h \times m^3$ reactor volume). A large amount of undesirable *trans* fatty acids is formed in the triglyceride molecule as a consequence of hydrogen mass-transfer control. The continuous hydrogenation with Deloxan AP II / 1 wt % Pd, palladium fixed-bed catalyst in the trickle-bed results in a higher space–time yield. However, the linoleate selectivity, a measure of the lack of formation of saturated fatty acids, is very low. Selectivities used to describe the hydrogenation of vegetable oils are defined in Figure 14.3. The increase in the hydrogen partial pressure results in a further increase in space–time yield. This result indicates that the reaction is strongly hydrogen mass-transfer controlled. In supercritical CO_2, hydrogenation activity is even further increased. Supercritical CO_2 lowers the viscosity of the reaction medium and increases mass transfer and diffusivity. Benefits include a higher hydrogenation activity and an increase in the linoleate selectivity. With a Deloxan immobilized palladium phosphine complex catalyst, Deloxan HK I / Pd, the linoleate selectivity is further increased. In comparison to the commercial batch hydrogenation with a supported catalyst, the Deloxan immobilized palladium phosphine complex catalyst in combination with supercritical CO_2 as a solvent gives higher space–time yields, a higher linoleate selectivity, and a significantly decreased *cis*/*trans* isomerization rate. Deloxan AP II / Pt supported platinum catalysts in supercritical CO_2 are less active than Deloxan AP II / Pd supported palladium catalysts, but they show an improved linoleate selectivity and a significantly lower *cis*/*trans* isomerization rate. The overall yield of undesirable *trans* fatty acids is 7.5 GC area % in the vegetable oil hardening using Deloxan AP II / 2% Pt, platinum catalyst. In batch hydrogenation with a commercial supported nickel catalyst, the undesirable *trans* fatty acid content was determined to

Reaction scheme:

$$\text{C18-3} \xrightarrow{k_3} \text{C18-2} \xrightarrow{k_2} \text{C18-1} \xrightarrow{k_1} \text{C18-0}$$

C18-3 (Triunsaturated fatty acids); C18-2 (Diunsaturated fatty acids); C18-1 (Monounsaturated fatty acids)

Selectivities:

Linolenate selectivity: $S_{Ln} = \frac{k_3}{k_2}$

Linoleate selectivity: $S_{Lo} = \frac{k_2}{k_1}$

Specific isomerization: $S_i = \left[\frac{\text{No. of trans double bonds formed}}{\text{No. of hydrogenated double bonds}}\right]$

Figure 14.3. Reaction scheme and selectivity definitions for the selective hydrogenation of vegetable oils.

be 40 area %. However, overall promising linoleate selectivities and low *trans* fatty acid contents have been observed so far only at higher iodine values (iodine value > 90 g iodine per 100 g product). Target iodine values are in the order of 60–90 g iodine per 100 g product. The iodine value of the applied raw material was determined to be 138.5 g iodine per 100 g product.

Propane or propane/CO_2 mixtures as liquid, near-critical, or supercritical fluids enhance the solubility of fats and oils (Harrod et al., 2000; Weidner and Richter, 1999). The decrease in viscosity and increase in diffusivity results in a higher hydrogenation rate (Figure 14.4). Harrod et al. (2000) have also demonstrated activity increases by reducing mass-transfer limitations in supercritical propane.

Complete Hardening of Free Fatty Acids

In the hardening of tallow free fatty acids in supercritical CO_2, we achieved iodine values (IV) below 1.0 g I_2 per 100 g product at space velocities up to 15 h^{-1}. A summary of various fixed-bed processes for the complete hardening of free fatty acids is given in Table 14.3. In comparison to trickle-bed hardening reactions using activated carbon- and titania-supported palladium fixed-bed catalysts (2% Pd), between 6 and 15 times higher space–time yields were obtained with a Deloxan AP II / 1% Pd supported palladium fixed-bed catalyst in supercritical CO_2. The hydrogen partial pressures in both processes were comparable (2.5 MPa H_2). Since the hardening is carried out at a lower temperature compared with the

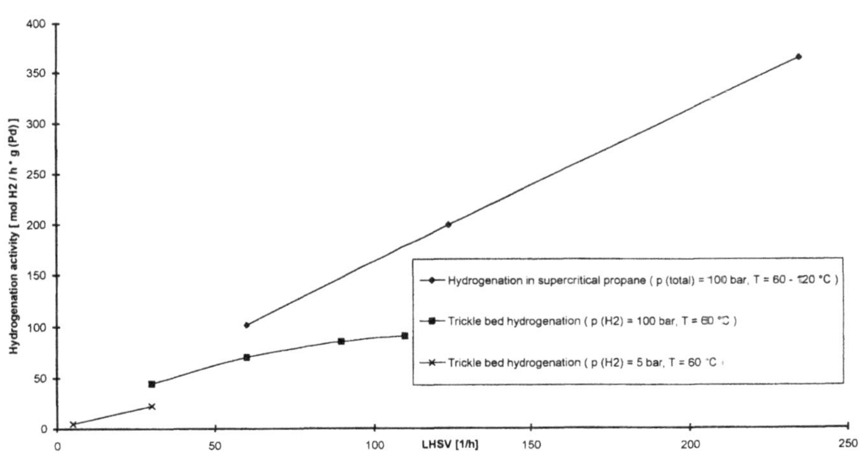

Figure 14.4. Hydrogenation activity in the selective hardening of fatty acid esters using Deloxan AP II / 1% Pd fixed-bed catalyst.

Table 14.3. Hardening of Free Fatty Acids with Precious-Metal Fixed-Bed Catalysts

Literature	Patent Holder	Catalyst	Temperature (°C)	H_2 Pressure (MPa)	Space–Time yield (m^3 oil/h × m^3 reactor vol.)	End Iodine Value (g I_2/100 g product)	Feedstock
CA 1 157 844 (1981)	Ruhr-Chemie	Pd/C	180	2.5	0.2	0.2–0.7	Dist. fatty acids
EP 0 505 863 (1991)	Henkel and Degussa	Pd/TiO_2	169	2.0	1.2	0.3	Raw fatty acids
EP 0 632 747 (1992)	Henkel	Pd/C	150–200	2.0	1–3.2	0.46–0.75	Dist. and raw fatty acids
DE 44 05 029 A1 (1994)	Degussa	Pd/ Deloxan	140–200	14.0 (H_2 + CO_2)	15	< 1.0	Dist. and raw fatty acids

conventional process, the acid value of the fatty acids—a measure for the selectivity of the reaction—remains at a very high level. In catalyst lifetime tests, we observed a three times higher catalyst productivity when using a Deloxan AP / 1% Pd supported palladium fixed-bed catalyst in supercritical CO_2 compared with the same catalyst in a trickle-bed hardening process (Table 14.4).

Complete Hardening of Free Fatty Acids Combined with in situ Extraction

Typically, free fatty acids are purified by distillation prior to and/or after the hydrogenation step to remove catalyst poisons and to achieve the necessary product quality. The distillation is done batchwise. The literature provides considerable information on the separation and purification of triglycerides, free fatty acids, and fatty acid esters based on continuous countercurrent extraction. We constructed a laboratory apparatus that combines catalytic reactions and countercurrent extraction, as shown in Figure 14.2. Carbon dioxide as well as carbon dioxide/acetone and carbon dioxide/propane mixtures could be used for the purification of raw fatty acids at temperatures between 60 °C and 100 °C and at a total pressure of 20.0 MPa, respectively. The used raw fatty acid is characterized by an acid number of 187.9 mg KOH/g and a sulfur content of 84 mg/kg. With a carbon dioxide/propane mixture, we were able to obtain a purified free fatty acid with a sulfur content of 19 mg/kg and an acid number of 193.1 mg KOH/g. These figures are comparable to distilled free fatty acids. The overall product yield of purified free fatty acid was 95.2%. These promising separation results led us to carry out hydrogenation experiments with an in situ countercurrent extraction step for the purification of a raw fatty acid. Catalyst lifetimes were compared with three other hydrogenation experiments: (1) trickle-bed hydrogenation at 2.5 MPa, (2) hydrogenation in supercritical carbon dioxide at 20.0 MPa, and (3) trickle-bed hydrogenation at 20.0 MPa. During all experiments, the same quality of the raw fatty acid was used. The hydrogenation reaction was started in all experiments at a reactor temperature of 140 °C and was increased stepwise during the experiment to compensate for the catalyst deactivation. The lowest iodine values and the longest catalyst lifetime was achieved in the hydrogenation with in situ countercurrent extraction in supercritical carbon dioxide/propane. Thus, we were able to demonstrate the feasibility of catalytic hydrogenation combined with an in situ countercurrent extraction step for the complete hydrogenation of raw fatty acids. This concept, however, still has to be significantly improved to compete with catalyst lifetime figures achieved upon use of fatty acids that have been purified by distillation prior to any trickle-bed or supercritical fluid hydrogenation process.

Table 14.4. Catalyst Productivity for the Hardening of Free Fatty Acids in Liquid Phase (Slurry, Trickle Bed) and in Supercritical CO_2

Catalyst	Process	Temperature (°C)	H_2 Pressure (MPa)	Catalyst Productivity (kg fatty acid/kg catalyst)
25% Ni-Silica (powder)	Batch/slurry/mass-transfer controlled	150–220	2.5	333.3
1% Pd/ Deloxan APII (fixed-bed)	Continuous fixed-bed process	150–200	2.5	2009.2
1% Pd/ Deloxan APII (fixed-bed)	New continuous fixed-bed process	140–200	14.0 (H_2 + CO_2)	6086

Conclusions

The combination of Deloxan-supported precious-metal fixed-bed catalysts and the use of liquid, near-critical, or supercritical CO_2 and/or propane mixtures creates new possibilities for continuous fixed-bed hydrogenations with significantly improved space–time yields and catalyst lifetimes. Short residence times and well-balanced diffusion and desorption of products and reactants results in a decrease in undesirable by-products and therefore higher selectivity. The characteristics of high-pressure hydrogenations in near-critical or supercritical fluids can be summarized as follows:

- Increased mass transfer and gas miscibility
  - high space–time yields
  - less undesirable by-products
- Continuous processing using stable fixed-bed catalysts
  - adjustable residence time of reactants
  - higher selectivities
  - higher product quality
- Increased heat transfer
- In situ purification of raw materials and catalyst reactivation can lead to longer catalyst life.

Currently, however, there is no continuous hydrogenation process design nor fixed-bed catalyst, working under either trickle-bed or supercritical fluid operation, available that will meet future edible oil market requirements in terms of low *trans* fatty acid content, and so on. It is likely that the future technology for the manufacture of edible products will be characterized by the use of other fats and oils, esterification/transesterification, fractionation, and the use of genetically modified oil seeds rather than by hydrogenation technology. With regard to the complete hydrogenation of free fatty acids in supercritical fluids, we have shown the process feasibility in a continuously operated bench-scale reactor with high catalyst lifetimes and very good product qualities in terms of iodine value, color, odor, and acid value. It is possible to increase catalyst lifetime in the complete hydrogenation of free fatty acids by in situ countercurrent extraction. This has to be optimized to compete with the traditional distillation process for raw fatty acids. At the moment, a supercritical fluid hydrogenation process design for the complete hardening of free fatty acids is not economically feasible. From the economical point of view, a trickle-bed process for the complete hydrogenation of purified free fatty acids seems to be more feasible (Buchold et al., 1999).

At the present time, catalytic reactions in supercritical carbon dioxide for the manufacture of fine chemicals appear to be more commercially feasible (McCoy, 1999). A large range of reactions have been investigated in cooperation with the University of Nottingham and Thomas Swan & Co., Ltd. (Consett, Co. Durham, U.K.) (Meehan et al., 1999). Very high space–time yields, selectivities, and catalyst lifetimes have been determined for some reactions (Tacke et al., 1998).

References

Baiker, A. *Chemical Reviews* **1999**, *99*, 453.

Buchold, H.; Tacke, T.; Beul, I.; Rehren, C.; Panster, P. New Palladium Fixed Bed Catalyst and Process for the Continuous Hardening of Unsaturated Fatty Acids. Poster presentation, AOCS Annual Meeting, Orlando, FL, 1999.

Harrod, M.; Moller, P. In *High Pressure Chemical Engineering*; von Rohr, R., Trepp, C.. Eds.; Elsevier: Amsterdam, 1996.

Harrod, M.; Macher, M.-B.; van den Hark, S.; Moller, P. *Proceedings of 5th International Symposium on Supercritical Fluids*; ISSF2000, Atlanta, GA, 2000.

Hitzler, M. G.; Poliakoff, M. *Chem. Comm.* **1997**, 1667.

Hitzler, M. G.; Smail, R. F.; Ross, S. K.; Poliakoff, M. *Organic Proc. Res. Dev.* **1998**, *2*, 137.

Hutchenson, K. W.; Foster, N. R., Eds. *Innovations in Supercritical Fluids: Science and Technology*; American Chemical Society: Washington, DC, 1995.

Jessop, P. G.; Ikariya, T.; Noyori, R. *Chem. Rev.* **1999**, *99*, 475.

Kiran, E.; Levelt Sengers, J. M. H., Eds. *Supercritical Fluids*; Kluwer Academic Publishers: Dordrecht, 1994.

McCoy, M. Industry Intrigued by CO_2 as Solvent. *Chemical Week* **1999**, *June 14*, 11–13.

Meehan, N. J.; Ross, S. K.; Poliakoff, M. *Chem. Ind.* **1999**, *19*, 750.

Meehan, N. J.; Gray, W. K.; Smail, F.; Ross, S.K.; Poliakoff, M. *Proceedings of 5th International Symposium on Supercritical Fluids*; ISSF2000, Atlanta, GA, 2000.

Perrut, M.; Brunner, G., Eds., *Proceedings of the Third International Symposium on Supercritical Fluids*; Strasbourg, France, 1994, Vol. 3.

Tacke, T.; Rehren, C.; Wieland, S.; Panster, P.; Ross, S. K.; Toler, J.; Hitzler, M. G.; Smail, F.; Poliakoff, M. Continuous Hydrogenation in Supercritical Fluids. In *Catalysis of Organic Reactions*; Herkes, F. E., Ed.; Marcel Dekker: New York, 1998; pp. 345–356.

van den Hark, S.; Harrod, M.; Moller, P. *J. Am. Oil Chem. Soc.* **1999**, *76*, 1363.

Veldsink, J. W.; Bouma, M. J.; Schoon, N. H.; Beenackers, A. A. C. M. *Catal. Rev. Sci. Eng.* **1997**, *39*, 253.

Weidner, E.; Richter, D. In *Proceedings of 6th Meeting on Supercritical Fluids*; Nottingham, U.K., 1999.

Wieland, S.; Panster, P. In *Catalysis of Organic Reactions*; Scaros, M. G., Prunier, M. L., Eds.; Marcel Dekker: New York, 1995.

15

Enhancing the Properties of Portland Cements Using Supercritical Carbon Dioxide

JAMES B. RUBIN
CRAIG M. V. TAYLOR
THOMAS HARTMANN
PATRICIA PAVIET-HARTMANN

Supercritical CO_2 (sc CO_2) is being used to accelerate the natural aging reactions (i.e., carbonation) of Portland cement. This treatment method alters the bulk properties of cement, producing profound changes in both structure and chemical composition. As a result of these changes, the mechanical and transport properties of these cements are also dramatically affected, and they display reduced porosity, permeability and pH, as well as increased density and compressive strength.

Two areas of application for the sc CO_2 treatment of portland cement have been undergoing investigation. Because the calcium carbonate ($CaCO_3$) formed during the accelerated carbonation reaction is found to have excellent cementing properties, it is possible to replace a large fraction of the relatively expensive Portland cement with industrial waste products, such as fly ash and kiln dusts, which have inherently inferior cementing properties. These modified Portland cements, incorporating significant volume fractions of industrial wastes, can be used as low-cost building materials. The second area of application deals with the enhancement of Portland cements used to encapsulate waste products. Portland cement is used as an immobilization matrix for low- and intermediate-level radioactive waste by both the U.S. federal government (Huang et al., 1994) and civilian nuclear power companies in the United States (Wilk, 1997) and abroad (Wilding, 1992). Transportation issues relating to water content, radiolysis, and radionuclide content often preclude the ultimate disposal of these cemented wasteforms (U.S. DOE, 1996). However, the structural and chemical changes produced by accelerated carbonation have been shown to address these problems satisfactorily (Hartmann et al., 1999).

Application to Unmodified Portland Cement

To produce cement, calcium carbonate, usually as calcite or limestone, is first calcined to produce lime:

$$3CaCO_3 \xrightarrow{\text{HEAT}} 3CaO + 3CO_2$$

The lime is then added to silica, usually obtained as an aluminosilicate pozzolan (clay or shale), and the mixture is heated in a kiln to produce tricalcium silicate (called alite):

$$3CaO + SiO_2 \xrightarrow{\text{HEAT}} 3CaO \cdot SiO_2 \text{ (or } Ca_3SiO_5 \text{ or } C_3S)$$

Tricalcium silicate is the major component of Portland cement. A cement is produced when the tricalcium silicate is mixed with water (or "hydrated"):

$$3CaO \cdot SiO_2 + xH_2O \rightarrow Ca(OH)_2 + \text{hydrated calcium silicates}$$

The natural curing reactions that occur in a standard Portland cement involve the formation of calcium hydroxide (portlandite) and $Ca(OH)_2$, as well as calcium silicate hydrate (CSH). Over time, the cement will absorb CO_2 from the air, converting the $Ca(OH)_2$, and some of the CSH, to calcium carbonate, $CaCO_3$, through reactions such as those shown in Figure 15.1 (Slegers and Rouxhet, 1976; Suzuki et al., 1985; Taylor, 1990). Supercritical CO_2 processing carbonates the calcium hydroxide, removes water from the

Manufacture of Cast Cement

$$2(3CaO{\cdot}SiO_2) + 6H_2O \rightarrow 3CaO{\cdot}2SiO_2{\cdot}3H_2O + 3Ca(OH)_2$$

$$2(2CaO{\cdot}SiO_2) + 4H_2O \rightarrow 3CaO{\cdot}2SiO_2{\cdot}3H_2O + Ca(OH)_2$$

$$3CaO{\cdot}Al_2O_3 + 6H_2O \rightarrow 3CaO{\cdot}Al_2O_3{\cdot}6H_2O$$

$$4CaO{\cdot}Al_2O_3{\cdot}Fe_2O_3 + 10H_2O + 2Ca(OH)_2 \rightarrow 3CaO{\cdot}Al_2O_3{\cdot}6H_2O + 3CaO{\cdot}Fe_2O_3{\cdot}6H_2O$$

Carbonation with Supercritical CO_2

$$Ca(OH)_2 + CO_2 \xrightarrow{\text{Supercritical } CO_2} CaCO_3 + H_2O$$

$$3CaO{\cdot}2SiO_2{\cdot}3H_2O + 3CO_2 \xrightarrow{\text{Supercritical } CO_2} 3CaCO_3 + 2SiO_2 + 3H_2O$$

$$3CaO{\cdot}Al_2O_3{\cdot}6H_2O + 3CO_2 \xrightarrow{\text{Supercritical } CO_2} 3CaCO_3 + Al_2O_3 + 6H_2O$$

$$3CaO{\cdot}Fe_2O_3{\cdot}6H_2O + 3CO_2 \xrightarrow{\text{Supercritical } CO_2} 3CaCO_3 + Fe_2O_3 + 6H_2O$$

Figure 15.1. Supercritical fluid carbonation of cast cement. (Top) Idealized chemical reactions occurring during the manufacture of a cast Portland cement. (Bottom) Idealized chemical reactions occurring during the accelerated carbonation using supercritical CO_2.

hydrated calcium silicates, and probably converts the calcium silicates (and calcium aluminates) to silica (and alumina):

$$Ca(OH)_2 + CO_2 \xrightarrow{sc\,CO_2} CaCO_3 + H_2O$$

$$\text{Hydrated calcium silicates} \xrightarrow{sc\,CO_2} \text{calcium silicates} + \text{water}$$

$$\text{Calcium silicates} \xrightarrow{sc\,CO_2} CaCO_3 + SiO_2$$

It turns out, however, that the molar volume of $CaCO_3$ (36.9 cm^3/mol) is larger than that of $Ca(OH)_2$, (33.1 cm^3/mol), resulting in the closure and/or blockage of pores, impeding the ingress of reactant, CO_2, and the egress of reaction product, H_2O (Parrott, 1987). The net result is that natural carbonation reactions are initially rapid, but slow dramatically with time. Consequently, the oldest, manufactured cement structures in Greece and China are still undergoing this carbonation reaction, even after several thousands of years. However, by exposing a Portland cement to supercritical CO_2, it is found that the carbonation reaction can be greatly accelerated. This acceleration is due to the ease of penetration of the supercritical fluid into the micropores of the cement, providing continuous availability of fresh reactant (CO_2), as well as the ability of the dense CO_2 to solubilize and facilitate removal of the reaction product (H_2O). Figure 15.2 shows an X-ray diffraction (XRD) pattern for a 6 in. × 12 in. cement cylinder that was treated with supercritical CO_2 for a period of time sufficient only to produce a reaction in the outer portion of the cylinder. It can be seen that the pattern for the outer portion of the cylinder indicates a nearly complete conversion of $Ca(OH)_2$ to $CaCO_3$, while the pattern for the inner portion shows minimal conversion. Further evidence can be obtained by simply treating the sample's cut surface with a pH-sensitive dye such as phe-

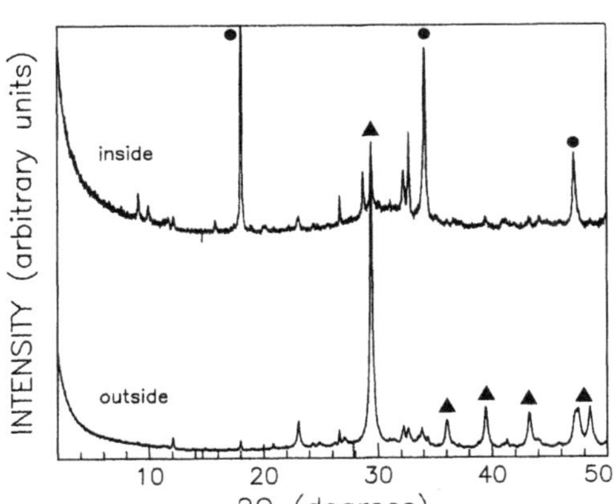

Figure 15.2. X-ray diffraction spectra of the inside and outside portion of a 6 in × 12 in Portland cement cylinder partially reacted with supercritical CO_2. The solid circles identify the peaks due to $Ca(OH)_2$, while the solid triangles show the diffraction peaks due to $CaCO_3$.

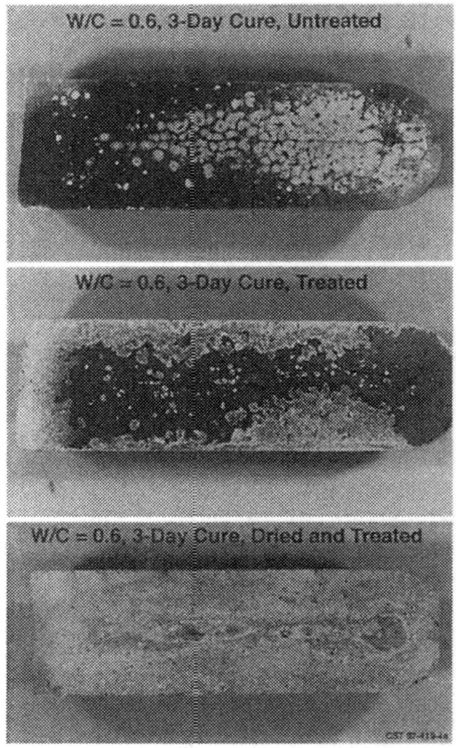

Figure 15.3. (a) Untreated, (b) partially converted, and (c) near-full-conversion cement cylinders bisected and dyed to show change of pH as a function of depth. Darker color indicates pH > 10.

nolphthalein. The high-pH hydroxide converts to a neutral-pH carbonate. Figure 15.3 utilizes this technique to show the depth of penetration of the CO_2 into samples with different initial free-water contents. We observe generally that the supercritical CO_2 treatment results in a carbonation, whose depth depends on the initial porosity, permeability, and free-water content of the cement, as well as on the temperature, pressure, and exposure time of the CO_2 (Hartmann et al., 1999). By suitable adjustment of these parameters, we have demonstrated complete carbonation in cement samples during a treatment time of several hours (cf. Figure 15.3).

The natural carbonation reactions within a Portland cement result in the formation of fine-grained $CaCO_3$ crystals, which degrade the structural integrity and reduce the overall strength of the cement (Young et al., 1974). However, for reasons that are still unclear, the accelerated carbonation reaction using sc CO_2 results in the formation of very large, well-consolidated $CaCO_3$ crystals, resulting in a significant increase in unconfined uniaxial compressive strength as shown in Table 15.1. The ingrowth of the crystals also results in an order of magnitude reduction in permeability (Rubin et al., 1998) as shown in Table 15.2. Figure 15.4 illustrates this mineralogical change within a Portland cement before and after sc CO_2

Table 15.1. Unconfined Uniaxial Compressive Strengths for Samples Treated with sc CO_2 Compared with Untreated Samples (water/cement ratio = 0.6 treated 2 h at 275.8 bar, 40 °C)

Treated samples	Applied Load (kg)	MOR (bar)	Control Samples	Applied Load (kg)	MOR (bar)
1	102.0	62.5	11	67.1	41.2
2	92.5	56.7	12	52.6	32.4
3	99.8	61.1	13	63.9	39.3
4	57.6	35.4	14	85.3	52.3
5	99.8	61.1	15	50.8	31.3
6	102	62.5	16	64.4	39.6
7	88.0	53.9	17	33.6	20.8
8	95.2	58.3	18	50.3	31.0
9	125.2	76.5	19	48.1	29.6

carbonation. We have found that the strength enhancement improves with the total curing time, indicating that the maximum strength enhancements will be achieved for cements that have been allowed to fully cure for a period of several weeks (Rubin et al., 1997). However, steam curing can accelerate the curing process itself. This can easily be accomplished in the same pressure vessel used for the sc CO_2 treatment.

Table 15.2. Permeability of Treated versus Untreated Samples

Sample	Curing Time (days)	Saturated K (cm/s)	Intrinsic Permeability, k (cm^2)
w/c = 0.6, untreated	3	2.09×10^{-7}	2.14×10^{-12}
		1.95×10^{-7}	2.00×10^{-12}
		1.85×10^{-7}	1.90×10^{-12}
		1.83×10^{-7}	1.88×10^{-12}
		1.74×10^{-7}	1.78×10^{-12}
w/c = 0.6, treated 2 h at 275.8 bar, 40 °C	3	3.43×10^{-9}	3.52×10^{-14}
		3.36×10^{-9}	3.44×10^{-14}
		3.38×10^{-9}	3.46×10^{-14}

The hydraulic conductivity, K, is equal to the intrinsic permeability (of the cement), k, multiplied by the fluidity of the migrating liquid, $\left(\frac{\rho g}{\eta}\right)$:

$$K = k \cdot \left(\frac{\rho g}{\eta}\right)$$

where ρ = fluid density (g/cm^3), g = acceleration due to gravity (= 980.07 cm/s^2), and η = fluid viscosity (poise = g/cm·s). For water, $\rho = 0.998$ g/cm^3, $\eta = 0.01007$ g/cm·s, and $g = 980.67$ cm/s^2, so that $K = k \cdot 9.7578 \times 10^4$.

(a)

(b)

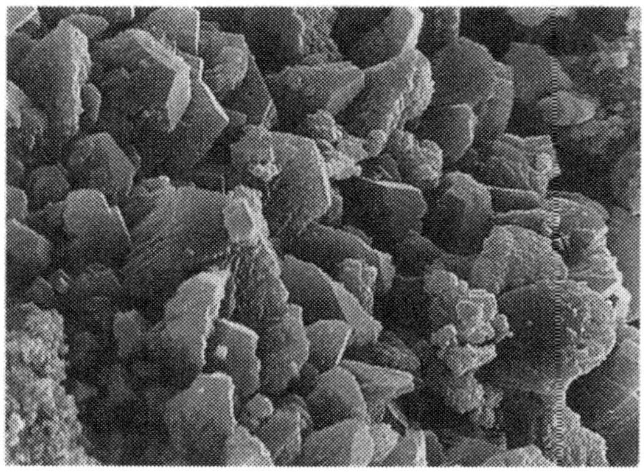

Figure 15.4. Micrograph of unmodified Portland cement (a) before supercritical CO_2 carbonation, (b) after supercritical CO_2 carbonation.

Supercritical CO_2 treatment affects the microstructure of the cement paste. In the first stage of the sc CO_2 treatment, free water in the cement pores is extracted. As a consequence of this dehydration process, channels of about 50-μm diameter develop. Dissolved calcium in the free water reacts with the CO_2 and crystallizes with the CO_2 as calcite along the channel walls. In the second stage, the structural water of the hydrated cement phases is extracted. The carbonation of the portlandite to form more calcite takes place. Water, bound to the CSH surrounding the partially hydrated cement clinker particles, is partially replaced by a carbonate formation. The short fibers of the CSH–cement framework, which are responsible for the physical properties of the cement, are not affected (Hartmann et al., 1999).

The sc CO_2 treatment leads to important chemical property changes as well. The resultant drop in pH due to the change from a hydroxide- to a carbonate-based material allows us to modify cements to form composite materials that were not possible in the past. This allows for the incorporation of polymers into the cement matrix. Polymers such as hydroxypropylmethylcellulose (HPMC), poly(acrylamide), or poly(vinylalcohol/acetate) can increase the plasticity of the mix so as to decrease the amount of water needed to make the cement workable and improve the microstructure. This will result in a more dense cement with significantly smaller micropores (approximately 15 mm).

The drop in pH will also allow the use of a number of reinforcing composite materials. Fiberglass is a common additive to increase the strength in cements, but it is a very expensive composite as one must use an alkali-resistant fiberglass or "e-glass". Normal borosilicate glasses break down in the cement matrix in a period of weeks to months. This is enough time to let the cement cure and then expose it to an appropriate sc CO_2 treatment, dropping the pH and preserving the normal glass fibers. Several polymer fibers such as high-tensile-strength nylons and poly(esters) will behave in much the same way.

In order to efficiently carbonate large pieces of Portland cement, the CO_2 must be supplied in large stoichiometric excess so as to dissolve the reaction product, H_2O. Further, the CO_2 must be supplied at a flow rate sufficiently high so that the water content in the CO_2 does not build up, thereby inhibiting the carbonation reaction. This is best achieved using a closed-loop, recirculating system, where the CO_2 is continuously scrubbed of the dissolved water. Figure 15.5 shows a schematic pressure–temperature phase diagram for CO_2, which also incorporates a process flow diagram for a closed-loop cement-treatment system. The treatment cycle begins with a liquid-CO_2 storage reservoir. The liquid is brought above its critical pressure during a pumping operation, which sends pressurized liquid to a heating unit. It is important to note that it is much easier and cheaper to pump liquids than it is to pump supercritical fluids owing to the much greater compressibility of the supercritical fluid. The heating unit warms the pressurized CO_2 above its critical temperature, forming a supercritical fluid. The supercritical fluid enters the treatment vessel and is brought into contact with the cement. During this time, the water is extracted from the cement and solubilized in the CO_2. There is a constant flow of CO_2 through the treatment vessel, so that clean, dry CO_2 is continuously available. In the supercritical state, the density of CO_2 is sufficient to dissolve up to several mole percent of water, so that a flowing system can remove large amounts of water in relatively short times. On exiting the treatment vessel, the supercritical CO_2—containing the dissolved water—is sent to a separation vessel, where the fluid is depressurized. As the pressure decreases to below the critical pressure, the CO_2 expands into a gas, with a large decrease in density. Since the solubility of the entrained water is proportional to the CO_2 density, the water falls out of solution and is deposited in the bottom of the separator. The clean, dry CO_2 gas exits the top of the separator, where it is

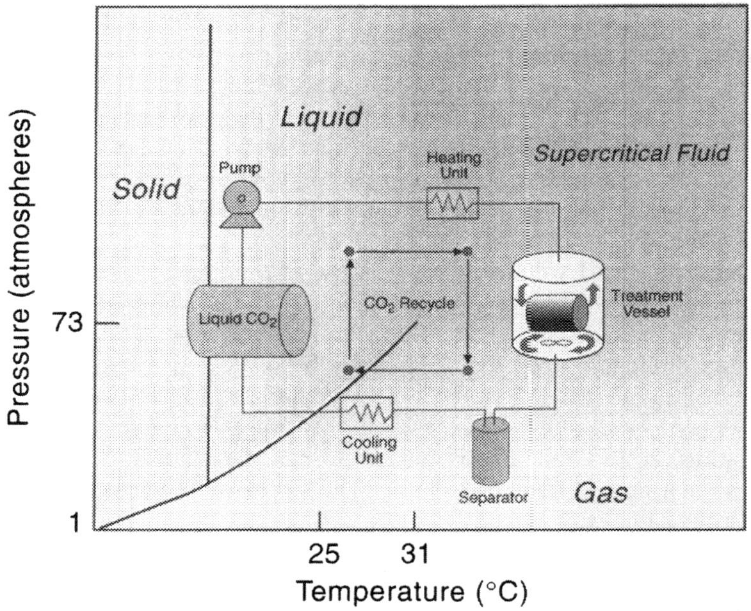

Figure 15.5. Pressure–temperature phase diagram of pure CO_2 with a superimposed flow diagram for a closed-loop supercritical fluid treatment process.

condensed to a liquid by a cooling unit before re-entering the storage vessel. This type of closed-loop extraction means that there are no uncontrolled waste streams exiting the extraction system. All of the extracted materials are retained in the separation vessel for subsequent analysis, treatment, recycle, and/or disposal. Further, the extracted materials are concentrated in the separator, reducing the volume of waste.

Application to Modified Portland Cement

We have been investigating two broad areas of application for Portland cement: incorporating industrial waste and the treatment of Portland cement using supercritical CO_2. The first is the utilization of solid wastes, generated as by-products of industrial processes, as additives to conventional cements to be used as low-cost building materials. One such waste material is fly ash from coal-fired power plants. Because the $CaCO_3$ formed during the carbonation reaction is itself chemically stable and an excellent cementing agent, it is possible to replace a large fraction of the Portland cement with fly ash, while maintaining similar levels of durability and mechanical strength. Our conception of a supercritical CO_2/fly ash plant

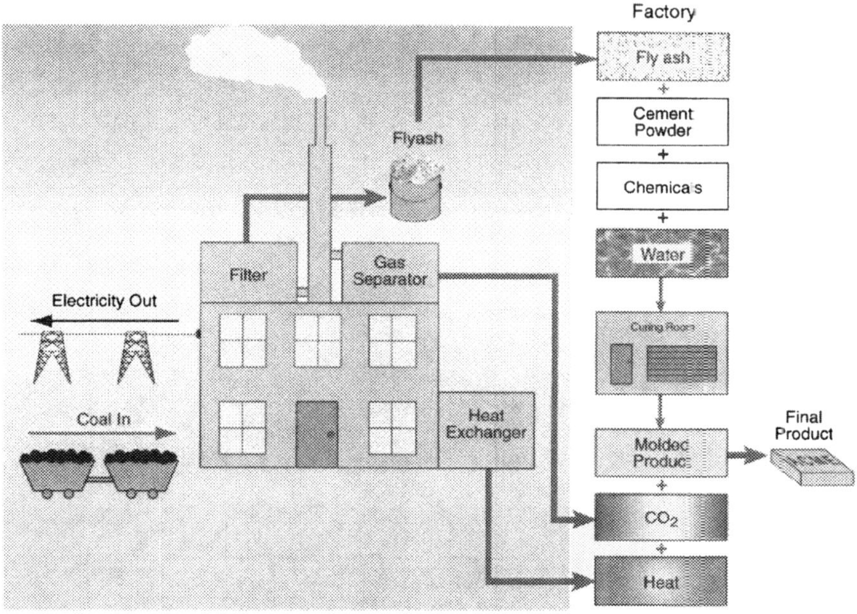

Figure 15.6. Schematic illustration of a supercritical CO_2/fly ash plant integrated with a coal-fired power plant.

integrated with a coal-fired power plant is shown in Figure 15.6. The fly ash would be filtered from the effluent gas stream, while the CO_2 would be separated from the stack gases downstream from the particulate filter. The technology to do this is available today. However, economics favor placement of such a separator where there is a cheap source of power and a local customer for the CO_2. The power plant can sell the electricity cheap to a colocated plant as there is no transmission loss. The cement plant would be colocated as well so the gas would not need to be transported (the major cost of any gas). In addition colocation provides a convenient means for transporting the product in the same way that the coal is delivered (generally by railcar). The fly ash is blended with a minor fraction of cement powder, water, and additional chemicals as necessary to activate the surface of the fly ash, to produce a paste, moldable into a desired product. Once formed into a final shape, the molded product is steam cured for a time, and then treated with supercritical CO_2, derived from the recovered stack gases and waste heat scavenged from heat exchangers. This fabrication process allows us to utilize the three main waste products of coal-derived power generation: fly ash, CO_2, and heat, combining them into a salable product to recover overall operating costs (and eliminating disposal costs for the fly ash—amounting to roughly \$14 per ton). As an example of the products that can be made with supercritical CO_2-processed, fly-ash modified cements, Figure 15.7

Figure 15.7. Roofing tiles, extruded from an 80 vol. % fly ash, 20 vol. % Portland cement slurry and treated with supercritical CO_2.

shows roofing tiles, produced by extrusion of an 80 vol. % fly ash, 20 vol. % Portland cement slurry and subsequently treated with supercritical CO_2. By colocating a supercritical CO_2 treatment facility with a fly ash source, such as a coal-fired power plant, we can close the CO_2 loop on the cement-manufacturing process, and at the same time produce salable products from an industrial solid waste. Supramics™ is commercializing this process on an industrial scale. Details can be found in Jones and Roger (1996) and will soon be used in construction products from ceiling tiles to wallboard to flooring.

The second general area of application is the enhancement of cement, which is itself an industrial waste. By this, we mean cements used to immobilize environmentally regulated metals and radioactive elements. For example, the immobilization of chromium in cement matrices was significantly improved by carbonation (Macias et. al., 1997). Since the 1960s, solidification of radioactive wastes utilizing Portland cement has been practiced worldwide, yet the scientific basis is still not clearly understood (U.S. DOE, 1992). For example, in order to meet present Department of Energy (DOE) transportation and storage requirements, cemented wasteforms must satisfy several requirements (U.S. DOE, 1992), including a maximum limit on the radioactive decay heat. Decay heat is defined as the heat produced by radioactive emissions that are absorbed in the surrounding materials. The current decay heat limit is based, in part, on the overall hydrogenous content of the cement. If the hydrogenous content of the cement (primarily water) can be reduced to less than 30% by weight, then the maximum allowable decay heat for this modified cement increases fourfold (U.S. DOE, 1996). The basis for this limit is the potential for radiolysis of the hydrogenous material, resulting in hydrogen gas generation, by the energetic radioactive decay products. Table 15.3 shows the relative weight percents of water, carbonate, and hydroxide for treated and untreated samples with different cure times. This data clearly shows the desired decrease in water content from greater than 30% to less. Clearly, the removal of a majority of the hydrogen-bearing material (water in the case of a Portland cement), would not only ensure compliance with transportation and storage regulations, but also would reduce the overall volume of cemented waste requiring disposal.

The accelerated carbonation of a cement wasteform using sc CO_2 not only removes the water, but also the formation of the large carbonate crystals results in an improved leaching resistance for the resulting matrix. Incorporation of certain radionuclides in cemented low-level radioactive wasteforms has been suggested to be controlled by carbonation through solid solution with calcite (Smith and Walfond, 1991). Apparently, the radionuclides become incorporated into the carbonates, rendering them unavailable for dissolution. Lange et al. (1997) demonstrated that natural carbonation in cement-type wasteforms is accompanied by an increase in calcite content, greater strength, and reduced leach ability of metals for all matrices examined. To demonstrate the beneficial effects of sc CO_2 treatment on radioactive cemented wastes, we have conducted leaching experiments using nonradioactive surrogate elements (Hartmann et al., 1999). The stable nuclide ^{232}Th was used as a surrogate for the tetravalent actinides such as ^{238}Pu, ^{239}Pu, and ^{241}Pu. Europium (^{151}Eu, ^{153}Eu) was used to substitute for the trivalent actinides ^{241}Am,

Table 15.3. Weight Percent of H_2O, $CaCO_3$, and $Ca(OH)_2$ in Untreated and sc CO_2-Treated Portland Cement (w/c = 0.6)

	3-Day Cure	7-Day Cure
Wt % H_2O	Untreated (TGA): 34.2 Treated (TGA): 22.4	Untreated (TGA): 34.3 Treated (TGA): 25.7
Wt % $CaCO_3$	Untreated (TGA): ≈ 0 Treated (TGA): 27.7	Untreated (TGA): ≈ 0 Treated (TGA): 25.3
	Untreated (XRD): 5.0 Treated (XRD): 18.3	Untreated (XRD): 3.4 Treated (XRD): 18.8
Wt % $Ca(OH)_2$	Untreated (XRD): 28.1 Treated (XRD): 4.6	Untreated (XRD): 36.3 Treated (XRD): 7.5

^{244}Cm, and ^{246}Cm. The results of this study showed that the treatment of the cement produced a fivefold decrease in the amount of the surrogate elements in the leachate.

The incorporation of fly ash into Portland cement has been identified as one of the treatment parameters of cement composition to be evaluated. There is already an extensive "experience database" on the performance of fly-ash-modified Portland cement for heavy-metal immobilization and the solidification/stabilization (S/S) of radioactive waste. The United Kingdom (Wilding, 1992) and the United States (Huang et al., 1994) have used these materials, in the form of cement grouts, for the S/S of low- and intermediate-level radioactive wastes. In this section, we will review the known benefits of fly-ash-modified Portland cement over unmodified Portland cement, along with the anticipated improvements expected by the supercritical CO_2 treatment of modified Portland.

- Fly ash increases the density, decreases the permeability, and increases the leaching resistance of Ordinary Portland Cement (OPC). It is a truism that "The leach resistance of solidified cement-waste systems can be improved by any process which accelerates curing, limits porosity, or chemically bonds fission product or actinide elements." (Jantzen et al., 1984). Supercritical CO_2 treatment of a modified Portland cement is expected to further increase the density over the untreated material, so that a reduced porosity and improved leachability should result. In addition, the high silica content of fly ash, with its well-known sorbent properties toward actinides and certain other radionuclides, enhances the immobilization characteristics.
- The presence of heavy metals is often observed to significantly interfere with proper setting of Portland cement. The incorporation of fly ash into Portland cement helps to partially compensate for this set-retardation, and a subsequent supercritical CO_2 treatment of such modified cement should perform even better.
- Fly ash incorporation improves the fluidity of a Portland cement mix, which improves workability. For solidification/stabilization applications, this increased fluidity may allow a reduced water/cement ratio to be used in the casting operation.
- The incorporation of fly ash lowers the initial heat evolution during setting, thereby reducing the incidence of cracking and spalling. It is desirable to maintain the modified cement in monolithic form for optimum leach resistance.
- Fly ash, as it is typically generated in a high free-carbon environment, contains iron in a reduced state, which helps to lower the redox potential in the cement. Maintaining both heavy metals, such as Cr, and radionuclides in lower oxidation states should result in lower solubilities and reduced leachabilities.
- Since the sc CO_2 treatment converts the alkaline phases into neutral-pH phases, a straight or fly-ash-modified Portland cement that has been carbonated will have a much reduced pH.

- In Portland cement, the highly alkaline environment precipitates actinide elements as hydroxides, rendering them immobile. We have recently begun experiments to evaluate the effect of supercritical CO_2 treatment, and the attendant pH reduction, on the leachability of actinide surrogates. We have shown that the treatment decreases the leachability by incorporating the surrogate elements within carbonate phases and in decalcified regions adjacent to these carbonate phases.
- While Portland cement is considered to be an inexpensive immobilization matrix, relative to other candidate materials, its cost is expected to rise in light of current and future projected shortages.

Fly ash, as it is a large-volume industrial waste, is both cheap and abundant, so that there is an economic incentive to use fly-ash-modified cements. In addition, CO_2 is also produced as a waste by-product of industrial processes (power generation, cement manufacture, etc.), and its permanent sequestration into cement is an added environmental benefit. A fully carbonated Portland cement permanently sequesters about 130 L of CO_2 per kilogram of cement. Figure 15.8 shows the structural and chemical modifications produced in cemented fly ash microspheres as a result of the supercritical CO_2 treatment. As is the case with fly ash, kiln dusts are primarily siliceous, so that the same benefits can be derived from their use as modifiers in immobilization and S/S matrices.

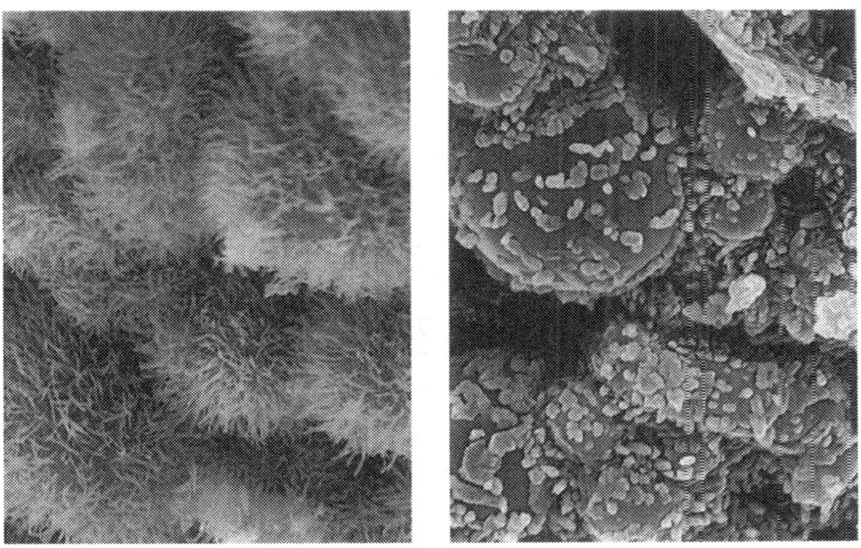

(a) (b)

Figure 15.8. Micrographs of cemented fly ash microspheres (a) before sc CO_2 treatment, (b) after sc CO_2 treatment.

Conclusions

We have demonstrated that supercritical CO_2 can be used to accelerate the natural carbonation reactions in unmodified Portland cements, and that this treatment enhances the physical properties of the cement. Further, it has been proven that the use of supercritical CO_2 allows the replacement of Portland cement powder with inexpensive, lower-grade pozzolans, such as fly ashes.

The accelerated carbonation reactions have also been shown to improve the immobilization properties of Portland cements used as radioactive wasteforms. The formation of carbonates eliminates the potential problem of radiolysis while at the same time improving the leaching resistance. The resulting carbonate matrix, which is more thermodynamically stable relative to the untreated cement, would also be preferred for the underground disposal of the wasteforms. Incorportion of fly ash into the cement-type wasteform further enhances the favorable properties of this waste type.

References

Hartmann, T.; Paviet-Hartmann, P.; Rubin, J. B.; Fitzsimmons, M. R.; Sickafus, K. E. The Effect of Supercritical Carbon Dioxide Treatment on the Leachability and Structure of Cemented Radioactive Waste-Forms. *Waste Manage.* **1999**, *19*, 355–361.

Huang, F. H.; Mitchell, D. E.; Conner, J. M. Low-Level Radioactive Hanford Wastes Immobilized by Cement-Based Grouts. *Nucl. Technol.* **1994**, *107*, 254–271.

Jantzen C. M.; Glasser, F. P.; Lachowski, E. E. Radioactive Waste-Portland Cement Systems: I, Radionuclide Distribution. *J. Am. Chem. Soc.* **1984**, *67*, 668–673.

Jones, R. H., Jr. Cement Treated with High-Pressure CO_2, U.S. Patent 5,518,540, 1996.

Lange, L.; Hills, C.; Poole A. (1997). Effects of Carbonation on Properties of Blended and Non-Blended Cement Solidified Wasteforms. *J. Hazard. Mater.* **1997**, *52*, 193–212.

Macias, A.; Kindness, A.; Glasser, F. Impact of Carbon-Dioxide on the Immobilization of Cemented Wastes. *Cem. Concr. Res.* **1997**, *27*, 215–25.

Parrott, L. J. A Review of Carbonation in Reinforced Concrete. British Cement Association, Berkshire, U.K., 1987.

Rubin, J. B.; Carey, J. W.; Taylor, C. M. V. Enhancement of Cemented Waste Forms by Supercritical CO_2 Carbonation of Standard Portland Cements. *Proceedings of the American Nuclear Society 1st Topical Meeting on Decommissioning, Decontamination, & Reutilization of Commercial & Government Facilities*; Knoxville, TN, Sept. 7–12, 1997.

Rubin, J. B.; Taylor, C. M. V.; Paviet-Hartmann, P.; Hartmann, T. Ash Cements Stabilized by Supercritical CO_2 Carbonation for Repository Overlayer. *Proceedings of NATO Advanced Study Workshop—Turning a Problem into a*

Resource: Remediation and Waste Management at the Sillamäe Site, Estonia; Tallinn, Estonia, Oct. 5–9, 1998.

Slegers, P.; Rouxhet, P. Carbonation of the Hydration Products of Tricalcium Silicate. *Cem. Concr. Res.* **1976**, *6*, 381–388.

Smith, R. W.; Walfond, C. The Effects of Calcite Solid Solution Formation on the Transient Release of Radionuclides from Concrete Barriers. *Scientific Basis for Nuclear Waste Management XIV*, **1991**, *212*, 403–409.

Suzuki, K.; Nishikawa, T.; Ito, S. Formation and Carbonation of CSH in Water. *Cem. Concr. Res.* **1985**, *15*, 213–224.

Taylor, H. *Cement Chemistry*; London: Academic Press, 1990.

U.S. DOE, TRUPAC-II User Requirements Document; Rev. February 2, 1992.

U.S. DOE, TRUPAC-II Content Codes (TRUCON), Rev. 10, DOE/WIPP 89–004, 1996.

Wilding, C. R. The Performance of Cement Based Systems. *Cem. Res.* **1992**, *22*, 299.

Wilk, C. M. Stabilization of Heavy Metals with Portland Cement: Research Synopsis; Portland Cement Association Report IS007; PCA: Skokie, IL, 1997.

Young, J. F.; Berger, R. L.; Breese, J. Accelerated Curing of Compacted Calcium Silicate Mortars on Exposure to CO_2. *J. Am. Chem. Soc.* **1974**, *57*, 394–397.

Index

absorption, 208, 222
  kinetics, 167, 168
adsorption, 86, 134, 135, 136, 146
aromatics, 21, 39, 51
asymmetric hydrogenation, 26, 27
atom, 66, 73, 74, 94, 127, 152

biocatalysis, 228, 103, 104, 116
blowing agents, 174, 176

capillary rheometer, 175, 176, 177, 187
carbohydrates, 50
carbonylation, 18, 23, 31, 33, 71
catalysis, 17, 18, 21, 30, 37, 39, 41, 43, 44, 49, 51, 52, 87, 92, 93, 135, 157
  oxidation, 22, 28, 30, 41
catalytic extraction using supercritical solvent (CESS), 21, 24, 28, 87, 88, 92, 95
catalytic hydrogenation, 15, 18, 25, 26, 51, 61, 237
chain, 59, 65, 66, 72, 75, 77, 93, 95, 103, 125–130, 150, 151, 152, 154, 155, 158, 160, 186
coatings
  free menscus, 203–206
  impregnation, 203, 206–208
  spin, 201–202
  spray, 194–200
copolymers, 128, 132, 137, 143, 151–156, 158, 159
cosolvent, 29, 35, 84, 109, 126, 193, 207

diffusion, 18, 25, 30, 31, 36, 66, 67, 74, 77, 91, 92, 97, 113, 114, 166, 176, 177, 239
dispersion polymerization, 135. 143, 145, 146, 150, 153, 154, 155, 156. 157, 160
dry cleaning equipment, 218, 224, 225, 226
dry cleaning solvents, 216, 217, 219, 221, 222

emulsion formation, 137
emulsions, 42, 43, 84, 134, 135, 137, 140, 141, 142, 143. 146, 152
environmental impact, 61, 64, 84
equation of state, 4, 5, 8, 9, 10, 11, 12, 13, 126, 175, 185, 186, 188, 199
equilibria, 7, 8, 9, 10, 13, 99
equilibrium, 10, 12
extraction, 12, 18, 28, 51, 64, 81, 87, 90, 115, 157, 230, 237, 239, 248
  in situ, 230, 237, 239
  supercritical fluid, 87, 157
  supercritical solvent catalytic (CESS), 21, 24, 28, 87, 88, 92, 95

fluid chromatography, 18
fluorous phase separation, 96
fractionation, 239
free-radical reactions, 65, 66, 67, 69, 78
free volume, 128, 130, 131, 132, 175, 184, 185, 186, 188
frequency vibrational spectroscopy, 61, 64, 84

glucose, 103–121
  acylation of, 108
  from cellulosic material, 107
  produced in nature, 50
  from starch, 106

green chemistry, 43, 64, 164

heavy metals, 87
high pressures, 12, 13, 71, 83, 130
homogeneous catalysis, 20, 30, 41, 49, 51, 92, 93
hydroformylation, 31, 32, 33, 36, 43, 89, 93, 94, 95
hydrogenation, 38–40, 42, 43, 61, 88, 89, 228, 229, 230, 231, 234, 235, 237, 239
  of carbon dioxide, 51–57
  catalytic, 25–28
  free fatty acids, 40, 228, 239, 229, 235, 237, 239
  of ketones, 23, 40, 116
  limitations, 36
  metal-catalyzed, 17, 18, 92, 93
  production of hydrogen peroxide from, 30–31
  suppressed side reaction of, 33
  of vegetable oils, 228, 229, 231, 234
hydrovinylation, 24

insolubility, 33, 34, 156
interaction parameters, 6, 8, 9, 10, 13
interface, 134, 135, 136, 137, 141, 145, 146, 200, 202, 203, 218, 219
ionic liquids, 44, 87

leachability, 252, 253
ligands, 19, 20, 21–24, 25, 26, 28, 32–35, 37–39, 41, 42, 52, 53, 56–59, 87, 93–97, 156
light-scattering, 135

metal-catalyzed reactions, 28, 57, 89
methyl formate, 55, 56
methyl methacrylate, 130, 151, 154, 155
microemulsions, 84, 134, 135, 136, 137, 138, 139, 140, 141, 142, 143, 146
mixtures, 8, 12, 17, 19, 42, 87, 88, 90, 96, 125, 128, 138, 151, 175, 181, 186, 188, 228, 230, 231, 235, 237, 239
molecular weight, 19, 20, 31, 43, 59, 87, 93, 127–130, 132, 141, 144, 150–154, 156–159, 165, 166, 169, 171, 179

organic compounds, 19, 20, 41, 48, 49, 72, 84, 85, 87, 194
oxidation, 29, 30, 31, 39, 41, 43, 72, 85, 97, 98, 103, 104, 116, 223, 229, 252
  catalysis, 22, 28, 30, 41

phase behavior, 87, 255
phase diagram, 3, 159, 247, 248
phase inversion, 137
phase separation, 19, 87, 90, 96, 99, 143, 165, 197
pollution prevention, 78, 160
polydimethylsiloxane, 137, 154, 155, 159, 175, 205
polyethylene, 128, 135, 165, 166, 167, 169, 170, 171, 173, 207, 208
polyethylene polystyrene blends, 171
polymer solubility, 4, 10, 12, 19, 27, 28, 30, 31, 35, 40, 53, 54, 64, 84, 85, 87, 90, 93, 97, 103, 109, 110, 116, 128, 158
  influence of free volume on, 128, 131, 132
  influence of polymer fluorine content, 127
  influence of pressure on, 126
  parameter, 19
  predicting, 130, 150
polymer solutions, 207, 125
polymerizations, 150, 151–155, 157, 160, 165
poly(methyl methacrylate), 130, 155
polystyrene, 129, 136, 138, 155, 165, 166, 168, 169, 170, 171, 172, 173, 175, 176, 177, 178, 179, 181, 182, 183, 184, 185, 187, 208
poly(vinylidene fluoride), 130, 131, 151
pressure, critical, 48, 56, 58, 71, 83, 202, 247
propane, 228–231, 235, 237, 239
propylene, 32, 39, 41, 90, 134, 137, 23
propylene oxide, 134, 137

rate constants, 66, 68, 72, 77, 78
regeneration, 99
reverse micelles, 44, 158
risk, 85, 216

scanning electron microscopy, 153
selectivity, 17, 18, 20–23, 26, 28, 29, 36, 37–44, 78, 81, 231, 233, 234, 237, 239
  enantiomer, 25, 26, 28, 73, 74, 92, 97, 114, 115
  regiospecific, 32, 33, 35, 94, 95
small-angle neutron scattering (SANS), 136, 158
solubility, 160, 165, 181, 195, 196, 201, 202, 205, 207, 208, 215, 235, 247
solvents
  conventional, 3, 26, 27, 33, 35, 48, 54, 69, 73, 74, 75, 82, 95, 113, 195, 200, 201, 207
  environmentally benign, 154, 195
  organic, 5, 10, 17, 18, 20, 25, 28, 29, 30, 32, 34, 58, 86, 90, 95, 103, 109, 112, 135, 149, 150, 195, 201, 203, 215
sorption, 164, 168
stabilization, 55, 95, 135, 141, 143, 145, 146, 160, 252
supercritical fluid extraction (SFE), 87

1,1,1,2-tetrafluoroethane, 174
temperature
  critical, 6, 8, 13, 48, 64, 83, 84, 116, 196, 247
  scaling, 139, 182, 184
  transition, 129, 130, 150, 166, 174, 179, 183, 223
thermodynamics, 199, 125
transmission electron microscopy (TEM), 142

vapor interface, 202, 218
viscoelastic scaling, 175, 179, 181–188

water
  in carbon dioxide microemulsions, 137–142
  to carbon dioxide ratio, 135, 136, 140, 141
  core, 142
  interphase, 135, 136, 145
  latex, 145, 150
  phase, 55, 137
  soluble, 41, 42, 43, 53, 54, 129

www.ingramcontent.com/pod-product-compliance
Lightning Source LLC
LaVergne TN
LVHW011753200926
848484LV00003B/692
* 9 7 8 0 1 9 5 1 5 4 8 3 2 *